Water Science and Technology Library

Volume 95

The aim of the *Water Science and Technology Library* is to provide a forum for dissemination of the state-of-the-art of topics of current interest in the area of water science and technology. This is accomplished through publication of reference books and monographs, authored or edited. Occasionally also proceedings volumes are accepted for publication in the series. *Water Science and Technology Library* encompasses a wide range of topics dealing with science as well as socio-economic aspects of water, environment, and ecology. Both the water quantity and quality issues are relevant and are embraced by *Water Science and Technology Library*. The emphasis may be on either the scientific content, or techniques of solution, or both. There is increasing emphasis these days on processes and *Water Science and Technology Library* is committed to promoting this emphasis by publishing books emphasizing scientific discussions of physical, chemical, and/or biological aspects of water resources. Likewise, current or emerging solution techniques receive high priority. Interdisciplinary coverage is encouraged. Case studies contributing to our knowledge of water science and technology are also embraced by the series. Innovative ideas and novel techniques are of particular interest.

Comments or suggestions for future volumes are welcomed.

Vijay P. Singh, Department of Biological and Agricultural Engineering & Zachry Department of Civil and Environmental Engineering, Texas A&M University, USA
Email: vsingh@tamu.edu

More information about this series at http://www.springer.com/series/6689

Ashish Pandey · S. K. Mishra ·
M. L. Kansal · R. D. Singh ·
V. P. Singh
Editors

Climate Impacts on Water Resources in India

Environment and Health

Springer

Editors
Ashish Pandey
Indian Institute of Technology Roorkee
Roorkee, India

S. K. Mishra
Indian Institute of Technology Roorkee
Roorkee, India

M. L. Kansal
Indian Institute of Technology Roorkee
Roorkee, India

R. D. Singh
Indian Institute of Technology Roorkee
Roorkee, India

V. P. Singh
Texas A&M University
College Station, TX, USA

ISSN 0921-092X ISSN 1872-4663 (electronic)
Water Science and Technology Library
ISBN 978-3-030-51426-6 ISBN 978-3-030-51427-3 (eBook)
https://doi.org/10.1007/978-3-030-51427-3

This Springer imprint is published by the registered company Springer Nature Switzerland AG
The registered company address is: Gewerbestrasse 11, 6330 Cham, Switzerland

Contents

About the Editors

Prof. Ashish Pandey did his B.Tech. in Agricultural Engineering from JNKVV, Jabalpur and obtained M. Tech. in Soil and Water Engineering from JNKVV, Jabalpur. He received Ph.D. in Agricultural Engineering from IIT Kharagpur. His research interests include Irrigation Water Management, Soil and Water Conservation Engineering, Hydrological Modeling of Watershed, Remote sensing and GIS Applications in Water Resources. He guided 10 Ph.D. and 78 M.Tech. students at IIT Roorkee. He published 155 research papers in peer reviewed and high impact international/ national journals/seminars/conferences/symposia. He has also co-authored a textbook on "Introductory Soil and Water Conservation Engineering".

Professor Pandey also served as Guest Editor for two issues of *Journal of Hydrologic Engineering* (ASCE). He is Editor, *Indian Journal of Soil Conservation*. His laurels include Eminent Engineers Award-2015 given by the Institute of Engineers (India), Uttarakhand State Centre, DAAD scholarship and ASPEE scholarship. He was offered prestigious, JSPS Postdoctoral Fellowship for Foreign Researchers, Japan and BOYSCAST fellowship of DST, GOI. He is a Fellow member of (1). Institution of Engineers (India); (2). Indian Association of Hydrologists (IAH) and (3). Indian Water Resources Society (IWRS).

Dr. S. K. Mishra is a 1984 Civil Engineering graduate of the then Moti Lal Nehru Regional Engineering College (presently MNNIT), Allahabad. He obtained his M.Tech. degree in Hydraulics and Water Resources from IIT Kanpur in 1986; and doctoral degree from the then University of Roorkee (presently IIT Roorkee) in 1999. He served the National Institute of Hydrology Roorkee at various scientific positions during 1987–2004. During the period, he also visited Louisiana State University, USA, as a post doctoral fellow during 2000–2001 and Department of Civil Engineering, IIT Bombay as a Visiting Faculty during 2002–2003.

Dr. Mishra joined IIT Roorkee in 2004 and is presently working as Professor. Besides having been the Head of the Department of Water Resources Development and Management, he is also presently holding Bharat Singh Chair of the Ministry of Jal Shakti, Government of India. He is specialized in the fields of hydraulics and water resources, environmental engineering, design of irrigation and drainage works, dam break analysis, surface water hydrology. He has published more 250 technical articles in various international/nationaljournals/seminars/conferences/symposia. His reference book on Soil Conservation Service Curve Number Methodology published by Kluver Academic Publishers, the Netherlands, has received a significant attention of the hydrologists/agriculturists/soil water conservationists around the globe. Of late, he has been associated with the *Journal of Hydrologic Engineering* as a Guest Editor of two special issues and the Executive Vice President and Editor of *Indian Water Resources Society*. He has visited several countries during the period and is on the role of several national/international professional bodies. Among several others, the Eminent Engineers Award and Dr. Rajendra Prasad Award are worth citing.

Prof. M. L. Kansal currently working as NEEPCO Chair Professor and Head in the department of Water Resources Development & Management at Indian Institute of Technology (IIT) Roorkee (India). He is a Civil Engineering graduate with post graduation in Water Resources Engineering. He obtained his Ph.D. from Delhi University (India) and holds the Post Graduate Diploma in Operations Management. Previously, he worked as Associate Professor in the department WRD&M, IIT Roorkee and served at Delhi Technical University, Delhi (erstwhile, Delhi College of Engineering, Delhi), NIT Kurukshetra, IIT Delhi and NIH, Rookree (India) at various levels. He has published more than 150 research papers and two books. He has got best paper awards from Indian Building Congress, Indian Water Works Association, and received star performer award from IIT Roorkee. He acted as reviewer for several international journals and research agencies. He has visited various countries as International Expert and Visiting Professor. Dr. Kansal acted as International expert for RCUWM of UNESCO and IUCN, etc. He is working as Executive Vice President of Indian Water Resources Society and served as expert panel member for All India Council for Technical Education (AICTE), India. He has contributed substantially by providing consultancy services to various national and international agencies of repute. He has worked in various administrative capacities such as Associate Dean of Students Welfare and as Chairman, Co-ordinating Committee of Bhawans at Indian Institute of Technology Roorkee and national expert for several bodies.

R. D. Singh did B.E. Civil Engineering and M.E. Civil Engineering with Specialization in Hydraulics & Irrigation Engineering from University of Roorkee (Now IIT Roorkee). He did M.Sc. in Hydrology from University College Galway, Ireland. Presently, he is working as Visiting Professor, Department of Water Resources Development & Management, IIT Roorkee. He worked as Director, National Institute of Hydrology (NIH), Roorkee for more than 9 years. Before taking over as Director NIH, he was holding the charge of Nodal Officer Hydrology Project-II, a World Bank Funded Project for Peninsular region of India completed by NIH during the year 2016. During his service at NIH, he had worked on more than eighty sponsored/ consultancy projects for solving the real-life problems in water sector. He had also worked as well as guided eleven International Collaborative projects at NIH.

He has research & development experience of more than 40 years in different areas of Hydrology and Water Resources. He has an extensive experience in flood estimation, flood management, drought management, hydrological modeling, environmental impact assessment and climate change & its impact on water resources etc. He has published more than 316 research papers in the reputed International & National Journals, International & National Seminar/Symposia, Workshops, etc. He has received C.B.I.P. Medal, Institution of Engineers Certificate of merit, Union Ministry of Irrigation award and best Scientist award form from NIH, Roorkee. He guided three Ph.D. and 14 M.E. & M.Tech. and one M.Phil. dissertations. He has widely travelled abroad for different assignments.

Prof. V. P. Singh is a University Distinguished Professor, a Regents Professor, and Caroline and William N. Lehrer Distinguished Chair in Water Engineering at Texas A&M University. He received his B.S., M.S., Ph.D. and D.Sc. degrees in engineering. He is a registered professional engineer, a registered professional hydrologist, and an Honorary diplomate of ASCE-AAWRE. He is a Distinguished member of ASCE, a Distinguished Fellow of AGGS, and an Honorary Member of AWRA. He has published extensively in the areas of hydrology, irrigation engineering, hydraulics, groundwater, water quality, and water resources (more than 1290 journal articles; 30 textbooks; 70 edited reference books; 105 book chapters; and 315 conference papers). He has received more than 95 national and international awards, including three honorary doctorates. He is a member of 11 international science/engineering academies. He has served as President of the American Institute of Hydrology (AIH), Chair of Watershed Council of American Society of Civil Engineers, and is currently President of American Academy of Water Resources Engineers. He has served/serves as editor-in-chief of three journals and two book series and serves on editorial boards of more than 25 journals and three book series.

1.1 Introduction

Microbial contamination of ambient waters such as rivers and lakes are a serious issue (USEPA 2012). Increased pathogen and pathogen indicator concentrations such as *E. coli* in ambient water increases the risk to public and animal health. In the event of increasing frequencies of extreme weather, how water borne pathogen levels will change in ambient water bodies is a matter of a great concern. Due to the fact that many rivers and streams are used for recreational activities as well as for supplying drinking water for humans and animals, any level of increase in the microbial load in ambient water may increase the health risk, and deriving strategies to cope with the increased risk of microbial contamination in water resources requires improving our existing understanding of the possible impacts of global warming/climate change on waterborne pathogens. Previous studies (Vezzulli et al. 2012; Harvell et al. 2002) have shown that there is a possibility that increase in water temperature may accelerate the growth and transmission of bacteria.

Climate, increased temperature, and water contamination have been found to be associated with many previous global outbreaks (Harvell et al. 2002; Daszak et al. 2000; Epstein 1999). An example is the Rift Valley fever outbreaks in East Africa during the years 1950–1998. The rift valley fever is a viral disease, common in domesticated animals in sub-Saharan Africa. The origin of the outbreak was in sub-Saharan Africa, and in 1977, it spread in Egypt through infected livestock trade. Rift Valley fever can also pose potential health risks to humans. The direct and indirect contact with infected animals was the main reason for transmission from livestock to humans (WHO 2018). Multiple studies have been conducted to understand the potential linkages between climate and outbreaks, and in this case, the relationship between temperature and outbreaks was noticeable (Epstein 1999). During warm seasonal weather increased cased of the fever were observed. This pattern of increased fever cases and warm weather continued for many decades.

A similar phenomenon with regards to temperature and the spread of cholera in Bangladesh has also been reported. Cholera-like disease outbreaks became common in Bangladesh for many years, which initially started around 1992. Outbreak of cholera is linked with poor water quality. At the early stages, the outbreak was localized, the epidemic began in the southern part of Bangladesh, however, it spread throughout the country. Each year, more than 100,000 cases and 4500 deaths in Bangladesh are reported due to the cholera-like diseases (Islam et al. 2018). Disease outbreaks such as cholera and diarrheal diseases are mainly due to the microbial contaminated water exposure, unsafe drinking water and lack of basic hygiene (Pandey et al. 2014; Islam et al. 2018). Improving access to safe water has profound impacts on reducing the public health risks. The long-term analysis of weather pattern and cholera outbreaks in Bangladesh signifies the linkages between the outbreaks and warm temperature. Specifically, the increase in Ocean water temperature is linked with cholera outbreaks in Bangladesh (Colwell 1996).

Substantial studies have attempted to use the reported outbreaks and the climate data of an affected area to define the relationships between ambient conditions and disease outbreaks (Vezzulli et al. 2012; Harvell et al. 1999). When establishing the relationship between water quality and the disease outbreaks, often the lack of long-term water quality data is a major challenge. Having an improved knowledge of climate, and water borne contamination can certainly helps in planning and implementation of control measures to reduce the public and animal health risk (Pandey et al. 2014, 2016).

In this work, we attempted to use the data of waterborne pathogens/pathogen indicator, which was measured in streams over many years. In streams, *E. coli* is often used as an indicator organism to assess water quality. We quantified the *E. coli* concentrations in stream water, and retrieved the data of air temperature, stream flow, and precipitation of the study area to understand the linkages between waterborne pathogens and climate conditions.

1.2 Study Area

Figure 1.1 shows Squaw Creek Watershed, streams, and sampling locations. The Squaw Creek Watershed is located in Iowa, USA. University town (Ames) is located downstream of Squaw Creek (Fig. 1.2). Two sampling locations are shown in the figure, which were used for collecting the stream water samples, and monitoring the stream flow. The sampling location near the City of Ames records precipitation, and stream flow continuously. Water quality measurements were carried out between May 2009 and October 2011. In winter, the watershed receives snow, and the weather is cold, thus limiting sample collection during this period. During spring and summer, snow melt and rainfall results in increased stream flow conditions. In the watershed, agricultural land is the dominant use. Details of the watershed are given in Table 1.1. More than 87% of the soil in Iowa is fine sand and clay. Only 8% of the soil is sandy (NRCS 2011). Loamy Wisconsin glacial till and clayey deposits (loam, silt, clay loam) are the major soil type in Squaw Creek Watershed. The watershed data were retrieved from Natural Resources Geographic Information System (NRGIS).

1.3 Sample Collection and *E. coli* Enumeration

Water sampler (Forestry Suppliers Inc., Mississippi, U.S.) was used to collect water samples. Between sampling locations, the sampler was cleaned and rinsed to avoid possible cross contamination. To collect water samples, the sampler was lowered from the bridge to the center of the stream. Each stream sample was collected from the top layer (top 12 cm). After collection of samples, all stream water samples were transported and stored at 4 °C prior to analysis. Samples were analysed within 24 h of sample collections. Standard modified mTEC agar and membrane filtration technique (EPA, method 1603) was used for enumerating *E. coli* (pathogen indicator) concentrations in water (APHA 1999).

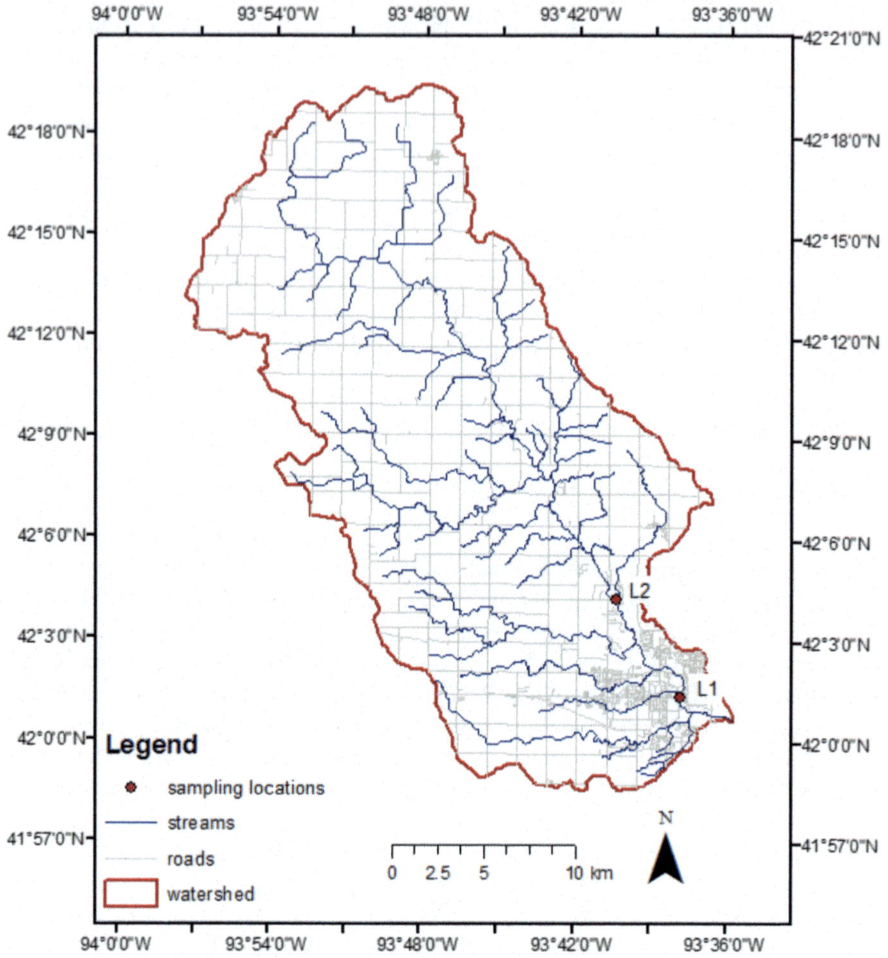

Fig. 1.1 Squaw Creek Watershed, sampling locations, streams and roads. Watershed is upstream of Ames, Iowa. Blue lines indicate stream lines, and light-gray lines indicate road networks. Red dots at the lower end of the watershed indicate sampling location

1.4 Climate Data

Data base of Iowa Environmental Mesonet (IEM Agronomy Department, Iowa State University, was used to obtain the air temperature, and precipitation data. Climate data (air temperature/precipitation) included in the database was recorded using a Campbell HMP 45 instrument, which was mounted on a radiation grill at 2 m height (IEM 2012).

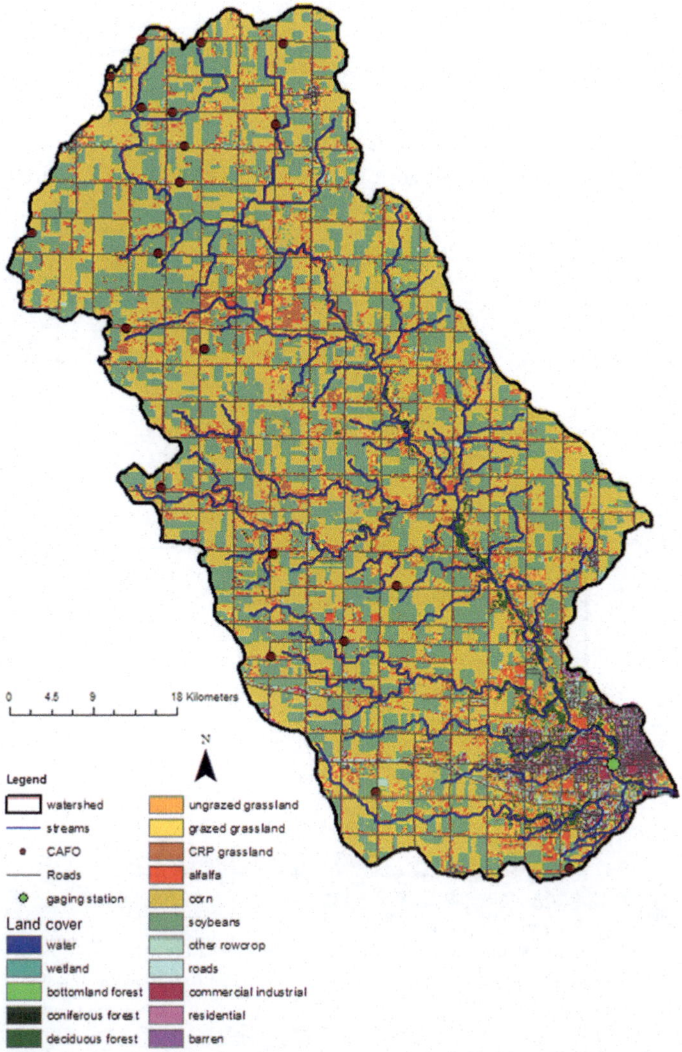

Fig. 1.2 Watershed land cover map of Squaw Creek Watershed, Iowa [green colour indicates soybean crops; yellow colour indicates corn crop (corn and soybean are under crop rotation); dark blue lines indicate streams; and lower end dense area indicate the city of Ames (urban land)]

1.5 Results and Discussion

Stream flow data is shown in Fig. 1.3. Stream flow increased with precipitation (Fig. 1.3) as a result of runoff and direct precipitation on the stream surface. The observations of pathogen indicator are shown in Fig. 1.4. Pathogen indicator (*E. coli*) concentrations at two locations are plotted in figure in time series. Results showed an

Table 1.1 Squaw Creek Watershed characteristics

Description	Values
Watershed area	592.39 km^2
Basin length	43.53 km
Basin perimeter	134.02 km
Land slope	2.01%
Main channel (squaw creek) length	60.46 km
Agricultural land use	74%
Corn and soybean under crop rotation	33 and 41%
Deciduous forest	2.71%
Ungrazed grass	10.87%
Grazed grass	2.52%
CRP grassland	1.70%
Alfalfa	1.84%

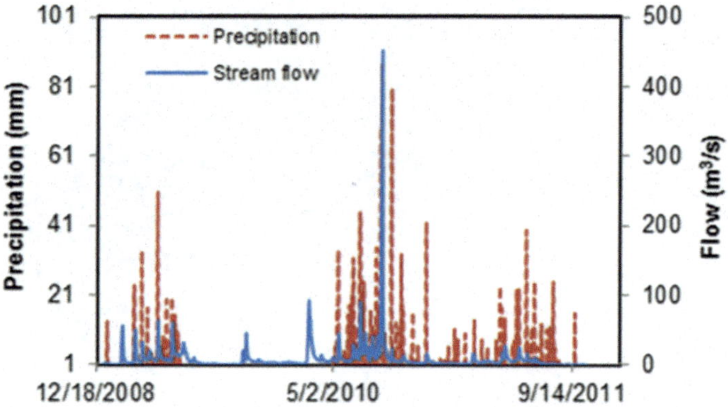

Fig. 1.3 Precipitation and stream flow in sampling locations of the watershed is shown. Red line indicates precipitation, and blue line indicate stream flow measured in sampling locations

increase and peaks of pathogen indicator among three seasons. To understand potential relationships between weather and pathogen indicator concentrations, we plotted pathogen indicator and air temperatures (2009–2011) (Fig. 1.5). The figure shows the variability pattern of the concentrations of pathogen indicator and air temperature. Results indicate that the pattern of waterborne pathogen/indicator organisms in the water column follows the pattern of air temperature (Fig. 1.5).

Relating Fig. 1.3 precipitation and stream flow data with *E. coli* concentrations (Fig. 1.4) in the water column at two locations, it is clear that the increase in *E. coli* concentrations occurs during elevated precipitation, air temperature and streamflow conditions. While relating air temperature data with waterborne pathogen indicator concentrations in the stream (Fig. 1.5), the air temperature peaks match the high pathogen indicator (*E. coli*) concentrations in stream water. When temperatures

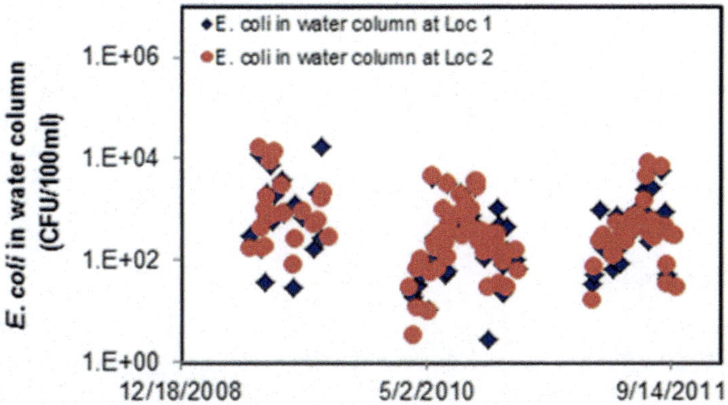

Fig. 1.4 Concentrations of pathogen indicator (*E. coli*) in water column in sampling locations [*E. coli* concentrations was measured at two locations (upstream and downstream and data of both locations are imposed in the plot)]

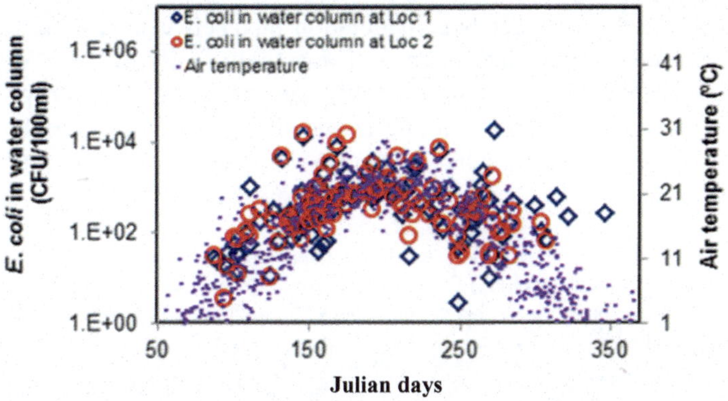

Fig. 1.5 Pattern of air temperature, and pathogen indicator concentrations in stream water

were lower, pathogen indicator concentrations in the water column were reduced. In addition to air temperature, stream flow and surface runoff caused by increase in precipitation resulted in high concentrations of the pathogen indicator in stream water, particularly during the first storm events (first part of the storm hydrograph shown in Fig. 1.3) (Pandey et al. 2016).

Our observations of waterborne pathogen indicator, *E. coli*, in stream water indicates a similar pattern between increasing air temperature and pathogen indicators in stream water. In general, warmer weather provides more conducive conditions for bacteria to grow. From these data of a single location, it can be challenging to understand how long-term increases in temperature/extreme weathers may have long term impacts on waterborne pathogens in streams, nevertheless, these observations suggest that during seasonal increase in temperature, the concentrations of waterborne pathogens are higher in the stream. Further, it can be inferred from these data that increased air temperature during climate warming may provide suitable temperature conditions for microbial populations to grow in ambient waterbodies, which may increase public and animal health risk through exposure to microbially contaminated water.

1.6 Conclusions

In this study, the change in waterborne pathogens over various seasons was analysed. The variability of rainfall, stream flow, and air temperature was studied. The change in stream flow, precipitation, and air temperature was related with the changes in waterbone pathogen concentrations in streams. During the summer seasons, when temperature was elevated, observations showed that waterborne pathogen concentrations increased. During low temperature seasons, waterborne pathogen concentrations were reduced. While additional long term studies from varous geographical locations are needed to understand the relationships between waterborne pathogens and ambient temperature, the findings of this study indicates that the waterborne pathogen concentrations were higher during summer (or higher temepraure conditions) that may increase risks to public and animal heath. This is also the time when humans and animals are more likely to enter waterbodies, and hence it increase the health risks.

Acknowledgement To conduct this research, financial supports were provided by the U.S. Environmental Protection Agency (EPA) Region 7, and the National Science Foundation (NSF). Data presentation and interpretation of this manuscript is of authors and do not reflect the views of the sponsoring agencies.

References

American Public Health Association (APHA) (1999) Standard methods for the examination of water and wastewater. AWWA, Water Environment Federation

Colwell RR (1996) Global climate and infectious disease: the cholera paradigm. Science 274:2025–2031

Daszak P, Cunningham AA, Hyatt AD (2000) Emerging infectious diseases of wildlife—threats to biodiversity and human health. Science 287:443–449

Epstein PR (1999) Climate and health. Science 285:248–347

Harvell CD, Kim K, Burkholder JM, Colwell RR, Epstein PR, Grimes DJ, Hofmann EE, Lipp EK, Osterhaus AD, Overstreet RM, Porter JW, Smith GW, Vasta GR (1999) Diseases in the ocean: emerging pathogens, climate links, and anthropogenic factors. Science 285:1505–1510

Harvell CD, Mitchell CE, Ward JR, Altizer S, Dobson AP, Ostfeld RS, Samuel MD (2002) Climate warming and disease risks for terrestrial and marine biota. Science 296:2158–2162

Iowa Natural Resources Conservation Service (NRCS) (2011) Information about soils. https://www.ia.nrcs.usda.gov/soils.html. Accessed 15 June 2011

Iowa Environment Mesonet (IEM) (2012) Iowa AG Climate Network. Iowa State University. https://mesonet.agron.iastate.edu/agclimate/hist/hourlyRequest.php. Accessed 26.05.12

Islam MT, Clemens JD, Qadri F (2018) Cholera control and prevention in Bangladesh: an evaluation of the situation and solutions. J Inf Dis 218(3):S171–S172. https://academic.oup.com/jid/article/218/suppl_3/S171/5085578

Pandey PK (2012) Modeling in-steram *E. coli* concentrations. Graduate Theses and Dissertations, Iowa State University, Ames, Iowa. https://lib.dr.iastate.edu/cgi/viewcontent.cgi?article=3862&context=etd

Pandey PK, Kass PH, Soupir ML, Biswas S, Singh VP (2014) Contamination of water resources by pathogenic bacteria. AMB Exp

Pandey PK, Soupir ML, Ikenberry CD, Rehmann CR (2016) Predicting streambed sediment and water column *Escherichia coli* levels at watershed scale. J Am Water Resour Assoc 52(1): 184–197

U.S. Environmental Protection Agency (USEPA) (2012) WATERS (Watershed Assessment, Tracking & Environmental ResultS). Washington, D.C.

U.S. Census Bureau (2010) Population estimates, American community survey. https://quickfacts.census.gov/qfd/states/19/1901855.html. Accessed 26.05.12

Vezzulli L, Brettar I, Pezzati E, Reid PC, Colwell RR, Hofle MG, Pruzzo C (2012) Long-term effects of ocean warming on the prokaryotic community: evidence from the vibrios. Int Soc Micro Ecol 6:21–30

World Health Organization (WHO) (2018) Rift valley fever. https://www.who.int/news-room/fact-sheets/detail/rift-valley-fever. Accessed on 5/3/2020

Chapter 2
Contamination in Drinking Water Supply: A Case Study of Shimla City, Himachal Pradesh, India

M. K. Sharma, Rajesh Singh, Omkar Singh, and D. G. Durbude

Abstract Shimla city is the capital of the hilly state, Himachal Pradesh. It is situated in the south of river Satluj. The drinking water supply to major portion of the city is met through Ashwani Khad and Dhalli water supply schemes. Ashwani Khad water treatment plant (WTP) receives water from a natural stream, Malyana sewage treatment plant (STP) treated water, and three open drains. During lean period, the treated water is supplemented with bore well water to fulfill the demand and is supplied to consumers after chlorination. Dhalli WTP receives water from Churat Nallah and Sayog catchment, and supplied to consumers after treatment. A mass level jaundice was reported in the Sanjauli-Malyana area of Shimla City during 2006–2007. In view of this, study of contamination in drinking water sources of Shimla City was carried out using hydrological and water quality data analysis. In the present paper, the causes of contamination in drinking water supply and options for ameliorative measures have been identified and discussed.

2.1 Introduction

Article 21 and 47 of the constitution of India prioritized the provision of clean drinking water for the citizens and advises the state governments to provide safe drinking water to improve the public health (Mishra 2018). The central, as well as state governments, has undertaken various schemes for providing safe drinking water

M. K. Sharma (✉) · R. Singh · O. Singh
National Institute of Hydrology, Roorkee, Uttarakhand 247667, India
e-mail: sharmamk.1967@gmail.com

R. Singh
e-mail: rsingh.nih@gmail.com

O. Singh
e-mail: omkar.nihr@gov.in

D. G. Durbude
WALMI, Aurangabad, India
e-mail: dgdurbude71@gmail.com

A. Pandey et al. (eds.), *Climate Impacts on Water Resources in India*, Water Science and Technology Library 95, https://doi.org/10.1007/978-3-030-51427-3_2

to the citizens since independence, however, due to the high population growth, the infrastructure for providing safe water and treatment of effluents to secure the water resources have always been lagging behind requirement. In addition, the average per capita water availability has reduced steadily due to the continuous growth of country's population, and it has been estimated that India will become water stressed country by year 2025 (IDSA Task Force Report 2010).

The health burden on the state exchequer due to poor water quality is enormous. Ingestion of contaminated water with pollutants results in water-borne diseases. Water gets contaminated either at source or while moving through poorly laid and maintained water transmission pipes, or in the homes due to improper storage and unhygienic practices. Mass level Jaundice has been reported due to influx of pollutants/bacteria in the drinking water of Shimla City during 2007. This study was undertaken to identify the sources of contamination and come up with options for avoiding the contaminant ingress into the drinking water sources of Shimla City.

2.2 Study Area

Shimla, the capital city of Himachal Pradesh, is located at 31°6′ North latitude and 77°13′ East longitude (Fig. 2.1) at 2130 m above mean sea level. Shimla City experiences four seasons, stretching almost equally in a year. First quarter (January–March) is snowy and stormy, second quarter (April–June) is dry and sunny, third quarter (July–September) is rainy and damp, and the last quarter (October–December) is bright and clear. Winters are very cold, and the chilly winds from the upper Himalayas make it colder. The city receives snowfall in the last week of December, around Christmas. Temperature of the city varies from 0 to 27 °C. In summers, the weather remains very pleasant with temperature ranging from 15 to 27 °C, and in winters, the temperature remains in the range of 0 to 17 °C. The geology of Shimla region comprises of 0.15 m thin soil layer on the ridges and around 7 m deep soil layer in valleys, an intervening layer of detritus, and hard bedrock. Main drinking water supply is made through Ashwani Khad and Dhalli water supply schemes. Ashwani Khad WTP receives water from Ashwani Khad, a natural stream, and from bore wells to full fill the demand. After different stages of treatment, the treated water is mixed with groundwater during the lean period, and supplied to consumers after chlorination. Dhalli WTP receives water from Churat Nallah and Sayog catchment and supplied to consumers after treatment.

2.3 Methodology

The methodology adopted for identifying the sources of drinking water contamination is described as below:

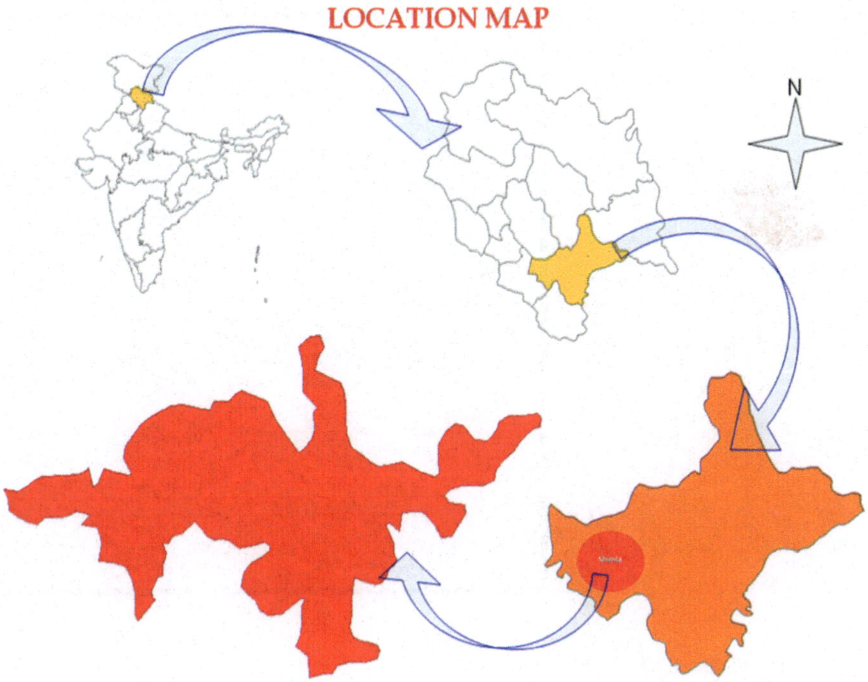

Fig. 2.1 Location map of Shimla city

(1) Generation of basin characteristics maps using ILWIS GIS/ERDAS, followed by hydrological and basin characterisation.
(2) Monitoring and analysis of drinking water sources, water supply lines, and sewage effluent on quarterly basis.
(3) Analysis of sewage influx in drinking water and efficacy of existing sewerage network in the part of Shimla city (Sanjauli Malyana) using SewerCAD Software (Bentley Systems 2008).
(4) Assessment of possible impact of sewage effluent/toxicants in drinking water sources based on water quality assessment using BIS/WHO standards.

The water samples, from the drinking water resources, natural drains, WTPs, water transmission break points, user points, and STP, were collected in the pre- and post-monsoon season of 2010–2011, and every month in 2011–12. The samples were preserved and analyze as per standard procedures (APHA 1995). All the chemical used during sampling and analysis were Merck make analytical grade.

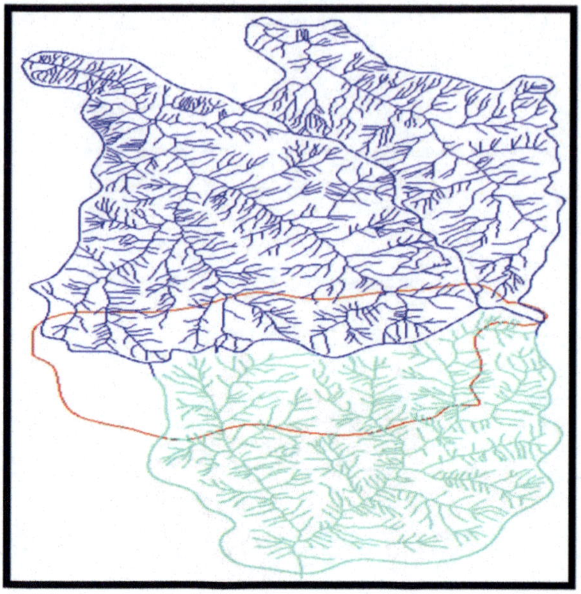

Fig. 2.2 Drainage map of Shimla city: partly in Satluj (above) and partly in Yamuna Basin (below)

2.4 Results and Implications

2.4.1 Basin Characteristics of Shimla City

The basin characteristics of the Shimla city were evaluated by digitizing the drainage area of Shimla city followed by preparing the digital elevation model (Figs. 2.2 and 2.3). The city lies partly in Satluj river basin and partly in Yamuna river basin, bifurcated by the mall road and the ridge road. The morphometric characteristics of stream (linear, aerial and relief aspect) of study area were analyzed and are presented in Table 2.1 (Tandon 2008). The Sanjali-Malyana region of the city lies in the Yamuna sub-basin with 91.97 km^2 drainage area with fifth order stream and drainage density 3.25.

2.4.2 Application of SewerCAD Software

SewerCAD is an analysis tool for modeling sanitary sewage collection and pumping systems. It is utilized for simulating and computing the sanitary loads during dry and wet-weather sources. The software was used in this study to investigate the efficacy of the existing sewerage system and find out the faults. It was observed from the runs that the existing sewerage network is adequate for designed sewage

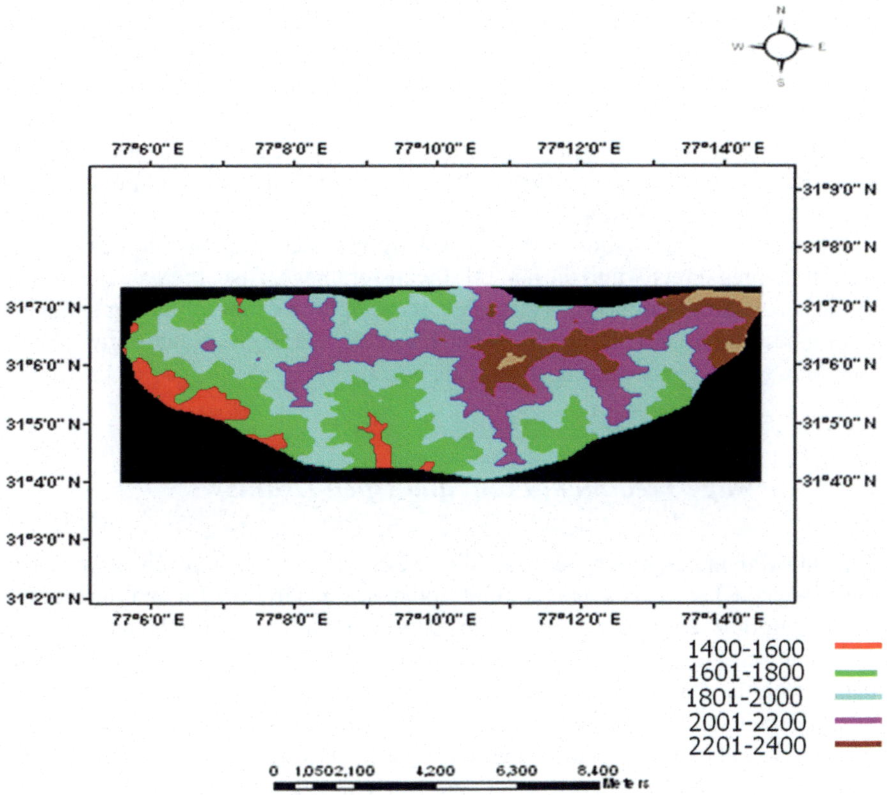

Fig. 2.3 Digital elevation model of Shimla city

Table 2.1 Morphometric characteristics of stream of study watersheds

Parameters	Watershed (Yamuna basin)	Sub-watershed (Sanjauli-Malyana)
Total no. of streams	465	166
Total length of streams (Km)	298.8	80.9
Watershed area (Km^2)	91.97	30
Drainage density (Km/Km^2)	3.25	2.67
Total relief (m)	1000	600

load and the elevation profile indicated smooth flow of sewage. Further, only 25–30% habitation was connected to the sewerage system ruling out any possibility of overflow. However, during field observations it was observed that some manholes were overflowing due to blockage by polybags/jute bags, crisscrossing of sewerage and drinking water lines, leakage in drinking water supply lines, and temporary provisions for arresting the leakages, which may lead to the ingress of polluted water into the drinking water supply.

About 70–75% habitation have primeval septic systems, the overflow from which was discharged directly into the natural streams/drainage. Also, the seepage from the septic tanks into groundwater is identified as one of the reasons for the contamination of groundwater as well as surface water resources, leading to water-borne diseases and more expensive treatment of drinking water.

2.4.3 Sewage Treatment Plant and Open Drains

Sanjauli-Malyana region, where Jaundice cases were reported, receives water from the Ashwani Khad water supply system, commissioned in 1992 for treating the water from a natural stream. However, over the period of time due to the change in the land use pattern in the upper stretches of the stream, the flow reduced and was contaminated with the domestic sewage. The WTP comprises of alum assisted entrapment of suspended solids, followed by sedimentation, filtration, and disinfection. During the lean period and high demand, groundwater is used to supplement the deficit by pumping directly in the treated water tank where it gets chlorinated. The disinfected water is pumped to the Kawalag storage tank and then to the Kusumpti tank where it is re-chlorinated before distribution.

The stream from which the water is pumped to Ashwani Khad WTP receives treated water from Sanjauli-Malyana STP and three natural drains namely housing board colony (Sanjauli) drain (polluted with sewage from the colony), Sanan open drain (polluted with sewage from Nawbhar, Chamyana, and Sanan), and Shivmandir (Malyana) drain (polluted with sewage). Accordingly, the samples were collected and analyzed from these sources to estimate the extent of pollution.

The installed STP is designed to treat 4.4 MLD sewage with 375 mg/l BOD, and produce treated water with BOD and TSS level less than 30 mg/l and 50 mg/l, respectively. It is based on the extended activated sludge process, comprising of bar screen, grit chamber, aeration tank, secondary clarifier, solids contact clarifier, and chlorine contact tank. The solids contact clarifier has been provided for removal of organics by adsorption on chemical sludge escaping from the biological reactor in the winter season when the activity of microorganisms is minimum. It was observed that the treatment units are adequate to treat the designed load, however, the treated water analysis results indicated partial treatment of sewage resulting in pollution of the natural stream. The reduced treatment efficiency of the STP was due to operational issues.

COD and nitrate concentration of open drains were observed in the range 100–400 mg/l and 15–40 mg/l respectively, indicating a substantial amount of sewage/human waste due to open defecation and absence of sewerage lines in the adjoining habitations like Dhingoo, Engine ghar etc. In addition, the kitchen and bathroom drains were not connected to sewerage lines and are discharged directly in the natural drains. It was also observed that the drains were flooded with garbage, which slowly degrades and provides media for microorganisms to flourish and ultimately polluting the stream. Further, despite a full-fledged solid waste management plant, it was observed that the solid wastes were dumped without any treatment near the plant without any engineered structure. It is expected that the leachates from this facility may pollute the water resources.

2.4.4 Water Treatment Plants and User Point

In order to trace the locations and causes of contamination during treatment and supply, samples were collected and analyzed from the inlet and outlet of WTPs, lifting stations, and user points. Dhalli and Ashwani Khad WTPs supply water to the affected areas and hence sampling was done from these two plants.

Dhalli WTP receives around 0.1 MLD water from Sayog catchment and 2 MLD water from Churat Nallah. The Sayog catchment is densely forested without any habitation due to which the water is expected to be pathogen free and without any contaminants, therefore, the water from this catchment is treated through slow sand gravity filters followed by chlorination. The water pumped from Churrat nallah is treated with the aid of alum and lime, flowed by sedimentation, filtration, and chlorination.

Ashwani Khad WTP is designed to treat 10 MLD water sourced from Ashwani Khad, which has six main tributaries namely—Malyana Nallah, Sanan Nallah, Housing Board Nallah, Jagroti Nallah, Koti Nallah, and Bharandi Nallah. Most of these nallah passes through densely populated habitations and one, Malayan Nallah, carries STP treated water. These nallah gets polluted due to the influx of greywater, outflow of numerous domestic septic tanks, and solid waste. The water is treated through flocculator and sedimentation tank, followed by filtration through rapid sand filter, and disinfection by chlorine. During the lean/high demand period, filtered water is blended with groundwater in the filtered/clear water tank. During the design phase, chlorine dosing was planned at inlet which later on shifted to the filter water tank. Based on the field survey and water quality analysis results following conclusions can be inferred:

(1) The input to both the WTPs, except Sayog catchment, is contaminated with organics as well as bacteria. The quality of input to Ashwani Khad deteriorates further in the lean period (summer season) due to the reduction in base flows. The presence of organics and fecal coliform requires continuous operator attention for proper disinfection. The low molecular weight organic compound and nutrients, as a result of chlorination, promote bacterial growth (Saeed et al. 2000),

and therefore, absence of residual chlorine at any moment, from treatment to supply, will lead to disaster.

(2) Dhalli and Ashwani Khad WTPs are designed for the treatment of surface water with practically little contamination and for removal of turbidity, however, the input quality is significantly contaminated, requiring upgradation of the WTPs.

(3) For such type of waters, solids contact clarifier with an option for in built sludge recirculation should be used. The coagulated and flocculated water is filtered through a metal hydroxide sludge blanket in the clarification zone of the clarifier resulting in efficient removal of organics as well as microbes, if operated properly.

(4) Installation of Ultra Filtration (UF) membranes for final polishing of treated water from Ashwani Khad WTP is recommended. UF membranes are a physical barrier for bacteria, viruses, and high molecular weight organic compounds. Reduction of these components will bring down the chlorine demand and also trihalomethanes in the treated water which are carcinogenic compounds. Installation of UF will ensure safe water for the citizens.

(5) Detailed study with ozone for oxidation of organics as well as microbes is suggested. Full scale system should be installed only after six months' field trials.

(6) Due to proper chlorination, in fact high chlorination, the quality of water samples was almost similar at municipal distribution tank and end user tap, free from bacteria.

Three user point samples were collected from each drinking water supply line of the affected area. The analysis of these samples confirmed the presence of free residual chlorine (FRC) and absence of microbial contamination up to the tail end. The treated water from Ashwani Khad WTP was observed to be high chlorine demanding due to the presence of organics in the treated water. The FRC reduced from 25 mg/l at WTP to 1 mg/l at the user point, in spite of re-chlorination at Kusumpti pumping station (Fig. 2.4).

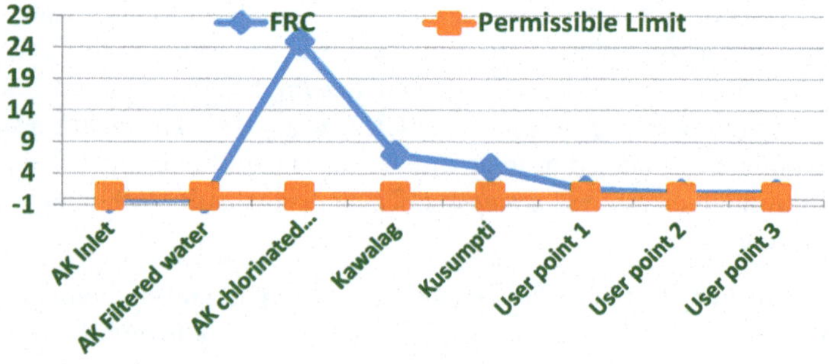

Fig. 2.4 Free residual chlorine profile from WTP to user point

From the reduction in COD, 35–50 mg/l in filtered water to 15–25 mg/l after chlorination, approximately 50 mg/l chlorine consumption was estimated at WTP. Moreover, 25 mg/l residual chlorine was detected in outlet of chlorine contact tank/storage tank, adding this to the chlorine consumed, the dosage comes out 65–75 mg/l. The reaction between organics and chlorine resulted in formation of organic halides, a known carcinogen. The presence was confirmed by FTIR Spectra. However, with present set-up, it is not possible to cut down on chlorine dosage to eliminate the organic halides formation, as it will result in elevated microbial growth.

2.5 Conclusions and Recommendations

(1) The sewerage system is adequate for the designated load and elevation profile indicates smooth flow of sewage.

(2) Contamination of natural stream supplying water to Ashwani Khad WTP is due to poor connectivity of habitation with the sewerage network, discharge of greywater to natural drains, open defecation, improper solid waste dumping, and reduced efficiency of Malyana STP.

(3) Ashwani Khad WTP is designed to remove suspended solids from the input water free of organics and pathogens. Since the inlet water quality has changes appreciably, the WTP needs to be upgraded by installing solids contact type Clarifier / Actiflow for removing the organics and microbes through coagulation, flocculation, and sedimentation process for providing safe water.
It is also recommended to install a physical barrier such as ultra-filtration (UF) membrane, which can remove almost 99.9% bacteria and viruses. Installation of the UF membrane will also help in reducing the organic content of the treated water.
It is further recommended to explore advanced oxidation techniques, like ozone-based oxidation, for the oxidation of organics which skip the Ashwani Khad water treatment scheme by converting them into inorganic forms. However, free residual chlorine should be dosed to maintain minimum concentration at the user end.

(4) Shimla Municipal Corporation should ensure connecting 100% habitation to the sewerage system for minimizing the contamination of the natural water resources.

(5) In absence of the transportation facility for the sludge in the septic tank, the content in the tank is discharged to natural drains in the monsoon season, therefore, the septic tank system needs to be discouraged and sewerage connections should be encouraged.

(6) It is recommended to incorporate an anoxic tank in the STP for reducing the nitrate concentration in drinking water supplied from Ashwani Khad WTP.

(7) The treated water from the secondary clarifier of STP should contain 1 mg/l dissolved oxygen, and this can be achieved by continuously recycling the settled sludge in the secondary clarifier along with periodic removal of biomass from

the system. This will help in reducing the suspended solids being escaping from the secondary clarifier and in turn the chlorine demand.

(8) Performance evaluation / audit of WTPs as well as STPs should be done in order to improve the performance.

Acknowledgement This work is a part of the Purpose Driven Study titled "Impact of sewage effluent on drinking water sources of Shimla city and suggesting ameliorative measures" sponsored by HP-II project and hereby acknowledged.

References

APHA (1995) Standard methods for the examination of water and wastewater, APHA, In: Clesceri LS, Greenberg AE, Trussel RR, Washington DC

Bentley Systems (2008) User manual on sanitary sewer design and modeling—applying SewerCAD, Version V8 XM, Watertown, CT 06795, USA

IDSA Task Force Report (2010) Water security for India: The external dynamics. Institute for Defense Studies and Analyses, New Delhi

Mishra A (2018) Right to water as a human right and Indian constitution: an analysis of various judgments of Apex Court of India. IOSR J Human Soc Sci 23(4):45–48

Saeed MO, Jamaluddin AT, Tisan IA, Lawrence DA, Al-Amri MM, andChida K. (2000) Biofouling in a seawater reverse osmosis plant on the red sea coast—Saudi Arabia. Desalination 128(2):177–190

Tandon M (2008) Catchment sensitive planning for sustainable cities. ITPI J 5(3):21–26

Chapter 3
Water Quality Status of Upper Ganga Canal

Rajesh Singh, Daniel Kanbienaa, and Ashish Pandey

Abstract Water pollution is a global challenge calling for attention. The Upper Ganga Canal (UGC) System which plays a significant role in Northern India, as a source of water for irrigation, drinking water, and spiritual upliftment, is also not spared. Due to the increasing population in the vicinity of canal, the quantity of untreated effluent being discharged into the canal is also increasing, leading to deterioration in the quality of canal water. The main objective of this study was to assess the status of UGC for agricultural application using Water Quality Index (WQI). Irrigation with poor quality water may cause salinity, specific ion toxicity, or infiltration problems in soils. In this regard, forty water samples were collected from five locations during July 2015 to February 2016. The samples were investigated for physico-chemical parameters and trace metals. The irrigation water quality parameters for deciding the suitability of water was computed from the analyzed parameters. The results were integrated in the calculation of the WQI of the UGC water for irrigation purposes, considering the permissible values recommended by Bureau of Indian Standards and Food and Agriculture Organization of United Nations. The calculated WQI varies from 37 to 42 during the study period, with an overall WQI of 38.22. The estimated WQI suggests that, the canal water quality is good for irrigation purposes throughout the study period.

R. Singh (✉)
Environmental Hydrology Division, National Institute of Hydrology, Roorkee 247 667, India
e-mail: rsingh.nihr@gov.in

D. Kanbienaa · A. Pandey
Department of Water Resources Development and Management, Indian Institute of Technology Roorkee, Roorkee 247 667, India
e-mail: dannykan110@yahoo.com

A. Pandey
e-mail: ashish.pandey@wr.iitr.ac.in

A. Pandey et al. (eds.), *Climate Impacts on Water Resources in India*, Water Science and Technology Library 95, https://doi.org/10.1007/978-3-030-51427-3_3

3.1 Introduction

Water pollution has become a global challenge in recent times; man has proven to be the foremost cause and victim. In India, the scenario is no different, most water bodies are polluted due to rapid population growth, urbanization, and anthropogenic activities (Goel 2009). The world is at a crossroad to device simple, but economic and efficient ways of monitoring, analyzing and communicating water quality information for understanding of the masses and policy makers. Water Quality Index (WQI) is proven to be one of the tools that could be used as a quality barometer in reporting and communicating water quality issues devoid of complex jargons (Horton 1965). Historically, Horton was the first to propose WQI in 1965 (Boyacioglu 2007), and since then several attempts have been made by many scholars to modify the brain child of Horton's and its application in the assessment of water quality (Bhargava 1983; Smith 1990; Chetana et al. 1997; Stambuk-Giljanovic 1999; CCMEWQI 2001; Cude 2001; Hebert 2005; Saeed et al. 2010; Singh et al. 2017). The weight of water quality parameter is generally dependent on personal opinion and discretion for many WQI models. Tiwari and Mishra (1985) propounded water quality index that employs the relative weight (RW) of a parameter as a function of its recommended permissible water quality standards; thus has the advantage of eliminating the uncertainty that will emanate from personal opinions and judgments. The UGC, an irrigation canal, is predisposed to pollution from human activities such as washing of clothes, swimming, dipping of ashes, dumping of idols, defecating, disposal of untreated domestic waste and cattle wading. Agriculture in this area is mainly dependent on irrigation, therefore, in-depth knowledge of the suitability and usability of the water for such purpose is imperative. Nonetheless, literature revealed limited and patchy information on the use of WQI in the assessment of water quality of the UGC water for irrigation purposes. In this study, the WQI model by Tiwari and Mishra (1985) has been applied for calculation of the WQI of the UGC system, in an attempt to assess the pollution status of the UGC. The findings would provide a simple, but better understanding of the quality status of the UGC water to farmers in decision making; and serve as a benchmark for future studies.

3.2 Study Area

Roorkee lies at a latitude of 29.8° N and longitude of 77.8° E and located in the District of Haridwar, Uttarakhand; with a population of 289,478 as of 2011 National Census of India. The UGC has its origin from the mythological Ganga River and it takes off from the right flank of Bhimgoda Barrage. The UGC flows from north to south and bisect Roorkee into two halves. The upstream of the canal for the purpose of this study is the Solani Park and Gaddajuda minor becomes the downstream. The midstream points are Asaf Nagar command area, Kurdi and Manglaur. ArcGIS 10.2 version has been used to present a synoptic view of the study area (Fig. 3.1).

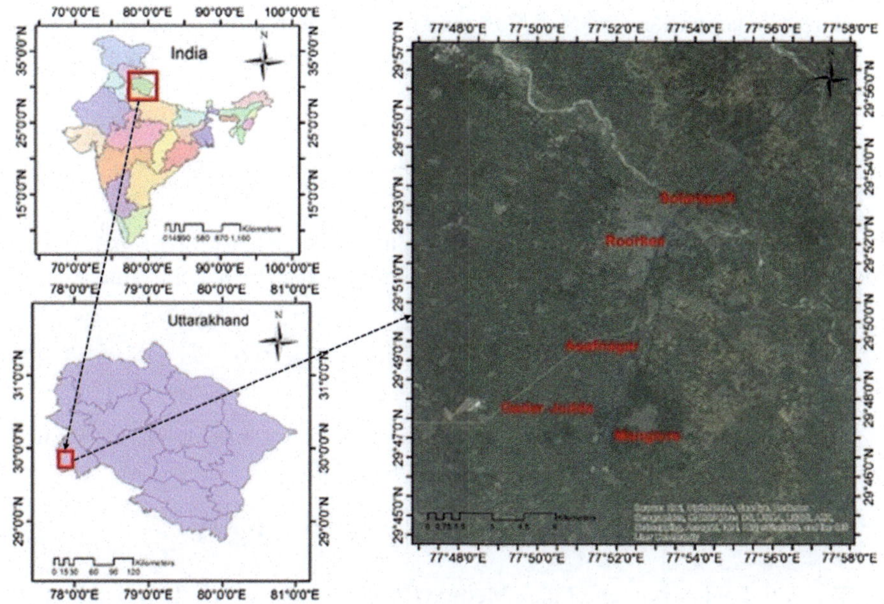

Fig. 3.1 Synoptic view of the study area and sampling location

3.3 Materials and Methods

To assess water quality status of the UGC, 40 water samples were collected and analyzed for physico-chemical parameters (pH, EC, TDS, Ca^{2+}, Na^{+}, Mg^{2+}, HCO_3^{-}, K^{+}, Cl^{-}, CO_3^{2-}) as per the standard methods for water quality analysis (APHA 1992). Samples were collected using DO sampler at a depth of 30 cm, and stored in a 1000 ml airtight, well labelled plastic containers, after 2 to 3 times rinsing with the water to curtail potential errors and heterogeneity (De 1989). Air and water temperature, EC, TDS, DO, and pH were measured in-situ using Portable Multi Parameters Meter, ORLAB MODEL. Ca^{2+}, Mg^{2+}, Cl^{-}, HCO_3^{-} and CO_3^{2-} were analyzed by titration; K^{+} and Na^{+} were determined by flame photometer −128. Samples were digested (APHA 1992) and trace metals (Aluminium (Al), Silver (Ag), Arsenic (As), Copper (Cu), Beryllium (Be), Lithium (Li), Cadmium (Cd), Chromium (Cr), Cobalt (Co), Lead (Pb), Manganese (Mn), Nickel (Ni), Selenium (Se), Vanadium (V), Iron (Fe), Zinc (Zn)) were analyzed using Inductively Coupled Plasma Mass Spectrometry (ICP-MS). Sodium Adsorption Ratio (SAR), the relative proportion of sodium to Calcium and Magnesium in irrigation water was calculated using the equation (Richards 1954).

$$\text{SAR} = \frac{\sqrt{2} \times \text{Na}^{+}}{\sqrt{\text{Ca}^{2+} + \text{Mg}^{2+}}} \tag{3.1}$$

Here, Concentrations for Na^{+}, Ca^{2+} and Mg^{2+} were in milliequivalent per litre.

Sodium Percentage (Na %) was calculated using Doreen expression by using the concentration of ions in me/l (Todd 1980).

$$Na\% = \left[\frac{Na^{+} + K^{+}}{Ca^{2+} + Mg^{2+} + Na^{+} + K^{+}}\right] \times 100 \tag{3.2}$$

Permeability Index (PI) determine sodium hazard/sodicity of irrigation water. Doneen (1964) equation was employed to determine the Permeability Index as

$$PI = \left[\frac{Na^{+} + HCO_3^{-}}{Ca^{2+} + Mg^{2+} + Na^{+}}\right] \times 100 \tag{3.3}$$

Residual Sodium Carbonate is used to ascertain the amount of carbonate in irrigation water by using the formula by Eaton (1950).

$$RSC = \left[CO_3^{2-} + HCO_3^{-}\right] - \left[Ca^{2+} + Mg^{2+}\right] \tag{3.4}$$

The Weighted Arithmetic Index Method (WAIM) proposed by Tiwari and Mishra (1985) was employed to determine the overall irrigation water quality index for the UGC water. Salinity hazard is represented in the WQI by electrical conductivity (EC); specific ion toxicity such as chloride and sodium ions are incorporated in the index. Permeability Index (PI), Residual sodium carbonate and percent sodium are also included in the Index. Miscellaneous hazards such as carbonate and bicarbonates, nitrate, and pH are also included in the index calculation. The weight of each parameter is dependent on its standard permissible limit for the designated use. The quality rating for each water quality parameter Qi is determined by the expression.

$$\text{Qi} = \text{qi} \times \left[\frac{\text{Xn} - \text{Xi}}{\text{Xs} - \text{Xi}}\right] \tag{3.5}$$

qi is the maximum value of Qi.

Xn is the analyzed/observed value for the parameter.

Xs is the recommended permissible standard value for the parameter (BIS 2001; Ayers and Wescot 1985).

Xi is the ideal value of the parameter, which is usually 0, except for pH and DO mg/l.

The Relative Weight (RW) is inversely proportional to the recommended water quality standard (Si) corresponding to the parameter-

$$RW = \frac{K}{Si} \tag{3.6}$$

K is a constant and equals to 1.

The overall Irrigation Water Quality Index (IWQI) was calculated by the expression

$$IWQI = \frac{\sum QI \times RW}{\sum RW} \tag{3.7}$$

3.4 Results and Discussion

The irrigation water quality and suitability of the study area was evaluated based on the pH, salinity hazards (EC), Sodium hazard (SAR), permeability index (PI), percent sodium (Na %), Residual Sodium Carbonate (RSC) and Water Quality Index (WQI). The pH values ranges from 7.3 to 8.2 with a mean value of 7.75 during monsoon season and 6.5 to 8.02 with a mean value of 7.26 in winter season and are presented in Figs. 3.2 and 3.3. The relatively high pH value in rainy season may be due to the influx of pollution due to runoff. However, the pH values in both seasons are within the permissible range of irrigated farming (BIS 2001). Sakthivel (2007) reported a pH range of 6.5 to 8.05 from Yamuna and Agra Canals; whiles pH value of 5.56 was observed for the Hindon river.

The EC values were increased from upstream to downstream of the canal, the lowest 181.4 μS/cm and highest 1020 μS/cm EC values were recorded at Solani (upstream) and Gaddajuda minor canal (downstream), respectively, with a mean value of 519.1 μS/cm in monsoon season. In winter season, EC varies from 168.4 μS/cm to 396 μS/cm with a mean value of 282.2 μS/cm. The significant difference in the EC value within and between the seasons is an indication of the amount of external input into the canal mainly at the downstream. This was also observed during the collection of water samples from the Gaddajuda minor canal. The results

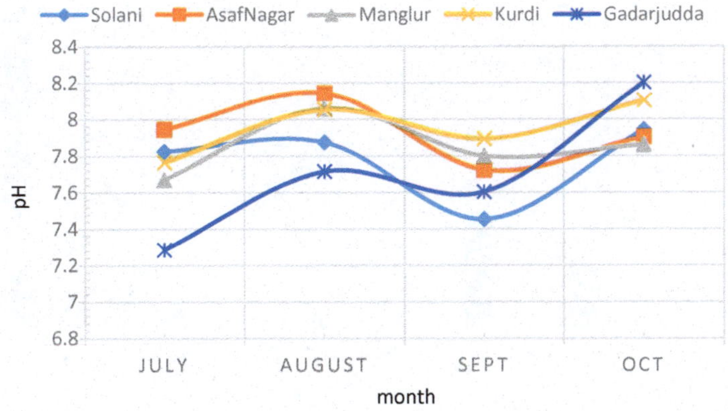

Fig. 3.2 Monthly Trend Fluctuation in pH (monsoon season)

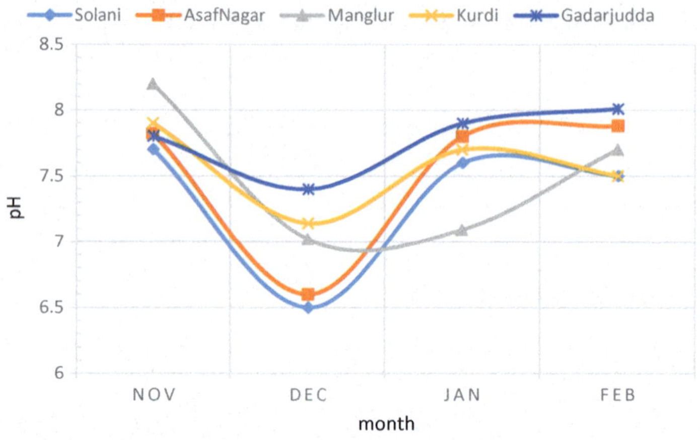

Fig. 3.3 Monthly Trend Fluctuation in pH (winter season)

revealed that based on EC values the canal water quality is suitable for irrigation in both seasons (BIS 2001). Matta et al. (2014) reported a maximum EC value of 145.4 μS/cm in monsoon season at Bahadrabad, while EC values of 38.0 to 170.0 μS/cm with a mean of 85.2 μS/cm was reported from Ganga River water in Rishikesh (Haritash et al. 2014). The Sodium Adsorption Ratio (SAR) was calculated using Eq. (3.1). The SAR values varies from 0.04 to 3.2 in monsoon season of the year 2015 and it varied from 0.14 to 1.2 during winter season, with 0.6106 as the mean value (Figs. 3.4 and 3.5). Percent Sodium (%Na) was computed in relations to cations present in the canal water and concentration of ions expressed in milliequivalents per litre. The results revealed that %Na ranges from 10 to 51 with a mean value

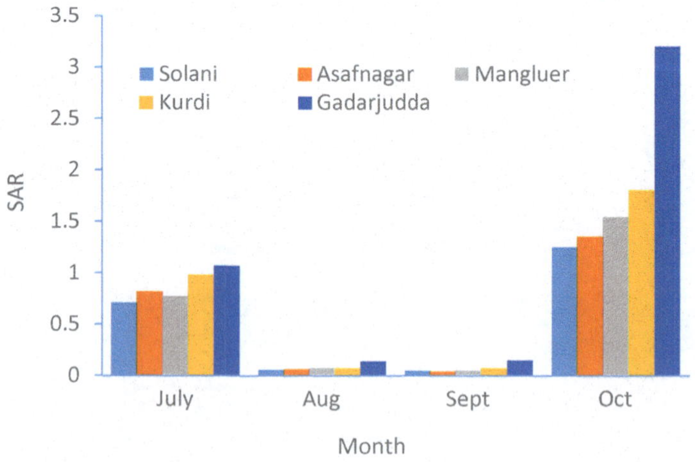

Fig. 3.4 Monthly trend fluctuation in SAR (monsoon season)

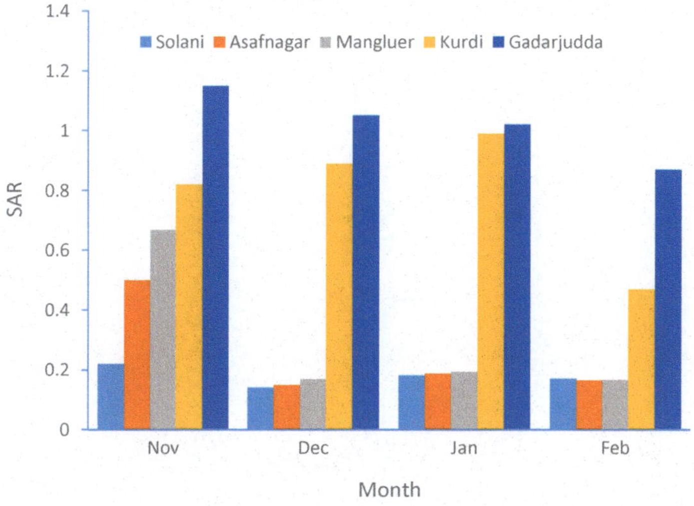

Fig. 3.5 Monthly trend fluctuation in SAR (winter season)

of 30.5 in rainy season compared to a range of 12–45 and mean of 28.5 in winter (Figs. 3.6 and 3.7). In monsoon, 30, 60 and 10% of the 20 samples show excellent, good and permissible classes respectively. However, in winter, 40, 50 and 10% of the 20 samples show excellent class, good class, and permissible accordingly. Joshi et al. (2009) reported %Na values between 23.56 and 52.35 from the Ganga River at

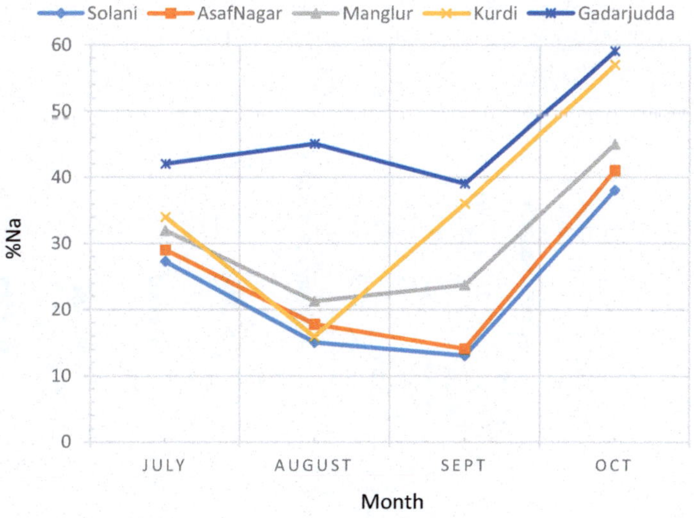

Fig. 3.6 Monthly trend fluctuation in %Na (monsoon season)

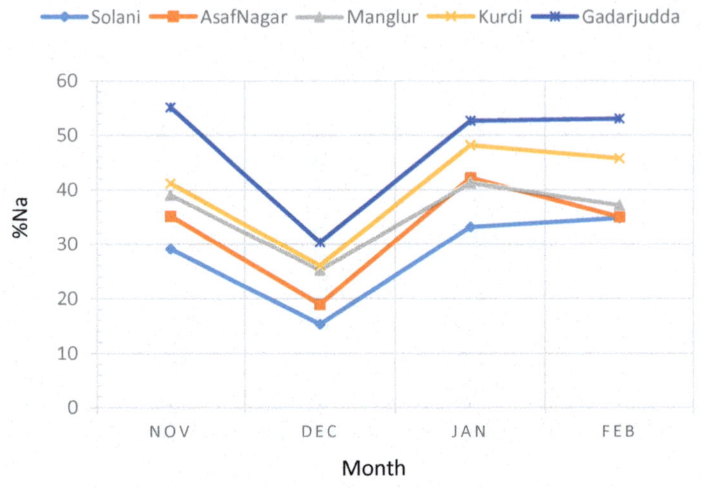

Fig. 3.7 Monthly trend fluctuation in %Na (winter season)

Haridwar. According to Richard (1954), the canal water is suitable for irrigation in both seasons.

Wilcox (1955) propounded an expression for evaluating residual sodium carbonates (RSC). Out of the 40 samples collected and analyzed for RSC in both seasons, 80% shows excellent, 15% reveals good and 5% depicts medium RSC hazard at the downstream of the canal in rainy season, whiles 90% of the sample analyzed can be described as excellent and 10% lie within the good class (BIS 2001). Joshi et al. (2009) observed RSC values less than 1.25 meq/lit from the Ganga River at Haridwar and concluded the water was suitable for irrigation purpose (Fig. 3.8).

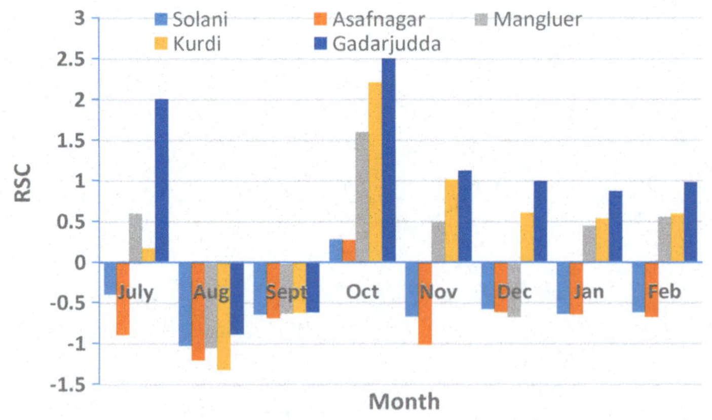

Fig. 3.8 Monthly trend fluctuation in RSC

Permeability Index (PI) was calculated using Eq. (3.3). The value of PI decreases from the upstream to downstream of the canal; which can be fathomed that there is some external pollutants that joins the flow from the Solani Park to Gaddajuda. The lowest and highest recorded PI values were 64 and 92, respectively (Figs. 3.9 and 3.10). The canal water quality falls under good to excellent category in relation to permeability Index (PI). Haritash et al. (2014) reported a similar trend of pollution

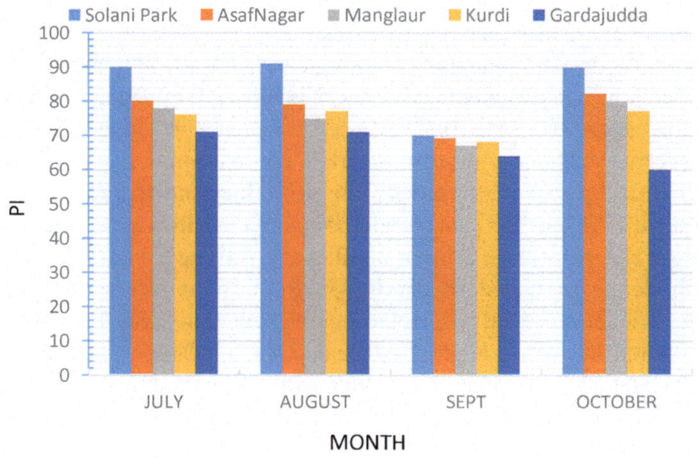

Fig. 3.9 Monthly trend fluctuation in PI (monsoon season)

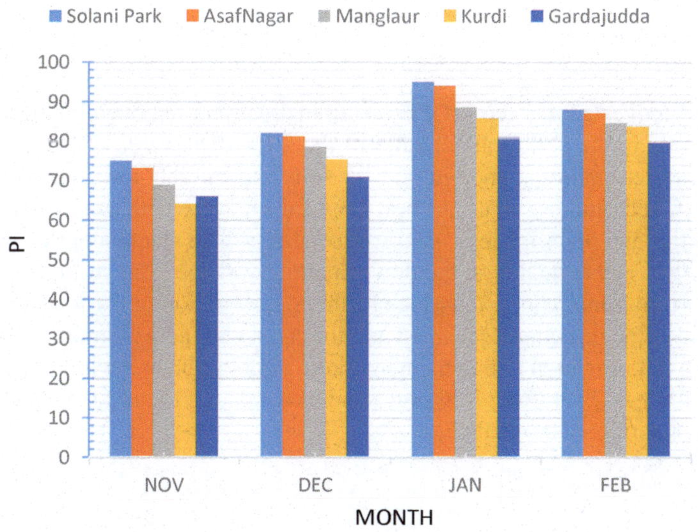

Fig. 3.10 Monthly trend fluctuation in PI (winter season)

in Ganga River, at Rishikesh. The minimum PI value of 62.57 was reported by Joshi et al. (2009), for Ganga River in Haridwar.

The Eqs. 3.5 and 3.6 were used to obtain water quality rating (Qi) and relative weight of parameter (RW) as provided in Table 3.1. Equation 3.7 was utilized to calculate the WQI. The values of WQI vary from 37.7 to 42 in both monsoon and winter seasons, with overall WQI of 38.22 (Fig. 3.11). The WQI shows that, the canal water quality is good for irrigation purposes in both the seasons. However, water at the Gaddajuda minor is recommended for use in soils with a light texture, permeable and on crops that are less sensitive to salinity .

The Pearson correlation analysis technique was employed to assess the degree of pollution and how one parameter existence depends and influences other parameters. A regression factors of 0.5 and -0.5 were used to represent positive and negative correlation respectively. This was done to ascertain the relation if, any among the various physico chemical parameters studied. The pH values showed a moderate negative correlation with TH and Ca^{2+}, but correlated positively with Mg^{2+} in monsoon season. The values of EC, TDS, and TH had a significant positive correlation with SAR, %Na, RSC but moderate correlated negatively with PI during monsoon season. Also, Ca^{2+} and Na^{+} has moderate negative correlation with Mg^{2+} and RSC, but showed a strong positive relationship with PI. The SAR and %Na values correlated positively with RSC but registered a negative association with PI in the rainy season (Tables 3.2 and 3.3). The strong relation among the various studied parameters show that these parameters do not only have the same input source but akin geochemical composition and behavior.

3.5 Conclusions

The study suggested that the WQI is useful in assessing the overall pollution status; and effective in the pollution zonation of the UGC. The value of WQI reveals that the canal water quality is good in both seasons and suitable for irrigation purposes. Water in Gaddarjudda canal can best be used on soils with good structure; high permeability and on crops less sensitive to salinity. The study suggests, continuous monitoring of water quality at the downstream. Moreover, the vegetation should be removed from minor canals to reduce further deterioration of the water quality. The household effluents being discharged into the canal will increase multifold in the near future due to population rise in the neighboring habitation and will in turn affect the quality of water. Therefore, it is suggested to discharge the effluents into the canal after removal of contaminants.

Table 3.1 Calculation of the WQI for the Upper Ganga canal water

Parameter	Mean tested result	Permissible value (BIS 2001; Ayers & Westcot 1985)	Relative weight (RW)	Water quality Rating (Qi)	Weighted Qi rating (WR × Qi)
pH	7.6	8.4	0.12	90.5	10.8
EC	338.2	2250	0.0004	15.0	6×10^{-3}
TDS	216.5	2100	0.0005	10.3	5.2×10^{-3}
Ca^+	24.3	200	0.005	12.2	0.1
Na^+	11.8	200	0.005	5.9	0.0
Mg^+	31.4	50	0.02	62.8	1.3
SAR	0.72	26	0.04	2.7	0.1
%Na	32.3	60	0.02	53.8	1.1
RSC	0.17	2.5	0.4	6.8	2.7
PI	91.8	25	0.04	367.2	14.6
Fe	0.32	5.0	0.2	6.4	1.3
Mg	4.7	75	0.0	10	0
Pb	0.0016	5.0	0.2	0.032	0.0
Al	0.17	5.0	0.2	3.4	0.7
As	0.0018	0.1	10	1.8	18.0
Cr	0.035	0.1	10	0.0035	0.0
Cd	0.0003	0.01	100	3	300.0
Ni	0.018	0.2	5	9	45.0
Hg	0.005	0.001	1000	500	50,000
Cs	0.0002	0.01	100	2	200
Mn	0.0088	0.2	5	4.4	22.0
Rb	0.013	0.2	5	6.5	32.5
Ag	0.0014	0.1	10	1.4	14
Co	0.0005	0.05	20	1	20.0
Li	0.0029	2.5	0.4	0.1	0.0
Se	0.0009	0.02	50	4.5	225.0
Be	0	0.1	10	0	0
V	0.0015	0.1	10	1.5	15
Zn	0.0085	2.0	0.5	0.43	0.2
	$\sum \boldsymbol{MR}=$ **760.1**		$\sum \boldsymbol{RW}=$ **1332.2**	$\sum \boldsymbol{QI}=$ **1182.0**	$\sum \boldsymbol{QI} \times \boldsymbol{RW}=$ **50,920.9**

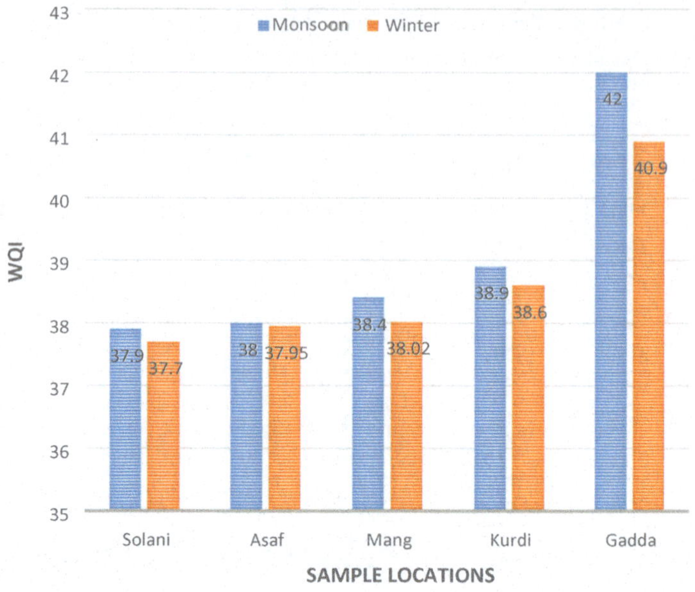

Fig. 3.11 Season—wise WQI's as per sample locations

Table 3.2 Correlation matrix of parameters in Monsoon Seasons

Para	pH	EC	TDS	TH	Ca^{2+}	Na^+	Mg^{2+}	SAR	%Na	RSC	PI
pH	1										
EC	−0.32	1									
TDS	−0.39	0.99	1								
TH	−0.63	0.84	0.89	1							
Ca^+	−0.52	0.09	0.09	0.03	1						
Na^+	−0.29	−0.27	−0.21	0.17	−0.55	1					
Mg^+	0.36	−0.01	0.01	0.18	−0.93	0.61	1				
SAR	−0.35	0.98	0.99	0.91	0.03	−0.17	0.10	1			
%Na	−0.23	0.98	0.97	0.79	0.13	−0.39	−0.03	0.97	1		
RSC	−0.38	0.91	0.91	0.74	0.44	−0.53	−0.30	0.89	0.95	1	
PI	−0.42	−0.57	−0.55	−0.32	0.14	0.57	−0.24	−0.59	−0.70	−0.61	1

Table 3.3 Correlation matrix of parameters in winter season

	pH	EC	TDS	TH	Ca^{2+}	Na^{+}	Mg^{2+}	SAR	%Na	RSC	PI
pH	1										
EC	−0.32	1									
TDS	−0.39	0.99	1								
TH	−0.63	0.84	0.89	1							
Ca^{+}	−0.52	0.09	0.09	0.03	1						
Na^{+}	−0.29	−0.27	−0.21	0.17	−0.55	1					
Mg^{+}	0.36	−0.01	0.01	0.18	−0.93	0.61	1				
SAR	−0.35	0.98	0.99	0.91	0.03	−0.17	0.10	1			
%Na	−0.23	0.98	0.97	0.79	0.13	−0.39	−0.03	0.97	1		
RSC	−0.38	0.91	0.91	0.74	0.44	−0.53	−0.30	0.89	0.95	1	
PI	−0.42	−0.57	−0.55	−0.32	0.14	0.57	−0.24	−0.59	−0.70	−0.61	1

References

APHA (1992) Standard methods for the examination of water and wastewater. American Public Health Association, Washington DC

Ayers RS, Westcot DW (1985) Water quality for agriculture. Food and Agriculture Organization of the United Nations, Rome, Italy

Bhargava DS (1983) Use of a water quality index for river classification and zoning of the Ganga River. Environ Pollut Series B 6(1):51–67

Boyacioglu H (2007) Development of a water quality index based on a European classification scheme. Water SA 33(1):101–106

BIS (2001) Irrigation water quality Guidelines and standards. Bureau of Indian Standards, India

CCME (2001) Canadian water quality guidelines for the protection of aquatic life. CCME WQI 1.0 User's Manual. https://www.ccme.ca/assets/pdf/wqi

Chetana SA, Somashekar RK (1997) sEvaluation of water quality index of the River Cauvery and its tributaries. Curr Sci 72(9):640–646

Cude CG (2001) Oregon water quality index; a tool for evaluating water quality management effectiveness. J Water Res Assoc 37(1):125–137

De AK (1989) Environmental chemistry. Wiley Eastern Ltd, New Delhi, India

Doneen L. D. (1964) Note on water quality in agriculture. Published as a Water Science and Engineering Paper 4001, Department of Water Science and Engineering, University of California.

Eaton FM (1950) Significance of carbonates in irrigation water. Springer, Berlin Publisher, Heidelberg

Goel PK (2009) Water pollution causes effects and control. New Age International Ltd., New Delhi, India

Haritash AK, Gaur S, Garg S (2014) Assessment of water quality and suitability analysis of River Ganga in Rishikesh India. Appl Water Sci 6(4):383–392

Hebert (2005) Comparison between the index of overall quality of water in Quebec, the water quality index CCMEWQI, for the protection of aquatic life, Quebic, Ministry of sustainable development, environment and parks Department in monitoring the state of environment. ISBN (2)-550-45900-8.

Horton RK (1965) An index number system for rating water quality. J Water Pollut Control Fed 37(3):300–306

Joshi DM, Kumar A, Agrawal N (2009) Assessment of the irrigation water quality of River Ganga in Hardwar district. Rasayan J Chem 2(2):285–290

Matta G (2014) A study on the physico chemical characteristics to assess the pollution status of river Ganga in Uttarakhand. J Chem Pharmaceutical Sci 7(3):210–217

Richards LA (1954) Diagnosis and improvement of saline and alkali soils. U.S Department of Agriculture Handbook, 60, Washington D.C., USA.

Saeed M, Abessi O, Sharifi F, Meragi H (2010) Development of groundwater quality index. Environ Monit Assess 163:327–335

Sakthivel R (2007) TDS levels for traditional crops, TDS—resistant crops submitted by India Water Portal on International Land Development Consultants New Delhi, India.

Singh S, Ghosh NC, Gurjar S, Krishan G, Kumar S, Berwal P (2017) Index-based assessment of suitability of water quality for irrigation purpose under Indian conditions. Environ Monit Assess 190(1):29

Smith DG (1990) A better water quality indexing system for Rivers and Streams. Water Res 24(10):1237–1244

Stambuk-Giljanovic N (1999) Water quality evaluation by index in Dalmatia. Water Res 33(16):3423–3440

Tiwari TN, Mishra M (1985) A preliminary assessment of water quality index on major Indian rivers. Indian J Environ Protect 5(4):276–279

Todd DK (1980) Groundwater Hydrology, 2nd edn. Wiley, New York

Wilcox LV (1955) Classification and use of irrigation waters (Circular No. 969). US Department of Agriculture, Washington D.C., USA.

Chapter 4
Rationalization of Water Quality Parameters for Krishna River Basin Using Multivariate Statistical Techniques and Water Quality Index

Vikas Varekar, Hasan Rameez, and Aditya Nanekar

Abstract Anthropogenic forces have led to the deterioration of water bodies over the course of time. This calls for implementation of an effective water quality management scheme. The design of an optimal water quality monitoring network (WQMN) is one of the earliest steps required towards the development of efficient water quality management scheme. One of the most important components of a WQMN is the rationalization of water quality parameters (WQPs). Since the water quality is stochastic in nature, multivariate statistical techniques are best suited to capture the variability. Krishna River is one of the major river basins of India, which has been influenced by various man-made activities. Hence, in present analysis rationalization of WQPs for the Krishna River is carried out by using factor analysis (FA)/principal component analysis (PCA), followed by estimation of water quality indices for pre and post rationalized WQPs. Water quality monitoring data for Krishna River was procured from Water Resources Information System, India, consisting 16 monitoring stations with 16 WQPs averaged over the period 2001–2010. The results showed that out of the 16 WQPs, 8 parameters were principal WQPs. The monitoring of the rationalized parameters may result in significant reduction in cost while capturing majority of the variation in water quality.

4.1 Introduction

Among the various surface water bodies like coastal waters, estuaries, rivers and lakes, rivers play a vital role in transporting or assimilating the wastewater from municipal and industrial sources along with agricultural runoff (Shrestha and Kazama 2007). Rivers are highly susceptible to pollution resulting from various natural (basin lithology, atmospheric conditions, climatic variations etc.) and anthropogenic activities (point and non-point sources of pollution, irrigation, damming etc.), leading to significant deterioration in the quality of water. This deterioration in the water quality

---

V. Varekar (✉) · H. Rameez · A. Nanekar
Veermata Jijabai Technological Institute, Mumbai 400019, India
e-mail: vbvarekar@ci.vjti.ac.in

A. Pandey et al. (eds.), *Climate Impacts on Water Resources in India*, Water Science and Technology Library 95, https://doi.org/10.1007/978-3-030-51427-3_4

may affect the socio-economic utility of the river as well as the aquatic ecosystem sustaining within. Hence, to maintain the socio-economic importance, public health and aquatic ecosystem of the river, the spatial and temporal variations in the quality of water is need to be monitored. This calls for the development of an effective water resource management system. One of the earliest necessary steps required for the development of such a water resource management system, is the establishment of an efficient water quality monitoring network (CPCB 2007). "Water quality monitoring network" means a systematic planning for collection, preservation and transportation, storage, analysis of water samples and dissemination of data for national water bodies restricted to surface and ground water in the country (MoEF 2005).

Water quality depends on various physical (temperature, colour, taste, odour, turbidity etc.), chemical (pH, alkalinity, hardness, electrical conductivity, SAR, BOD, COD, DO, major cations and anions etc.) and biological parameters (total coliform, fecal coliform, plankton, etc.). Monitoring the water quality for such large number of parameters is very tedious, time consuming and uneconomical, hence a need is felt for rationalizing the water quality parameters. Since the water quality is stochastic in nature, statistical approaches are widely considered for rationalization of water quality parameters. The use of various multivariate statistical techniques, like principal component analysis (PCA), factor analysis (FA), cluster analysis (CA) and discriminant analysis (DA), aids in the analysis of complex data matrices to understand the water quality and ecological status of the area under consideration. Further it allows the identification of possible factors that influence water systems and offers a valuable tool for reliable management of water resources (Shrestha and Kazama 2007). Hence multivariate statistical techniques have been extensively for effective rationalization of water quality parameters to capture the variability in the water quality. Among different multivariate statistical techniques available, the most widely used technique is Principal Component Analysis/Factor Analysis (PCA/FA) (Parinet et al. 2004; Singh et al. 2004; Ouyang 2005; Shrestha and Kazama 2007; Díaz and Lopez 2007; Strobl and Robillard 2008; Altın et al. 2009). Also, the interpretation of the water quality in terms of Water Quality Index (WQI) plays an important role in the management of water quality (Sahoo et al. 2015). WQI is a dimensionless quantity that considers different water quality parameters and reflects the composite influence of them. WQI is obtained by transforming and aggregating selected water quality parameters. Krishna River is one of the longest and significantly polluted rivers of India. Hence objective of the present study is to rationalize the water quality parameters for the Krishna River Basin using multivariate statistical techniques (PCA/FA), followed by estimation of water quality indices.

4.2 Study Area

In terms of the river basin area and water inflow, the Krishna River is the fourth largest river in India (Fig. 4.1). It is also the largest river in terms of storage capacity (41.80 km^3). It lies between 73°17″ to 81°9″ E and 13°10″ to 19°22″ N. The basin

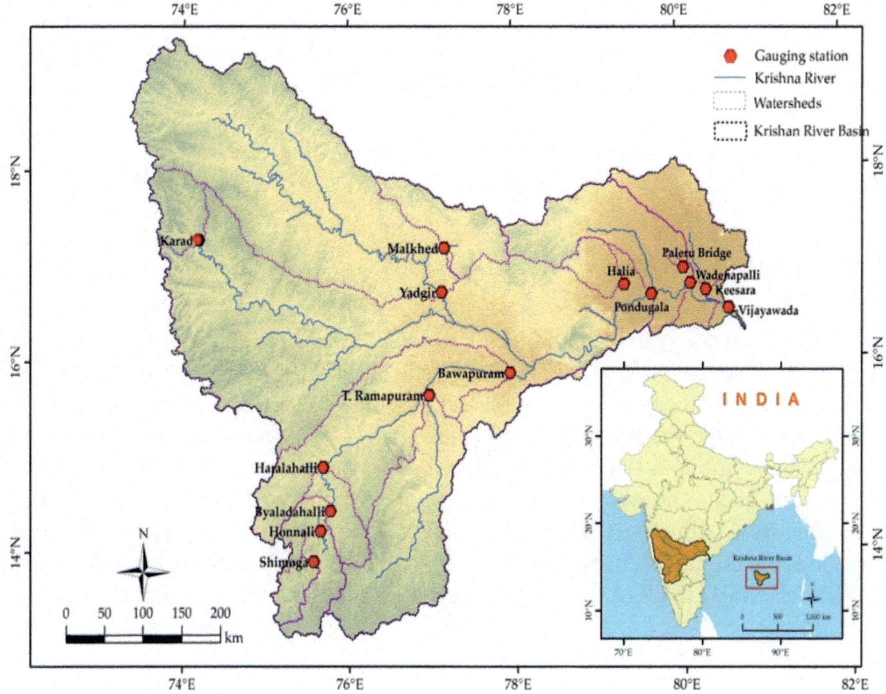

Fig. 4.1 Krishna river basin

is almost triangular shaped with periphery including Eastern Ghats on the south and the east, Balaghat range on the north and Western Ghats on the west. The river is 1400 km long, originating from the Western Ghats near Mahabaleshwar, in the state of Maharashtra and outfalls into the Bay of Bengal. It covers nearly 8% (258,948 sq. km) of the total geographical area of the country. The Krishna Basin extends over the states of Maharashtra, Karnataka, Telangana and Andhra Pradesh. "The overall catchment comprises of 7 sub-basins i.e. Krishna upper, middle and lower Basin, Bhima lower and upper basin and Tungabhadra lower and upper basin. Some of the major tributaries of Krishna river are Bhima, Musi, Tungabhadra, Malprabha etc. The majorly found soil types in the basin are black soils, red soils, laterite and lateritic soils, alluvium, mixed soils (red and black, red and yellow, etc.) and saline and alkaline soils. There are 16 multipurpose projects in the Krishna basin. Source of drinking water in the basin includes a total of 23,310 wells, 11,006 tube wells and 29,569 number of hand pumps" (WRIS, India). Also, there is a National Waterways—4 falls in this basin. Considering the development in this region, the river water quality is gradually degrading due to disposal of municipal wastewater and significant volume of agricultural runoff entering into the river (Jha et al. 2007; Varekar et al. 2015).

Krishna River and its tributaries are monitored at 16 water monitoring stations for 34 water quality parameters by government bodies (WRIS, India). As a measure towards development of an integrated water resources management program, Water Resources Information System (WRIS) was established by Indian Space Research Organization (ISRO), National Remote Sensing Centre (NRSC) and Central Water Commission (CWC). WRIS provides access to all water resources data and information in a standardized national GIS framework.

4.3 Methodology

In this study, water quality monitoring parameters are rationalized by using multivariate statistical technique (i.e. PCA/FA) and WQI. PCA/FA is implemented for the determination of the principal parameters based on the variations in water quality. A continuous water quality data for 10 years (2001–2010) was procured from WRIS, India. The continuous data was available for 16 monitoring stations, analyzed for 16 parameters each. The water quality parameters considered for the analysis were DO, pH, Electric Conductivity, BOD, NH3, NO2 + NO3, Total Alkalinity, Total Hardness, Ca, Na, Mg, SO4, F, B, K and Cl. The statistical analyses were carried out using MATLAB, STATISTICA (version 8) and SPSS (version 16) software packages.

4.3.1 Multivariate Statistical Analysis

"PCA is a multivariate statistical technique which can be used for reducing complexity of input variables when there is a large volume of information and it is intended to have a better interpretation of variables" (Noori et al. 2009, 2010). In this technique, principal components (PC) are formed by transforming the original variables into new, uncorrelated variables (axes) that give the directions of maximum variance. The "PCs are linear combinations of the original variables that provide information of the most meaningful parameters that describes a whole data set with minimum loss of original information" (Helena et al. 2000). First principal component captures maximum variation in the data set. Steps in PCA are as follows: (1) Standardization of the variables so that they have equal weights during analysis; (2) Calculation of covariance matrix; (3) Calculation of Eigen values and Eigen vectors; (4) Discarding Principal components having Eigen values below the decided criteria; (5) Developing factor loading (correlation coefficients between the factor and the variable) matrix and performing varimax rotation on this matrix so as to infer the principal stations. In this study Principal components having eigen values greater than zero were considered as significant. Final step of PCA is initial stage for FA.

FA is used to reduce the contribution of less significant variables to make the output of PCA even simpler. This can be achieved by developing new variables, called varifactors (VF). VF consists of unobservable, hypothetical and latent variables (Vega et al. 1998; Helena et al. 2000). PCA was performed on the normalized variables to extract PCs and further these were subjected to varimax rotation generating VFs

(Brumelis et al. 2000; Singh et al. 2004; Singh et al. 2005; Love et al. 2004; Abdul Wahab et al. 2005).

Mathematically PCA/FA can be represented as follows (Eq. 4.1):

$$Zij = a_{i1}X_{1j} + a_{i2}X_{2j} + a_{i3}X_{3j} + \cdots + a_{im}X_{mj} \tag{4.1}$$

FA is calculated as (Eq. 4.2):

$$Zji = af_1f_{1i} + af_2f_{2i} + af_3f_{3i} + \cdots + af_mf_{mi} + efi \tag{4.2}$$

where z, a, i, x, m, j, e and fare component score, factor loading, sample numbers, measured value of variable, total number of variables, other source of variation and the factor score, respectively.

4.3.2 *Water Quality Index*

WQI is a widely used measure for the evaluation and extent of water pollution. "WQI is a mechanism for presenting a cumulatively derived numerical expression defining a certain level of water quality" (Bordalo et al. 2006). In simple words, WQI is a representation of large amount of water quality data in the form of a single numerical figure corresponding to a water quality type (e.g., excellent, good, bad etc.).

In the present study, the method used to calculate WQI is Horton's Method (Horton 1965). The WQI is calculated using the following expressing (Eq. 4.3),

$$WQI = \frac{\sum q_n W_n}{\sum W_n} \tag{4.3}$$

where, q_n = sub-index or quality rating of n_{th} water quality parameter.

W_n = unit weight of n_{th} water quality parameter.

The quality rating (q_n) value can be calculated as follows (Eq. 4.4),

$$q_n = \frac{V_n - V_{id}}{S_n - V_{id}} * 100 \tag{4.4}$$

where, V_n = Estimated value of nth water quality parameter at a given sample location, V_{id} = Ideal value for nth parameter in pure water (V_{id} for pH = 7, DO = 14.6 mg/L and 0 for all other parameters), S_n = Standard permissible value of nth water quality parameter, and W_n can be calculated as follows (Eq. 4.5),

$$W_n = \frac{k}{Sn} \tag{4.5}$$

where, k = Constant of proportionality, which can be calculated as (Eq. 4.6):

$$k = \frac{1}{\sum \frac{1}{Sn}} \tag{4.6}$$

4.4 Results and Discussion

The results of PCA/FA analysis for the 16 water quality parameters' data procured from the monitoring stations under consideration are as shown in Fig. 4.2. As the Krishna river system is dynamic, highly non-linear and large, a conservative criterion of factor loading (correlation coefficient between factors and variables) of 0.90 was considered significant. Out of the sixteen parameters considered, eight were found to be principal as their factor loadings were greater than the threshold value of 0.9. The principal water quality parameters were: DO, pH, NH_3, BOD, F, B, Ca and Mg. As stated earlier, multivariate statistics is a data reduction technique. The data quantum with sixteen parameters was considerably reduced to eight parameters. The result of the analysis showed that these eight parameters would be sufficient to capture most of the variance of the system.

Water quality index was calculated for 16 monitoring stations before and after rationalization of WQPs (Table 4.1). The results showed that the WQI for each station considering the 8 rationalized parameters was around 98% of the WQI calculated by considering all the 16 WQPs.

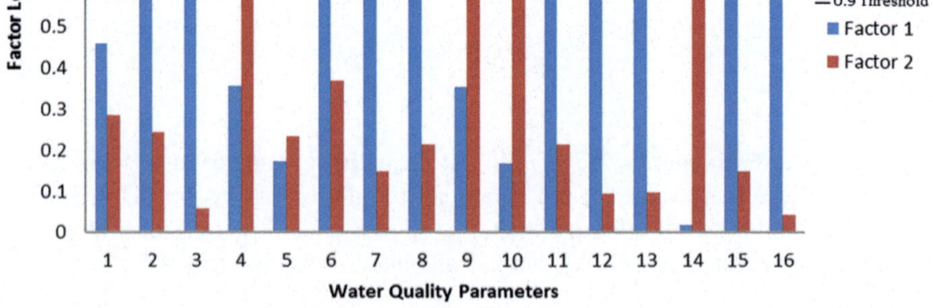

Fig. 4.2 Rationalization of water quality parameters for Krishna River Basin

Table 4.1 WQI of stations before and after rationalization of WQPs

Stations	WQI before rationalization	WQI after rationalization
Bawapuram	26.4	26.0
Cholachguda	38.0	37.5
Damercharla	27.8	27.5
Halia	31.1	30.8
Haralahalli	17.8	17.5
Honnali	16.3	16.1
Karad	25.1	24.9
Keesara	25.6	25.3
Malkhed	20.7	20.5
Paleru Bridge	28.8	28.5
Shimoga	28.9	28.7
T. Ramapuram	24.8	24.3
Takli	39.8	39.3
Vijayawada	18.7	18.5
Wadenapalli	18.9	18.7
Yadgir	18.7	18.4

4.5 Conclusions

In this analysis an attempt was made to rationalize water quality parameters being monitored at Krishna River basin using multivariate statistical techniques i.e. FA/PCA. Further, estimation of WQI showed that the eight rationalized WQPs accounted for 98% of the WQI before rationalization of parameters, indicating the significance of these eight parameters towards the water quality of the Krishna River. The results illustrated that there is a potential for augmenting the economy and efficiency of the monitoring network by reducing the number of parameters from sixteen to eight. The proposed rationalization technique may result in effective cost reduction without any significant loss in information of water quality. Rationalized water quality monitoring network could prove valuable if budget constraint demands for the refinement of the existing monitoring network. Furthermore, the seasonality factor should be considered for effective temporal rationalization of monitoring network. The cost of parameter monitoring can be considerably reduced by rationalization using FA/PCA.

Acknowledgments The authors thank Veermata Jijabai Technological Institute, Matunga, Mumbai, 400019, for providing infrastructural facilities and support for successful completion of their research work.

References

Abdul Wahab SA, Bakheit CS, Al-Alawi SM (2005) Principal component and multiple regression analysis in modelling of ground—level ozone and factors affecting its concentrations. Environ Modell Softw 20(10):1263–1271

Altın A, Filiz Z, Iscen CF (2009) Assessment of seasonal variations of surface water quality characteristics for Porsuk Stream. Environ Monit Assess 158:51–65

Bordalo AA, Teixeira R, Wiebe WJ (2006) A water quality index applied to an international shared river basin: the case of the Douro River. Environ Manage 38:910–920

Brumelis G, Lapina L, Nikodemus O, Tabors G (2000) Use of an artificial model of monitoring data to aid interpretation of principal component analysis. Environ Modell Softw 15(8):755–763

Central pollution control board, India (CPCB) (2007) Guidelines for water quality monitoring. https://www.mpcb.gov.in/sites/default/files/water-quality/standards-protocols/GuidelinesforWQMonitoring%5B1%5D.pdf. Accessed on 17 May 2017

Helena B, Pardo R, Vega M, Barrado E, Ferna´ndez JM, Ferna´ndez L (2000) Temporal evolution of groundwater composition in an alluvial aquifer (Pisuerga river, Spain) by principal component analysis. Water Res 34:807–816

Horton RK (1965) An index number system for rating water quality. J Water Pollut Control Fed 37(3):300–306

Jha R, Ojha CSP, Bhatia KKS (2007) Development of refined BOD and DO models for highly polluted Kali River in India. J Environ Eng 133(8):839–852

Love D, Hallbauer D, Amos A, Hranova R (2004) Factor analysis as a tool in groundwater quality management: two southern African case studies. Phys Chem Earth 29:1135–1143

Ministry of Environment and Forests, India (MoEF) (2005) https://www.indiawaterportal.org/sites/indiawaterportal.org/files/Uniform%20Protocol_Water%20Quality%20Monitoring_MoEF_2005.pdf. Accessed on 17 May 2017

Noori R, Abdoli MA, Ghasrodashti A, Ghazizade M (2009) Prediction of municipal solid waste generation with combination of support vector machine and principal component analysis: a case study of Mashhad. Environ Prog Sustain Energy 249–258

Noori R, Sabahi MS, Karbassi AR, Baghvand A, Zadeh HT (2010) Multivariate statistical analysis of surface water quality based on correlations and variations in the data set. Desalination, Elsevier B.V, 260(1–3): 129–136

Ouyang Y (2005) Evaluation of river water quality monitoring stations by principal component analysis. Water Res 39(12):2621–2635

Parinet B, Lhote A, Legube B (2004) Principal component analysis: an appropriate tool for water quality evaluation and management—application to a tropical lake system. Ecol Model 178(3–4):295–311

Sahoo MM, Patra KC, Khatua KK (2015) Inference of water quality index using ANFIA and PCA. Aquatic Procedia 4(2015):1099–1106

Sedeño-Díaz JE, López-López E (2007) Water quality in the Río Lerma, Mexico: An overview of the last quarter of the twentieth century. Water Resour Manage 21(10):1797–1812

Shrestha S, Kazama F (2007) Assessment of surface water quality using multivariate statistical techniques: A case study of the Fuji river basin Japan. Environmental Modelling and Software 22(4):464–475

Singh KP, Malik A, Mohan D, Sinha S (2004) Multivariate statistical techniques for the evaluation of spatial and temporal variations in water quality of Gomti River (India)—a case study. Water Res 38(18):3980–3992

Singh KP, Malik A, Sinha S (2005) Water quality assessment and apportionment of pollution sources of Gomti river (India) using multivariate statistical techniques: a case study. Anal Chim Acta 538:355–374

Strobl RO, Robillard PD (2008) Network design for water quality monitoring of surface freshwaters: a review. J Environ Manage 87(4):639–648

Varekar V, Karmakar S, Jha R (2015) Seasonal rationalization of river water quality sampling locations: a comparative study of the modified Sanders and multivariate statistical approaches. Environ Sci Pollut Res 23(3):2308–2328

Vega M, Pardo R, Barrado E, Deban L (1998) Assessment of seasonal and polluting effects on the quality of river water by exploratory data analysis. Water Res 32:3581–3592

Water Resource Information System, India. https://www.india-wris.nrsc.gov.in. Accessed on 24 May 2017

Chapter 5
Analysis and Mapping of Groundwater Quality in Vicinity of Kala Sanghian Drain

Kirti Goyal and Bhanu Magotra

Abstract In India, one of the major sources of water is the groundwater. Currently, its quality is being degraded due to many reasons such as intrusion of waste from various chemical factories and other industries. The objective of the study was to analyze the groundwater quality in the vicinity of the Kala Sanghian drain, Jalandhar. The drain carries waste from various industries in Jalandhar and is unlined which might affect the quality of the nearby groundwater. The physiochemical and heavy analysis was carried out on 30 samples collected from different locations of varying depths along the drain. It was observed that the values for some of the parameters are exceeding the permissible limit. The reason for dwindling quality of water in the area may be due to the discharge of the industrial waste into the Kala Sanghian drain. Hence, quick and reliable monitoring measures are essential for safeguarding the water quality. Geographic Information System was used to map the parameters values in terms of WQI in spatial domain which would help in better visualization of the problem and improved decision making.

5.1 Introduction

Groundwater is one of the widely distributed water resources and an invaluable source of water supply globally. The groundwater quality can be degraded either through natural sources or by the various human activities due to progressive industrialization and urbanization (Singh et al. 2013). The pollution of groundwater has become a major concern and has led to many severe environmental and ecological problems.

In India, due to improper facilities, domestic sewage often gets mixed with the effluents coming from the industries or factories which in turn causes many problems such as contamination of crops, groundwater contamination and soil degradation

K. Goyal (✉)
Hydro and Renewable Energy, IIT Roorkee, Roorkee, Uttrakhand, India
e-mail: goyalkirti2012@gmail.com

B. Magotra
Galgotias College of Engineering & Technology, Greater Noida, Uttar Pradesh, India

A. Pandey et al. (eds.), *Climate Impacts on Water Resources in India*, Water Science and Technology Library 95, https://doi.org/10.1007/978-3-030-51427-3_5

etc. Mixing of toxic chemicals, wastes from disposal sites and industrial sites are some of the reasons for the contamination of ground water. Generally, wastewater is disposed off through infiltration ponds, laying onto the ground surface and discharge to streams which may leach into the underlying, shallow aquifers (Chilton 1996). The drain carrying the waste from different sources can also cause groundwater pollution to its nearby areas. Some of the examples of such kind of pollution are Najafagarh drain (Delhi), Buddhah-Nala (Ludhiana), Mirja tannery drain (Unnao) etc. (Shekhar and Sarkar 2013; Verma et al. 2013; Jain et al. 2013).

According to a WHO report, about 80% of the diseases are water-borne. Once the groundwater becomes contaminated, it would be very hard to restore back its quality. For human consumption, the groundwater must be free from any chemical substances and the limits must comply with the standards of IS 10500-2012.

In this study, the groundwater quality has been analyzed in the nearby areas of Kala Sanghian drain, Jalandhar and pollution maps for the water quality index have been prepared using GIS. Various types of industries like leather and tannery, electroplating, textile etc. lie near that drain and discharge their waste into the drain and pollute it. The quality of drain has been deteriorated way too far and it might pollute the nearby groundwater and the surrounding environment in many ways. The pollution mapping using GIS is very useful as it provides efficient capacity to visualize the spatial data.

Various researchers have done many studies on the above discussed issue and found that the groundwater deterioration due to intrusion of wastewater from industries and other sources is one of the major problems being faced these days which needs proper management in terms of treatment and disposal.

5.2 Experimental Work

5.2.1 Study Area

The research is focused in the industrial hub of the state of Punjab i.e. Jalandhar. The district is located between 30°59′: 31°37′ North latitudes and 75°04′: 75°57′ East longitudes. It is the home to some of the major manufacturing industries in the world such as leather, surgical tools, textiles, pulp and paper, hand tools, sanitary fittings etc. (Source: Central Ground Water Board Report, Govt. of India, 2007). The main industrial estates in Jalandhar are Industrial Area, Industrial Estate, Focal Point, Focal Point Extension, Leather Complex and Sports & Surgical Complex. Most of the industries are concentrated on the outskirts of the city (Singh et al. 2007).

Table 5.1 Number of tube wells in the region

Items	Quantity (No.)	Remark
Total no. of tube wells	320	Depth between 180 and 460 ft.
Deep	305	Depth 400 ft. and above
Shallow	15	Depth 180 ft. and above

5.2.2 *Status of Sources of Water Supply*

The water sources in the city include both surface and ground water. There are two main sources of surface water in the city viz., Beas River, situated on the western side of the city and the Kali Bein River originating from the surrounding village. The Beas river remains dry for most part of the year and is not a reliable sources of water supply whereas, the Kali Bein river has actually converted into a drain because of the disposal of the toxic effluents by the industries situated at its banks.

Thus, the city has to depend mostly on groundwater sources for the municipal and industrial needs. At present, there are 320 tube wells extended throughout the city at various locations which are pumped to supply water by the Jalandhar Municipal Corporation. The details regarding the tube wells is given in Table 5.1.

5.2.3 *Site Details*

Kala Sanghian drain originates from Raowali village, Jalandhar and flows through Bulandpur, focal point, leather complex in Jalandhar city and goes to Chiti Bein through many villages in the vicinity, and east (Chiti) Bein finally connecting with the Satluj. The total length of the drain from the point of origin to the point where it meets east (Chiti) Bein is 4408 m. Many industries e.g. leather and tannery, electroplating, textile etc. located in the vicinity of the Kala Sanghian drain dumps their waste into it which is the major cause of pollution in the drain (Singh et al. 2007).

The quality of drain has been deteriorated way too far. This may pollute the nearby groundwater sources as well causing problems to the health of the people staying in the surrounding area. Thus, studies have been performed in order to check the quality of groundwater in the surroundings of the Kala Sanghian Drain.

A total of 30 samples have been drawn at different depths from various locations in the vicinity of the drain. The samples were analysed for physico-chemical parameters and heavy metals to check the effect of drain on the groundwater. After that, GIS mapping of the same has been done in order to obtain the quality of water in the whole area.

The use of a GIS mapping software often helps to visualize the study area better and allows for better planning of sample collection as well as mapping of acquired results. ArcGIS software has been used for the purpose. Using a base map from Bing Maps and longitudes and latitudes of sample points, collected by handheld GPS device, the drain and corresponding points are plotted. In this way, the study area can be very well visualized as apparent from Fig. 5.1.

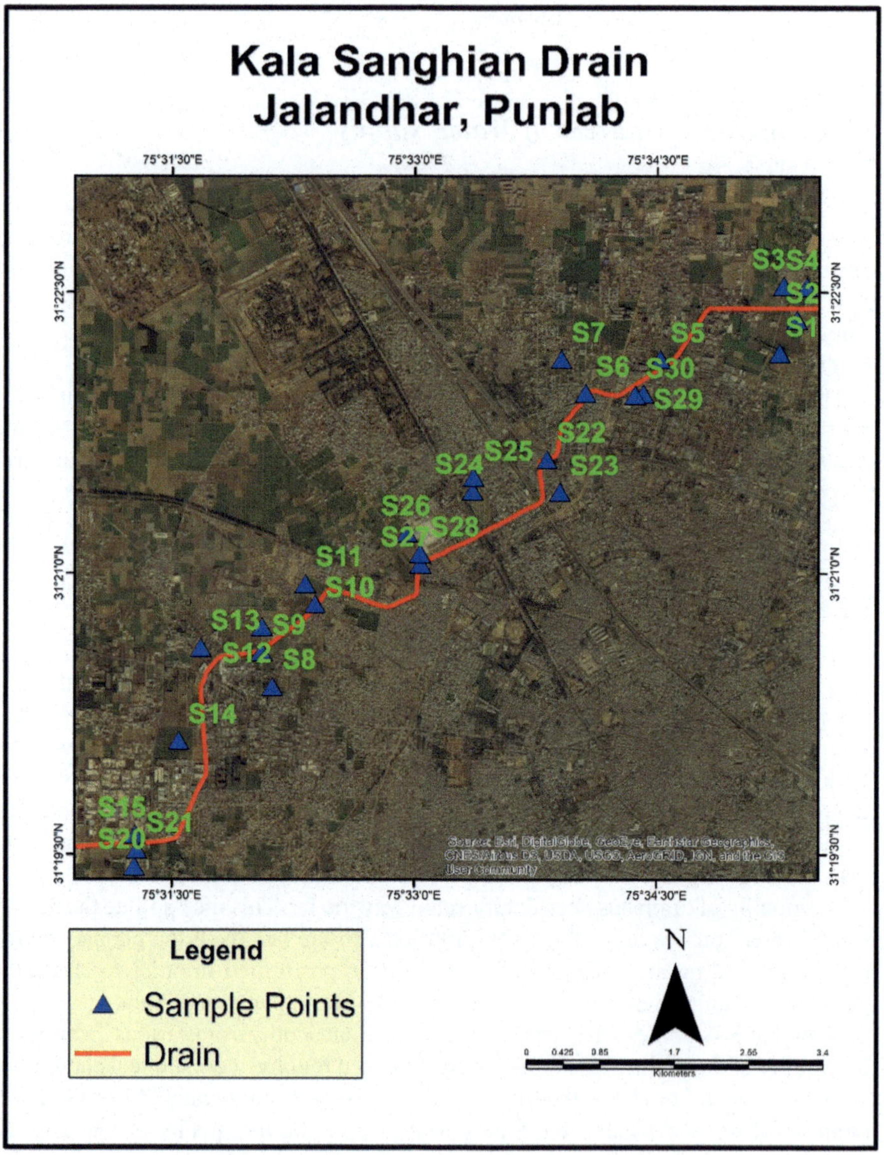

Fig. 5.1 A GIS Map of the drain and location of sample points

Table 5.2 Equipment/methods used for the analysis

S. no	Parameter	Method	Instrument
1	pH	Potentiometric method	pH meter
2	Total hardness as $CaCO_3$	Titrimetric method	Titration
3	Chlorides	Titrimetric method	Titration
4	Total dissolved solids	Gravimetric method	Oven
5	Alkalinity	Titrimetric method	Titration
6	Sulphates	Turbidimetric method	Turbidity meter
7	Turbidity	Turbidimetric method	Turbidity meter
8	Iron	Colorimetric method	Colorimeter
9	Heavy metals—Arsenic, Manganese, Chromium, Copper, Zinc, Lead, Cadmium	Inductively coupled plasma method	iCAP 6000 series ICP emission spectrometer

5.2.4 Analytical Methods and Equipment Used

The various methods and equipment used for the physico-chemical and heavy metal analysis are as per the required standards and have been listed in Table 5.2.

5.2.5 Formulation of WQI

WQI is a mathematical tool which combines numerous water quality data into a single number in a comprehensive manner which aids in conveying information regarding the pollution status of the river (Table 5.3).

It is an excellent management and administrative tool that may be used for communicating water quality information to general public as well as the stakeholders (Horton 1965).

Table 5.3 WQI and corresponding water quality status

S. no	WQI	Status	Possible usage
1	0–25	Excellent	Drinking, irrigation and industrial
2	26–50	Good	Domestic, irrigation and industrial
3	51–75	Fair	Irrigation and industrial
4	76–100	Poor	Irrigation
5	101–150	Very poor	Restricted use for irrigation
6	Above 150	Unfit for drinking	Proper treatment required before use

It is generally evaluated considering the physical, chemical and biological parameters concerned with the water quality. It shows the overall water quality at a specific location and time.

The WQI is calculated by Eq. 5.1.

$$WQI = \frac{\sum Q_i W_i}{\sum W_i} \tag{5.1}$$

The quality rating scale (Qi) for each parameter is calculated by Eq. 5.2.

$$Q_i = 100[(V_i - V_o)/(S_i - V_o)] \tag{5.2}$$

where,

Vi is estimated concentration of ith parameter in the analysed water
Vo is the ideal value of this parameter in pure water Vo = 0 (except pH = 7.0 and DO = 14.6 mg/l)
Si is recommended standard value of ith parameter (Tyagi et al. 2013).

The unit weight (Wi) for each water quality parameter is calculated by using the following formula:

$$Wi = \frac{\mathrm{K}}{\mathrm{Si}}$$

where, K = proportionality constant and can also be calculated by using the Eq. 5.3.

$$K = \frac{1}{\sum \frac{1}{S_i}} \tag{5.3}$$

Table 5.4 Standard values of water quality parameters and their ideal values and unit weights

S. no	Parameters	S_i	Recommending agency for S_i	Ideal value, V_o	K value	W_i
1	TDS	500	IS	0	0.286	0.000572
2	pH	8.5	IS	7	0.286	0.033647
3	Total hardness	300	ICMR	0	0.286	0.000953
4	Total Alkalinity	120	ICMR	0	0.286	0.002383
5	Chlorides	250	IS	0	0.286	0.001144
6	Iron	0.3	IS	0	0.286	0.953333
7	Sulphates	200	IS	0	0.286	0.00143

For the calculation of quality rating (Qi) and unit weights (Wi) (Table 5.4) the drinking water standards as recommended by the Indian Council of Medical Research (ICMR) and Indian Standards Institution (ISI) have been considered.

For calculation of the WQI, 7 parameters have been selected, namely, TDS, pH, Alkalinity, Hardness, Iron, Chlorides and Sulphates. Turbidity has not been selected for the calculation of WQI as it is related to other parameters and is more of aesthetic importance rather than health concerns.

5.3 Results and Discussion

5.3.1 Physico-Chemical Parameters

The analysis of the various physico-chemical parameters as per the standard methodology has been done and the results are displayed in Table 5.5. As per the results, most of the parameters have values within the permissible limit, though, the values for Alkalinity are high in few of the samples which may be due to contamination from natural sources. The values of iron are quite alarming in some of the samples which may be due to intrusion of industrial wastewater.

Also, the WQI for different samples has been calculated. In most of the places, the quality of the water is ranging from excellent to fair. But, in few samples, the value is quite high which makes the water unfit for drinking.

5.3.2 Heavy Metal Analysis

The samples have also been analyzed for the heavy metals such as Zinc, Manganese, Copper, Arsenic, Cadmium, Lead and Chromium as shown in Table 5.6. As per the analysis, all the values are coming well within the permissible limits. The values of the concentration in the samples for Zinc varies between 0.01 and 2.41 mg/L. The values of Manganese are exceeding above the permissible limit for drinking water. The disposal of industrials effluents is expected to be the reason for the same. The maximum concentration of copper is coming out to be 0.02 mg/L. The concentration of Chromium lies in the range of 0.00406–0.012 mg/L which is well within the permissible limits. The values of arsenic, cadmium and lead in all the samples are coming out to be non-detectable or in traces only.

Table 5.5 Results of physic-chemical analysis

Sample no.	pH	Hardness	Chlorides	TDS	Alkalinity	Sulphates	Turbidity	Iron	WQI
S1	7.1	156	68	192	104	73	4.5	0.10	33
S2	7.2	162	53	237	152	177	4	0.23	75
S3	7.1	99	39	177	89	63	3	0.18	58
S4	7.1	121	37	201	97	92	3.6	0.13	42
S5	7.3	113	48	149	88	101	4.2	0.13	43
S6	7.8	226	104	251	189	180	3.9	0.70	226
S7	7.4	184	83	289	154	69	3.6	0.69	222
S8	7.4	179	84	139	149	73	4.2	0.23	75
S9	7.5	182	96	224	138	76	2.9	0.22	72
S10	7.6	169	75	199	134	83	3.7	0.31	101
S11	7.3	154	86	192	118	175	4	0.17	56
S12	7.1	149	88	298	103	87	3.6	1.01	324
S13	7.1	136	42	256	97	78	3.5	0.01	4
S14	7.3	163	47	273	113	89	4.7	0.11	36
S15	7.2	136	39	146	86	76	2.7	0.40	129
S16	7.2	116	38	252	77	177	3.8	0.13	42
S17	7.3	124	44	172	82	82	4	0.27	87
S18	7.3	129	42	192	92	104	3.7	0.26	84
S19	7.3	127	41	149	89	116	4.6	0.34	110
S20	7.3	109	104	245	68	85	3.5	0.18	59
S21	7.4	184	97	287	136	169	3	0.50	161
S22	7.6	179	89	135	129	78	4.2	0.98	315
S23	7.5	186	98	286	135	66	3.7	0.29	94
S24	7.2	179	97	286	127	105	3.8	0.12	39
S25	7.2	98	51	142	54	133	4.8	0.09	30
S26	7.2	119	108	250	67	129	3.1	0.05	17
S27	7.2	146	103	238	86	90	2.9	1.05	337
S28	7.3	109	95	280	59	100	4	0.81	260
S29	7.4	113	88	275	64	75	3.6	1.15	369
S30	7.2	97	49	138	59	80	4.3	0.25	81

5.3.3 Mapping of WQI Using GIS

Sample wise WQI data is mapped with point shapefile containing the sample points using "join" command in attribute table. Theissen polygons are developed for all 30 sample points within the extent of 800 m buffer around the drain (Jordan 2017). The

Table 5.6 Results for heavy metal analysis

Sample no.	Arsenic	Manganese	Chromium	Copper	Zinc	Lead	Cadmium
S1	Trace	0.01	0.00497	N. D	0.02	N. D	N. D
S2	Trace	0.03	0.00566	N. D	0.05	N. D	N. D
S3	Trace	0.03	0.00492	N. D	0.18	N. D	N. D
S4	Trace	0.01	0.00406	N. D	0.07	N. D	N. D
S5	Trace	0.01	0.00491	N. D	0.01	N. D	N. D
S6	Trace	1.20	0.01141	0.02	1.33	Trace	N. D
S7	Trace	0.14	0.00702	0.01	0.82	N. D	N. D
S8	Trace	0.20	0.00491	Trace	0.91	N. D	N. D
S9	Trace	0.09	0.00521	Trace	0.54	N. D	N. D
S10	Trace	1.69	0.00780	0.02	0.92	Trace	N. D
S11	Trace	0.08	0.00446	N. D	0.08	N. D	N. D
S12	Trace	0.62	0.00551	N. D	2.41	N. D	N. D
S13	Trace	0.07	0.00265	N. D	0.06	N. D	N. D
S14	Trace	0.05	0.00423	N. D	0.08	N. D	N. D
S15	Trace	0.04	0.00422	N. D	0.11	N. D	N. D
S16	Trace	0.01	0.00477	N. D	0.44	N. D	N. D
S17	Trace	0.04	0.00650	N. D	0.53	Trace	N. D
S18	Trace	0.04	0.00646	N. D	0.58	N. D	N. D
S19	Trace	0.05	0.00606	N. D	0.68	N. D	N. D
S20	Trace	0.03	0.00968	N. D	0.61	N. D	N. D
S21	Trace	0.11	0.01201	N. D	0.71	Trace	N. D
S22	Trace	0.89	0.00899	0.02	0.90	Trace	N. D
S23	Trace	0.03	0.00786	N. D	0.27	Trace	N. D
S24	Trace	0.05	0.00742	N. D	0.72	N. D	N. D
S25	Trace	0.06	0.00621	N. D	1.50	N. D	N. D
S26	Trace	0.02	0.00667	N. D	0.10	N. D	N. D
S27	Trace	0.70	0.00866	N. D	0.24	N. D	N. D
S28	Trace	0.07	0.00808	Trace	0.14	Trace	N. D
S29	Trace	1.25	0.01002	0.04	0.46	Trace	N. D
S30	Trace	0.01	0.00576	N. D	0.09	N. D	N. D

N. D. (Not Detectable)

WQI value for each of the points, thus, gets mapped to the corresponding polygon covering specific localities. A classification of polygons is done based on Table 5.3 to identify the area with unfavorable drinking water quality (Fig. 5.2).

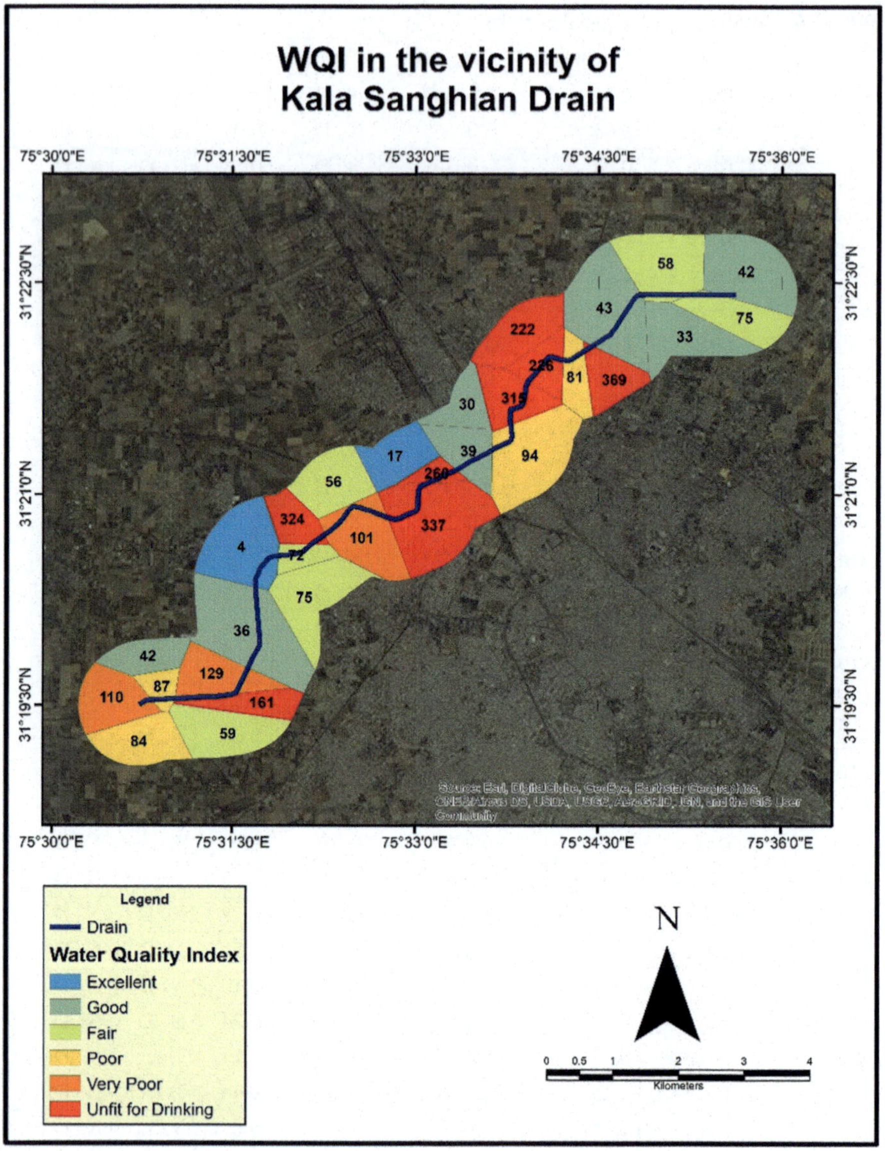

Fig. 5.2 Map showing WQI values and its classification in vicinity of drain

5.4 Conclusion

The groundwater quality in the nearby regions of the Kala Sanghian drain in the Jalandhar city has been studied and analyzed. Multiple samples have been collected from 30 different locations throughout the area. On the basis of these analytical

Table 5.7 Summary of WQI values

WQI	No. of samples
0–25	2
26–50	7
51–75	6
76–100	4
101–150	3
Above 150	8

findings, it can be said that most of the parameters are well within permissible limits except for a few which are ranging above the desired range. Iron ion concentration is coming out to be higher than the permissible limit in the samples that are near/on the polluted drain, especially, in the shallow wells. It may be due to the percolation of contaminants from the drain to the aquifer. Electroplating industrial effluents may also be the cause of ferrous pollution in the samples. Manganese ion concentration is higher than the permissible limits which may be because of the presence of leather and tannery industrial waste in the drain.

Water Quality Index, being a single value, has further helped in reflecting the overall quality of water taking into account, the composite influence of different quality parameters. The summarized results can be seen as per Table 5.7.

As it has been observed, 50% samples have water quality poor or worse which shows that the industrial discharge is leading to the declining quality of groundwater in the vicinity of the Kala Sanghian drain. Hence, there is a strong need for quick and consistent real time monitoring techniques to maintain the water quality so as to avoid problems associated with the health and environment.

References

Chilton J (1996) Water quality assessments—a guide to use of biota, sediments and water in environmental monitoring, 2nd edn. E & FN Spon

Horton R (1965) An index number system for rating water quality. J Water Pollut Cont Fed 37(3):300–305

Jain C (2013) Report of the committee on pollution caused by leather tanning industry to the water bodies/groundwater in Unnao district of Uttar Pradesh

Jordan L (2017) Applying Thiessen polygon catchment areas and gridded population weights to estimate conflict-driven population changes in South Sudan. ISPRS Ann Photogramm Remote Sens Spatial Inf Sci IV-4/W2:23–30

Shekhar S, Sarkar A (2013) Hydro geological characterization and assessment of groundwater quality in shallow aquifers in vicinity of Najafgarh drain of NCT Delhi. J Earth Syst Sci 122(1):43–54

Singh G, Singh D, Sharma S (2013) Effect of polluted surface water on groundwater: a case study of Budha Nullah. IOSR J Mech Civ Eng (IOSR-JMCE) 5(5):01–08

Singh S, Sharma A, Marwaha S, Gupta S (2007) Report by Central Groundwater Board, Ministry of Water Resources, Government of India, North Western Region, Chandigarh

Tyagi S, Sharma B, Singh P, Dobhal R (2013) Water quality assessment in terms of water quality index. Am J Water Resour 1(3):34–38

Verma VK, Setia RK, Khurana MPS (2013) Ugly face of urbanization and industrialization: a study of water pollution in Buddha Nala of Ludhiana City, India. J Environ Conserv Res 1(1):6–11

Chapter 6
Geospatial Analysis Coupled with Logarithmic Method for Water Quality Assessment in Part of Pindrawan Tank Command Area in Raipur District of Chhattisgarh

Purushottam Agrawal, Alok Sinha, Srinivas Pasupuleti, Rajesh Nune, and Sarbani Saha

Abstract Water Quality Index is an effective tool to measure the quality of water based on different parameters. In this study, a part of Pindrawan tank command area under Mahanadi river basin in Raipur district of Chhattisgarh was selected to assess the suitability of water for drinking purpose. 25 water samples were collected from various surface and groundwater sources of the study area for comprehensive water quality analysis and 11 physio-chemical parameters such as: pH, fluoride, chloride, total hardness, total dissolved solids, iron, calcium, sodium, potassium, electrical conductivity and alkalinity were considered for calculating Water Quality Index (WQI). WQI was computed using logarithmic method. Spatial distribution maps for different parameters were prepared using IDW technique in GIS interface. The study area was categorised from excellent to unsuitable for drinking water zones according to the computed WQI values. Out of 25 water samples analysed for drinking water quality, 2 samples were found to be unsuitable for drinking, 10 samples fall under good/excellent and 13 samples quality were poor. Water in the poor zone can be utilised for drinking after proper treatment. This study demonstrates the efficiency of GIS for analysing the complex spatial data for water quality mapping.

P. Agrawal (✉) · A. Sinha
Department of Environmental Science & Engineering, Indian Institute of Technology (ISM), Dhanbad, Dhanbad 826004, India
e-mail: agrawal.purushottam@gmail.com

S. Pasupuleti · S. Saha
Department of Civil Engineering, Indian Institute of Technology (ISM), Dhanbad, Dhanbad 826004, India

R. Nune
IDC, ICRISAT, Patancheru 502324, Hyderabad, India

A. Pandey et al. (eds.), *Climate Impacts on Water Resources in India*, Water Science and Technology Library 95, https://doi.org/10.1007/978-3-030-51427-3_6

6.1 Introduction

Water quality assessment in developing nations is very essential due to the fresh water scarcity. The quality of water is a critical issue because improper management of water resources affect the entire population of any country. According to the WHO reports most of the diseases in human are caused by drinking water. Water quality assessment is required to maintain the quality of water resource. It aids to the feasibility for development analysis and planning for water resources. One of the major techniques of water management is water quality mapping. Water Quality Index is very efficient method for evaluating the suitability of water for drinking purpose. In this research study, physical and chemical parameters were used to determine the water quality index.

Geographical Information System (GIS) was used to map the spatial distribution of different parameters using IDW technique. This interpolation technique is very useful for mapping the sample locations in a well-defined manner. This tool helped to explore the spatial data by analysing the hydrochemistry of water by indicating spatial variability. Before introduction of GIS only laboratory investigations were used to analyse the quality of water. After introducing GIS it is very easy to plot the data according to their spatial variability. Many research studies were conducted GIS based mapping for water quality studies (Nas and Berktey 2010; Chatterjee and Raiziuddin 2002; Srinivasamoorthy et al. 2009; Rupal et al. 2012; Naik and Puohit 1996; Singh et al. 2013; Srikanthan et al. 2013; Shomar et al. 2010; Selvam et al. 2013).

The main objective of this study is to use the complex data in an easy and comprehensive manner so that it will be useful to obtain the necessary information in a very lucid manner. It will also help the policy makers to manage the water quality for the entire area. GIS also helps to correlate the laboratory-based data with geographical data accurately. In this study water quality index was calculated using the logarithmic method proposed by Tiwari and Mishra (1985).

6.2 Study Area

The present study area is a part of Pindrawan tank command area which is located in Raipur district of Chattisgarh, India. The present study area geographically located between 81°45′, 81°50′ E longitudes, and 21°20′, 21°25′ N latitudes which is shown as Fig. 6.1. This area consists of few villages such as Pauni, Khauna, Amlitalab, Deogaon, Dhansuli, Kurra, Bangoli, Nilja and Baraonda. It is situated at the south eastern part of the upper Mahanadi River valley. Raipur has tropical wet and dry climate. Temperature remains moderate throughout the year. March–June are having the highest temperature in the year.

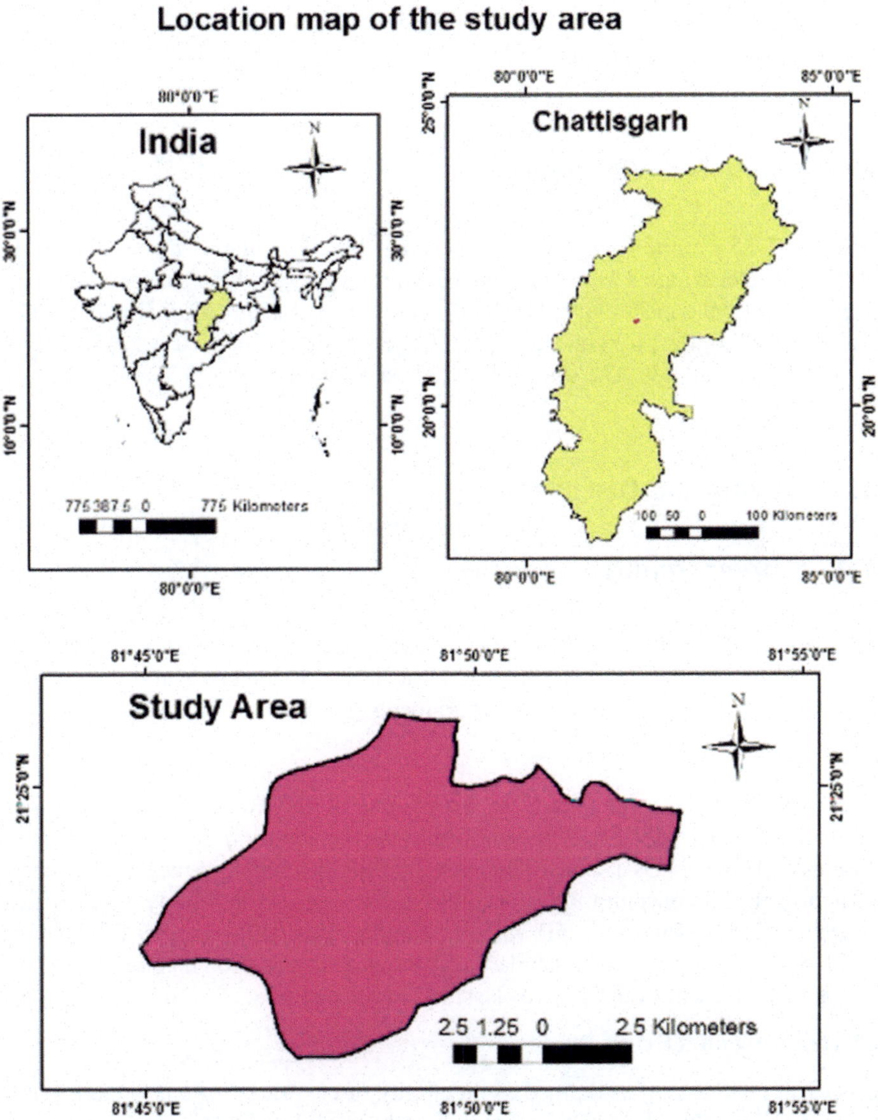

Fig. 6.1 Location map of the study area

6.3 Materials and Method

The methodology adopted in this research work includes study area demarcation from toposheets, water sample collection from field, laboratory testing of samples, map analysis using Arc GIS, preparation of spatial distribution maps for different parameters and preparation of WQI using IDW techniques in GIS. Methodology flow chart is shown as Fig. 6.2.

6.3.1 Sample Collection

A total of 25 sample locations were selected based on the study area map. The collected water samples were analysed for different physio-chemical parameters such as pH, Electrical conductivity, Alkalinity, Sodium, Potassium, TDS, Total Hardness, Chlorides, Fluorides, Iron and Calcium. The analysis was carried out following the standard BIS methods. The sample locations are shown in Fig. 6.3.

6.4 Results and Discussion

6.4.1 Water Quality Parameters

To compute the water quality index 11 physio-chemical parameters were considered. The maximum and minimum value of these parameters were obtained from Table 6.1 and significance of various parameters are briefly discussed.

PH

The pH of water is considered as a measure of the equilibrium of acid-base. pH is usually controlled by the equilibrium between carbon dioxide–bicarbonate–carbonate. Lower pH is a result of increased carbon-di-oxide. The negative logarithm of hydrogen ion concentration of a solution is the pH. In this study area pH value ranges 7.12–8.3. Therefore, pH value of samples lies within the permissible limit 6.5–8.5 according to the BIS standards. The spatial distribution map of pH is shown as Fig. 6.4 where pH values were classified into 5 classes.

Electrical Conductivity

Electrical conductivity is defined as the ability of electrical current to pass through the water. Total dissolved salts in water directly related to the electrical conductivity of water. Conductivity of electricity depends on the salts which are dissolved in positively and negatively charged ions. In this study area EC value ranges from 154 to 1777 (μS). The EC value of the samples lies within the described permissible limit. The spatial distribution map of EC is shown as Fig. 6.5 and EC values were classified into 5 classes.

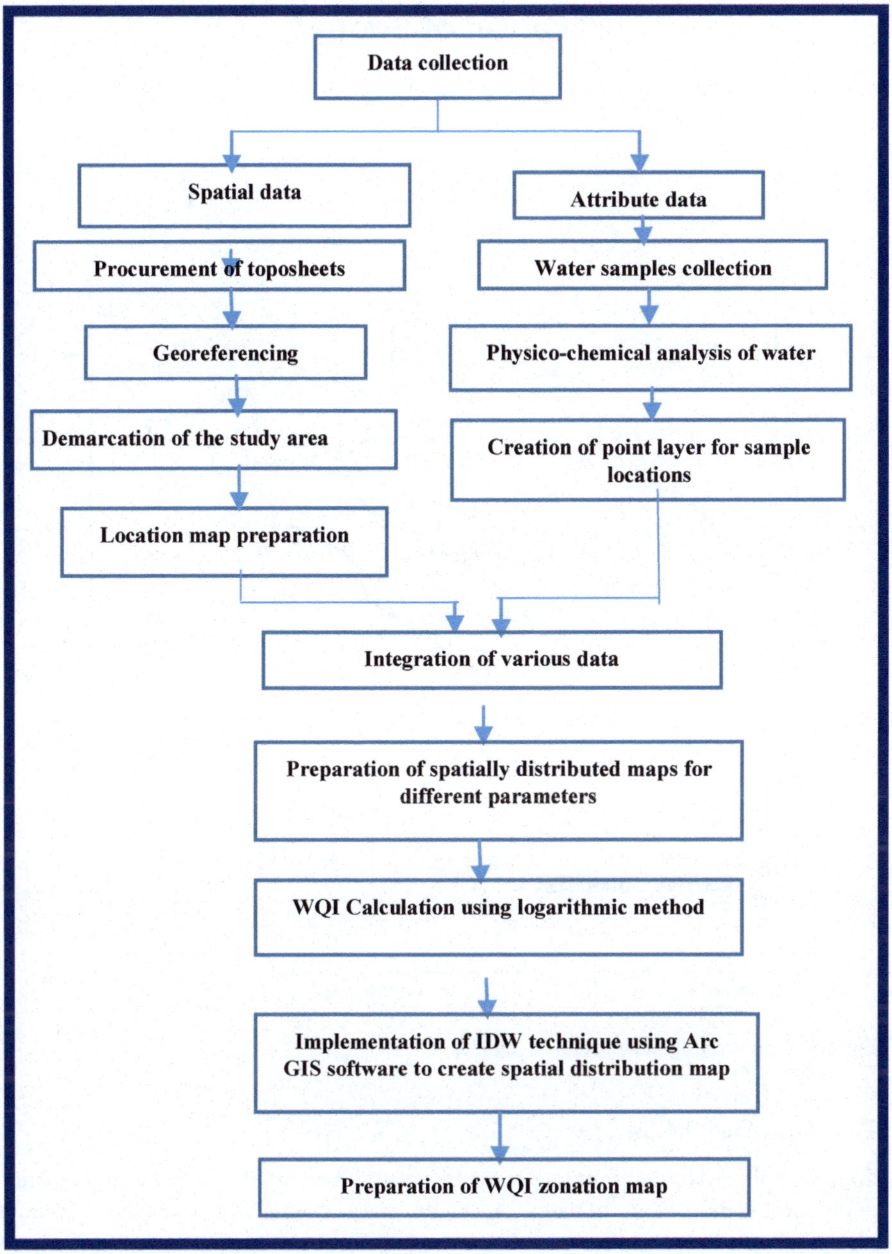

Fig. 6.2 Detailed methodology flow chart

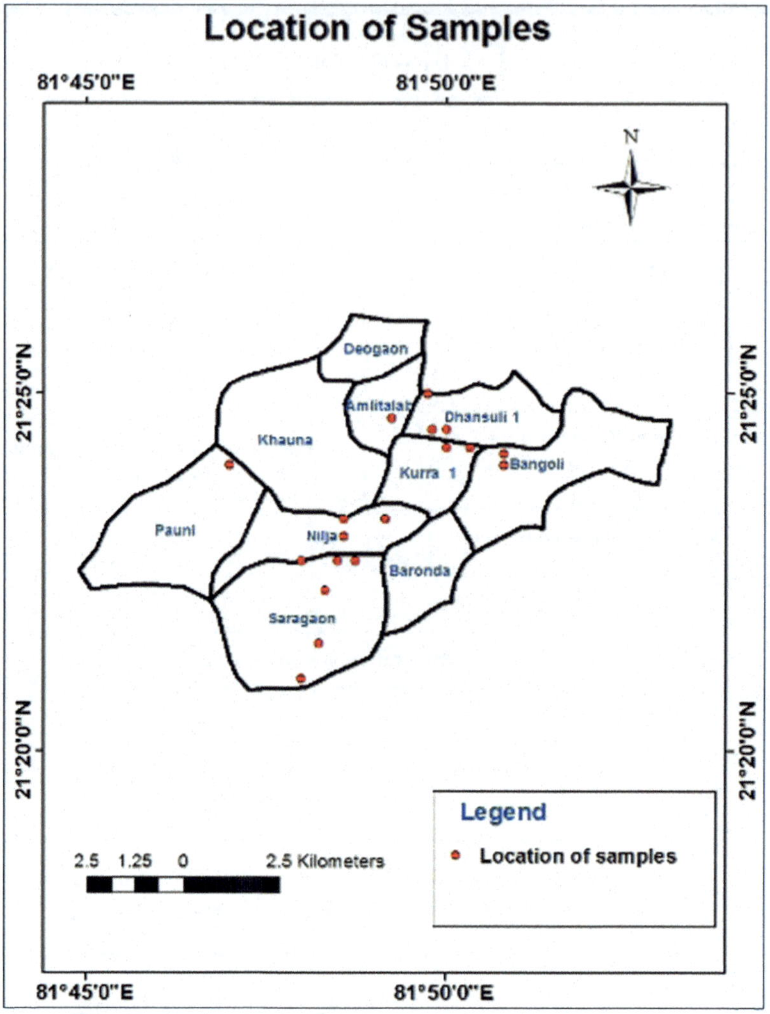

Fig. 6.3 Map showing the location of samples

Total Dissolved Solids

Total dissolved solids comprise inorganic salts (principally calcium, magnesium, potassium, sodium, bicarbonates, chlorides, and sulfates) and some small amounts of organic matter that are dissolved in water. TDS in water originates due to discharge from sewage treatment plant, leaching of contaminated soil urban and agricultural runoff. In this study area TDS value ranges 231–1137 (mg/l). BIS fixes the upper limit of TDS in drinking water at 2000 mg/l. The spatial distribution map of TDS is shown as Fig. 6.6 and TDS values were classified into 5 classes.

Table 6.1 Water Quality parameters

No. of samples	pH	EC (μS)	Alkalinity (mg/l)	Total hardness (mg/l)	Chloride (mg/l)	TDS (mg/l)	Fluoride (mg/l)	Iron (mg/l)	Calcium (mg/l)	Sodium (mg/l)	Potassium (mg/l)
1	7.36	518	110	100	80	348	0.420	0.1322	31.8432	62	713
2	7.26	444	130	110	60	360	0.313	0.0825	30.3460	88	714
3	7.34	712	100	200	110	492	0	0.0930	73.4310	6.8	410
4	7.32	356	130	70	40	288	0.449	0.2366	30.3504	20.4	1170
5	7.14	327	115	100	30	294	0.337	0.0907	42.6316	6.1	826
6	7.12	525	150	110	60	264	0.524	0.1380	36.0732	40	4087
7	7.42	724	180	100	100	456	0.405	0.0565	30.8989	6.2	699
8	7.52	260	110	110	30	300	0.405	0.1241	41.4451	15	634
9	8.03	616	160	130	70	432	0.421	0.1088	41.3761	110	684
10	7.71	695	130	110	100	408	0.220	0.1601	30.3398	96	672
11	7.47	837	170	100	140	612	0.441	0.2304	37.4391	288	688
12	7.32	502	180	130	60	444	0.437	0.2438	44.5299	236	704
13	7.46	553	190	220	40	540	0.081	0.0570	78.6501	26	667
14	7.46	313	120	125	30	330	1.801	0.1308	60.0217	6.8	447
15	7.62	801	180	260	80	624	1.817	0.1492	124.254	86	144
16	7.63	347	120	130	40	348	0.161	0.0493	76.1683	20.4	523
17	7.8	356	120	150	40	372	0	0.0675	56.0277	7	482
18	8.16	375	150	50	20	264	0.449	0.0860	71.9124	20.4	1
19	7.59	1379	260	370	150	936	0.275	0.4106	173.8369	198	679
20	7.9	584	190	190	50	516	0.314	0.0723	101.5050	122	596

(continued)

Table 6.1 (continued)

No. of samples	pH	EC (μS)	Alkalinity (mg/l)	Total hardness (mg/l)	Chloride (mg/l)	TDS (mg/l)	Fluoride (mg/l)	Iron (mg/l)	Calcium (mg/l)	Sodium (mg/l)	Potassium (mg/l)
21	7.54	1777	339	389	219	1137	0.172	0.0772	152.0580	320.9	490
22	7.62	836	210	280	70	684	0.057	0.0355	105.2680	114	688
23	7.57	1139	250	370	110	876	0	0.1030	60.1712	144	642
24	7.96	421	150	160	40	420	0.170	0.0472	26.6998	28	599
25	8.3	154	61	80	50	231	0.077	0.0414	26.9431	40	698

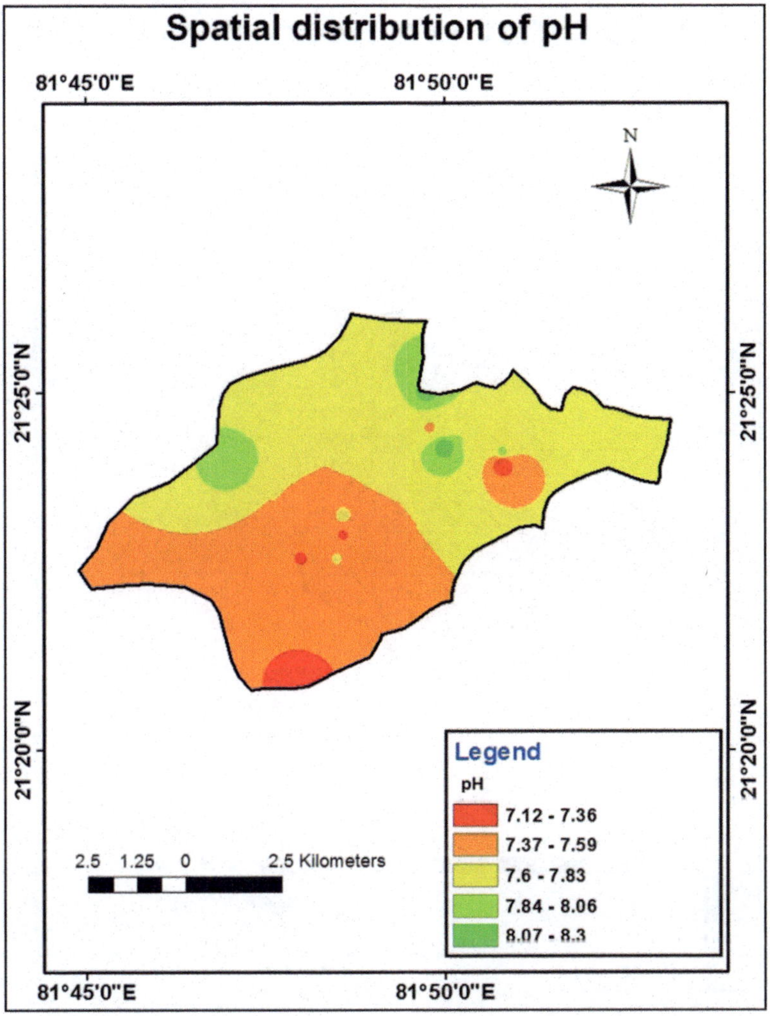

Fig. 6.4 Spatial distribution map of pH

Alkalinity

Alkalinity is capacity of measuring the water's ability to neutralize acidity. It is used to measure the bicarbonates, carbonates, and hydroxides level in water. The results of alkalinity tests are generally expressed as "ppm of calcium carbonate ($CaCO_3$)".

Alkalinity value ranges 61–339 (mg/l) in this study area. Carbon dioxide, carbonates and bicarbonates in dissolved state in water cause alkalinity. Alkalinity values of the samples lies within the permissible limit 600 mg/l. The spatial distribution map of alkalinity is shown as Fig. 6.7 which were classified into 5 classes.

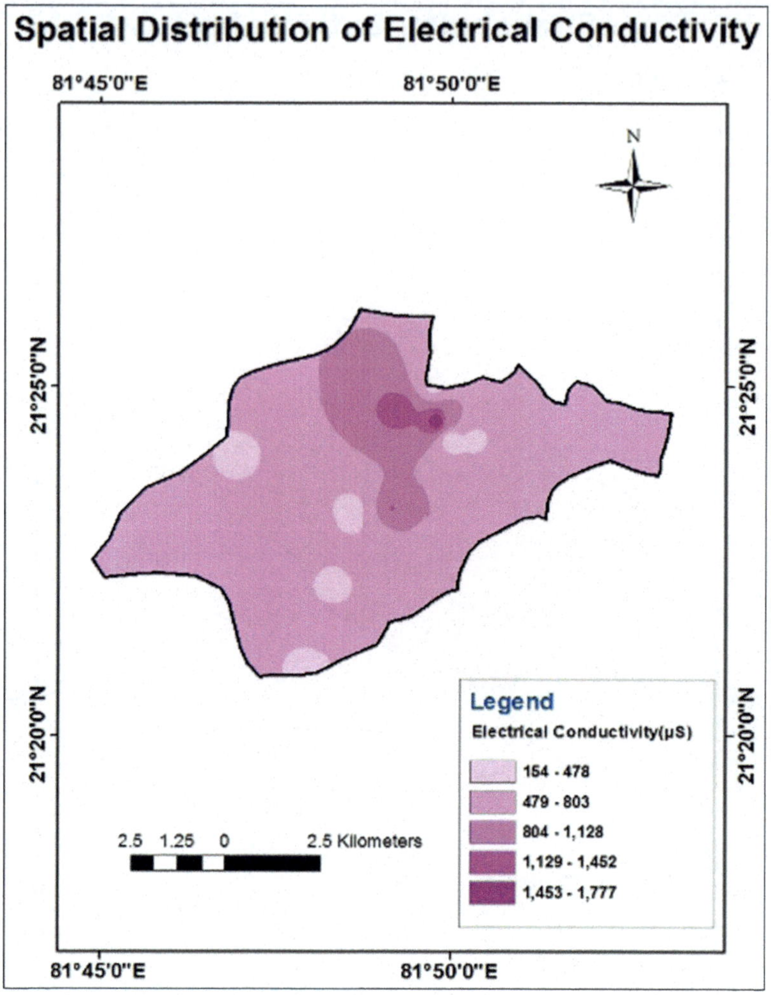

Fig. 6.5 Spatial distribution map of EC

Chloride

Chloride originates from the use of inorganic fertilizers, landfill leachates, septic tank effluents, industrial effluents in surface and groundwater. BIS prescribes 250 mg/l as desirable limit and 1000 mg/l as permissible limit. Chloride value ranges 20–219 (mg/l) in this study area. Chloride values of the samples lies within the permissible limit. The spatial distribution map of chloride is shown as Fig. 6.8 and chloride values were classified into 5 classes.

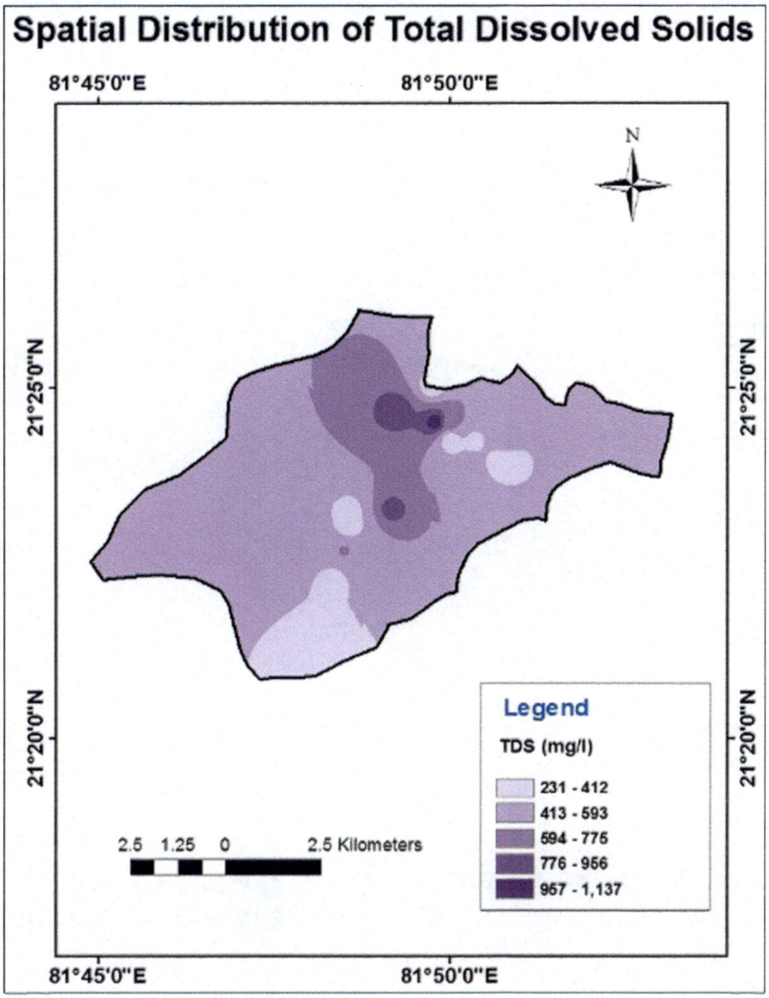

Fig. 6.6 Spatial distribution map of TDS

Fluoride

Fluoride is usually determined by means of an ion-selective electrode, which makes it possible to measure the total amount of free and complex-bound fluoride dissolved in water. According to the BIS water having more than 1.5 mg/l concentration of fluoride causes fluorosis, dental mottling and bone related diseases. The range should be between 0.6 and 1.0 mg/l. Fluoride value ranges from 0 to 1.817 (mg/l) in this study area. Fluoride values of the samples in some areas did not lie within the permissible limit. The spatial distribution map of fluoride is shown as Fig. 6.9 and fluoride values were classified into 5 classes.

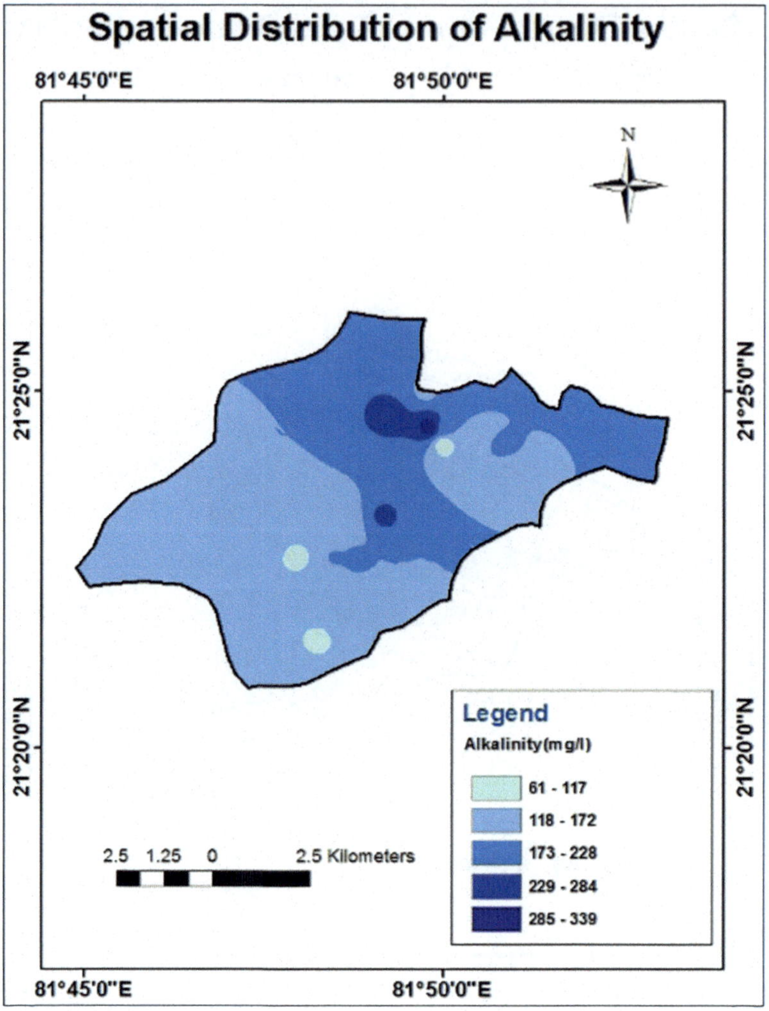

Fig. 6.7 Spatial distribution map of Alkalinity

Total Hardness

Total hardness is defined as the sum of calcium and magnesium hardness, in mg/L as $CaCO_3$. Total hardness is determined by the concentration of multivalent cations present in water sample. Total hardness value ranges 50–389 (mg/l) in this study area. Total hardness values of the samples lie within 600 mg/l according to BIS permissible limit. The spatial distribution map of Total hardness is shown as Fig. 6.10 and total hardness values were classified into 5 classes.

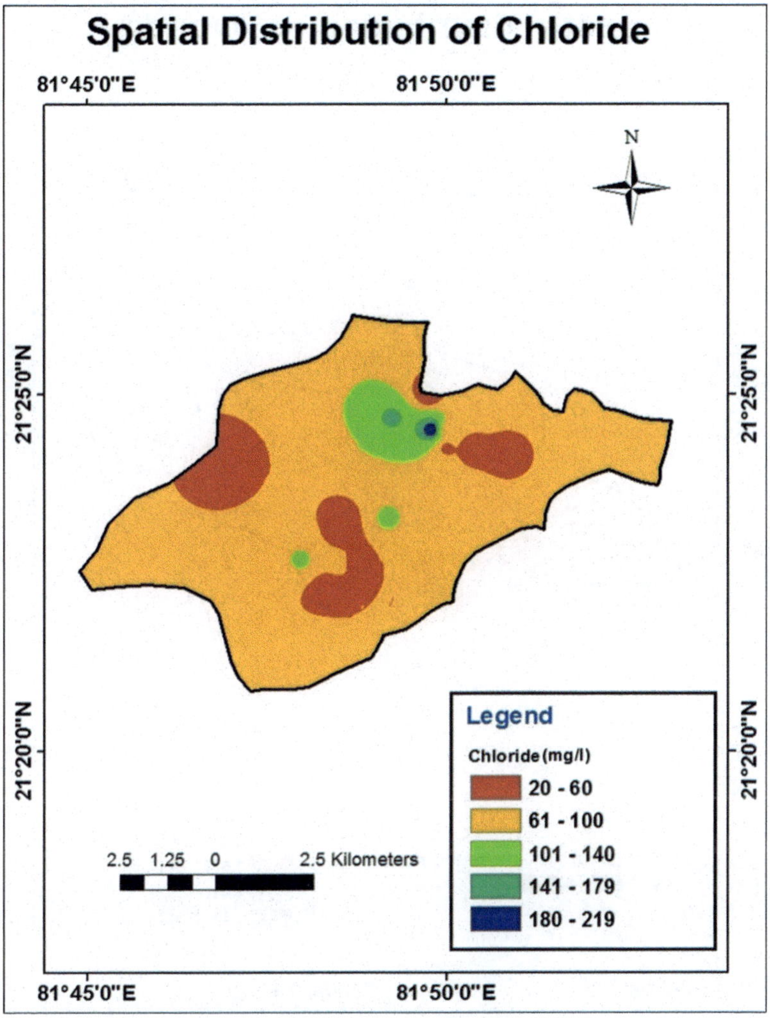

Fig. 6.8 Spatial distribution map of Chloride

Calcium

Calcium occurs in water mainly due to the presence of lime stone, gypsum, dolomite and gypsy-ferrous minerals. Excessive calcium leads to formations of concretions in the body and may cause gastro-intestinal disease. The desirable limit for calcium is 75 mg/l and permissible limit is 200 mg/l. Calcium value ranges 27–174 (mg/l) in this study area. Calcium values of the samples lies within the described permissible limit. The spatial distribution map of Calcium is shown as Fig. 6.11 and the values were classified into 5 classes.

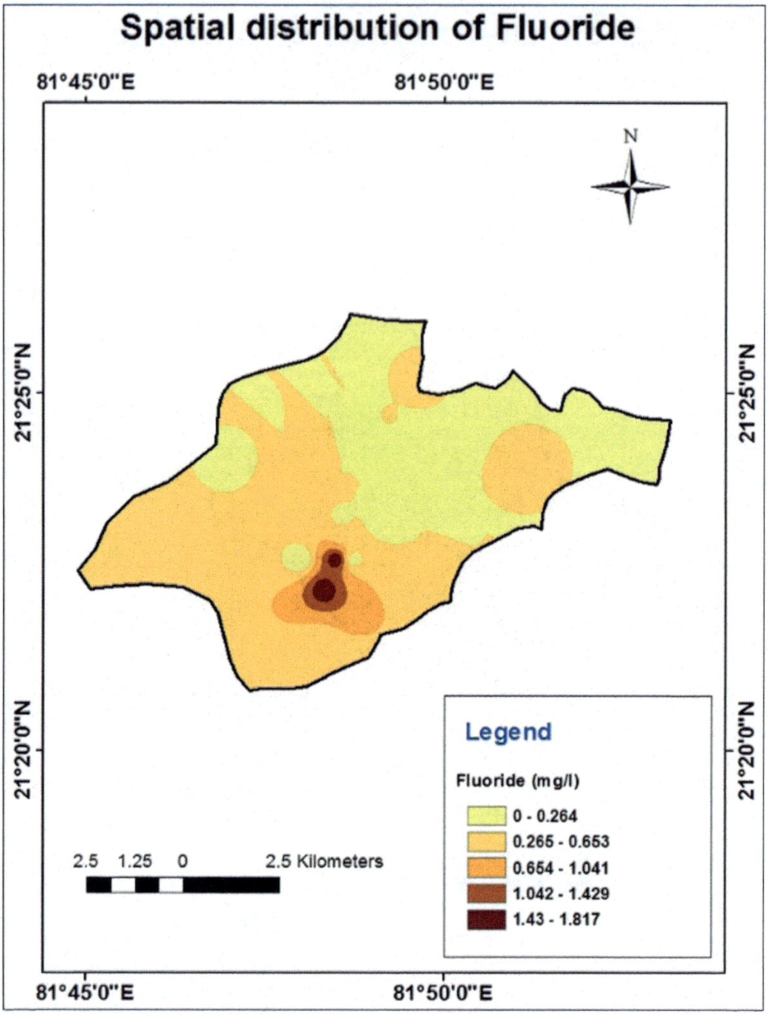

Fig. 6.9 Spatial distribution map of Fluoride

Iron, Sodium and Potassium

Iron value ranges from 0.0355 to 0.4105 (mg/l), Sodium value ranges 6.1–320.9 (mg/l) and Potassium value ranges from 1 to 4087 (mg/l) in this study area. Iron value of the samples lies within the described permissible limit 1.0 mg/l whereas the values of sodium and potassium exceeds the permissible limit value. The spatial distribution map of iron, sodium and potassium are shown as Figs. 6.12, 6.13 and 6.14 respectively.

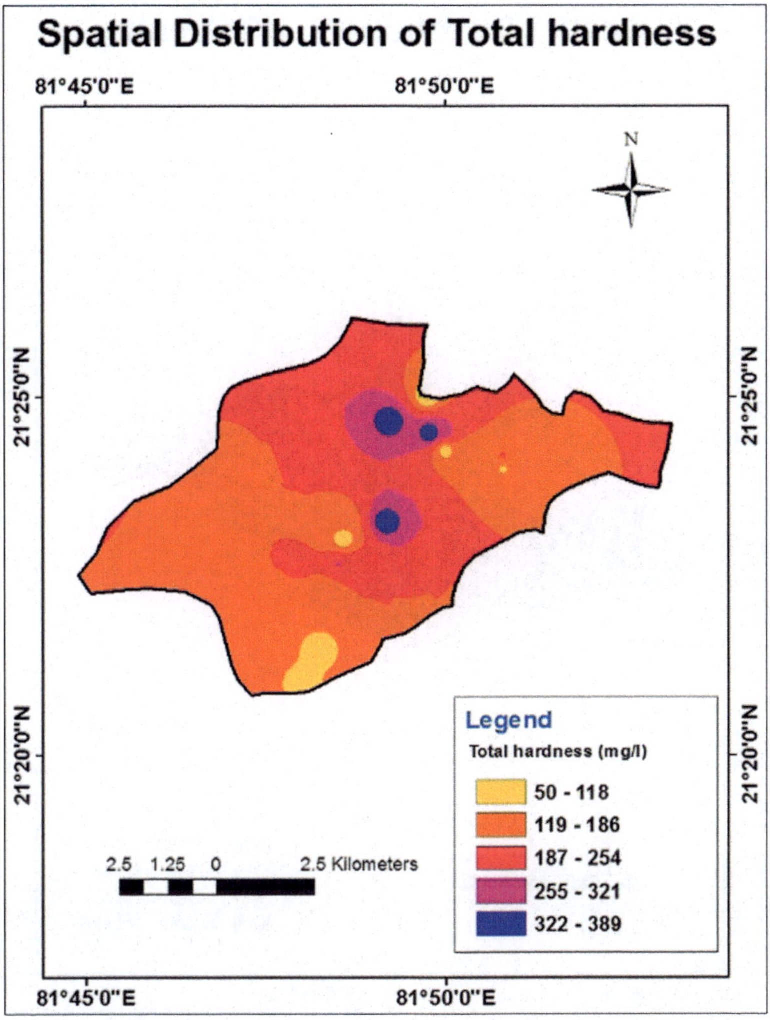

Fig. 6.10 Spatial distribution map of Total Hardness

6.4.2 Water Quality Index

Water quality index is a rating reflecting the composite influence of different water quality parameters on the overall quality of water. WQI is calculated using the logarithmic formula proposed by Tiwari and Mishra (1985). The WQI spatial distribution map was prepared using IDW technique. Inverse distance weighted interpolation is an exact interpolator, where the minimum and maximum values in the interpolated

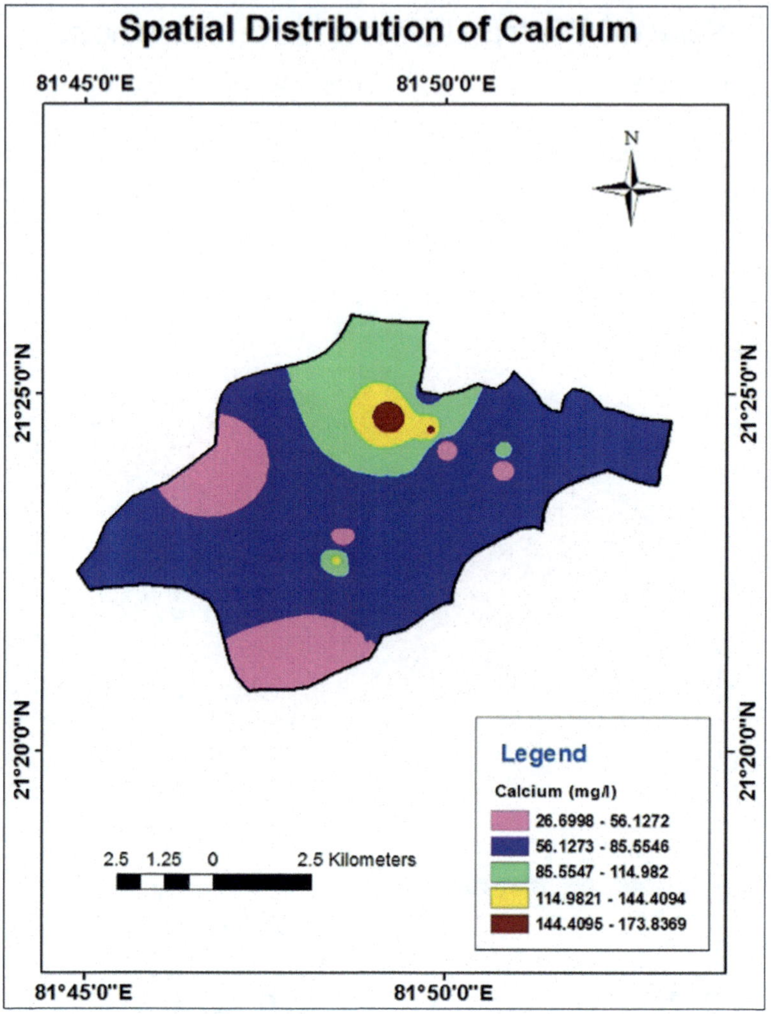

Fig. 6.11 Spatial distribution map of Calcium

surface can only occur at sample points. IDW assumes that the phenomenon being model is driven by local variation. The map for water quality index classified water samples into 5 classes namely Excellent to Unsuitable for drinking.

The steps involved in computing WQI using logarithmic method are as follows.

$$WQI = \text{anti log}\{[Wi \times \log(q_i)]\} \tag{6.1}$$

where, W_i is weightage factor and q_i is quality rating for the ith water quality parameter and are defined as,

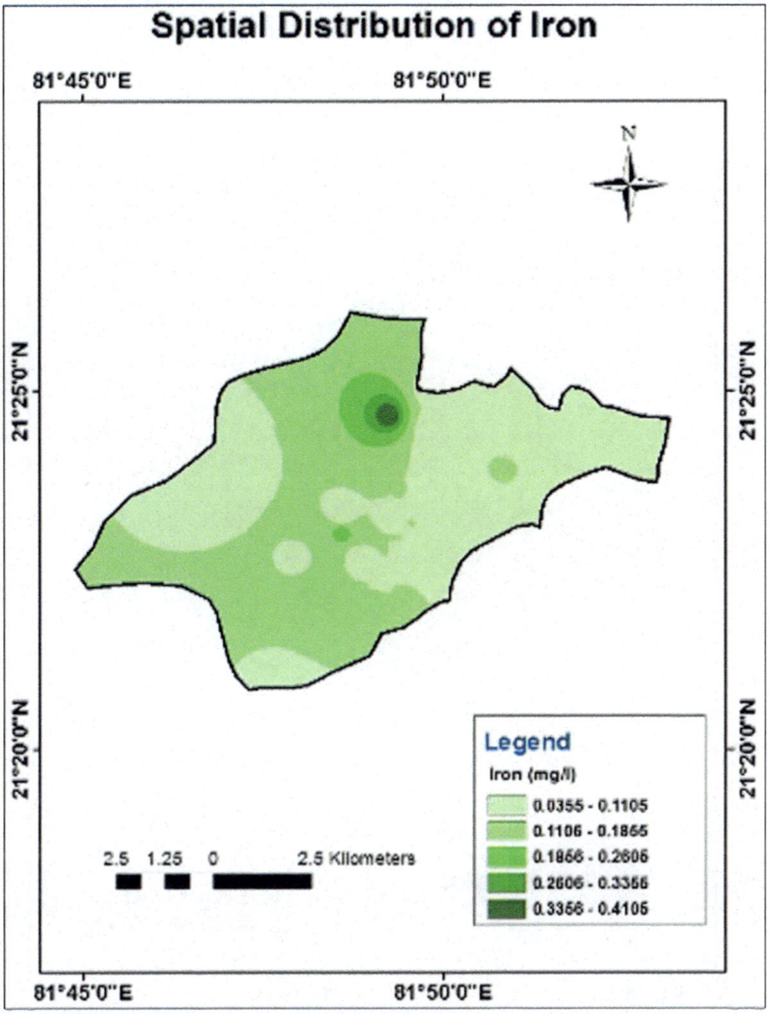

Fig. 6.12 Spatial distribution map of Iron

$$W_i = \frac{K}{P_i} \text{ and } q_i = 100 \times \frac{C_i}{P_i}$$

K is proportionality constant given as

$$K = \frac{1}{\sum (1/P_i)} \tag{6.2}$$

where, C_i is observed concentration of ith parameter and P_i is permissible limit of ith parameter. The water quality index map is prepared for the study area

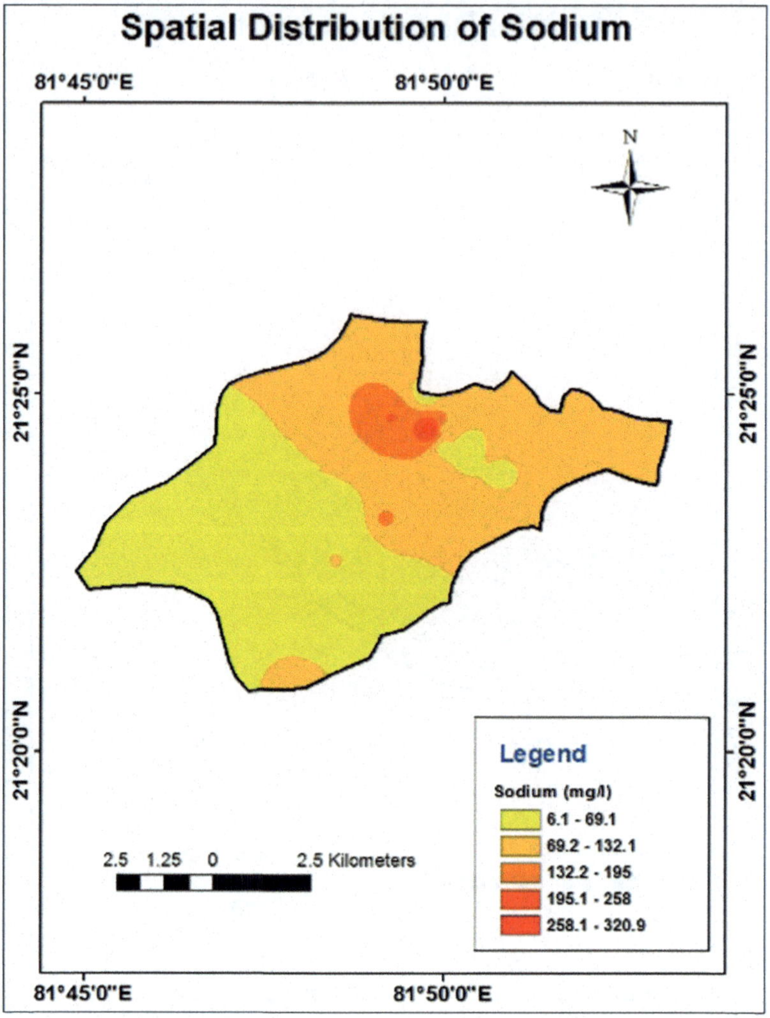

Fig. 6.13 Spatial distribution map of Sodium

shown as Fig. 6.15. Classification was done according to the BIS standards as shown in Table 6.2. The WQI values for the samples at various locations are given in Table 6.3.

6.5 Conclusions

Out of 25 water samples analysed for drinking water quality, 2 samples are found to be unsuitable for drinking, 10 samples fall under good and excellent zones and 13 samples quality were poor and very poor. Water in the poor and very poor zones

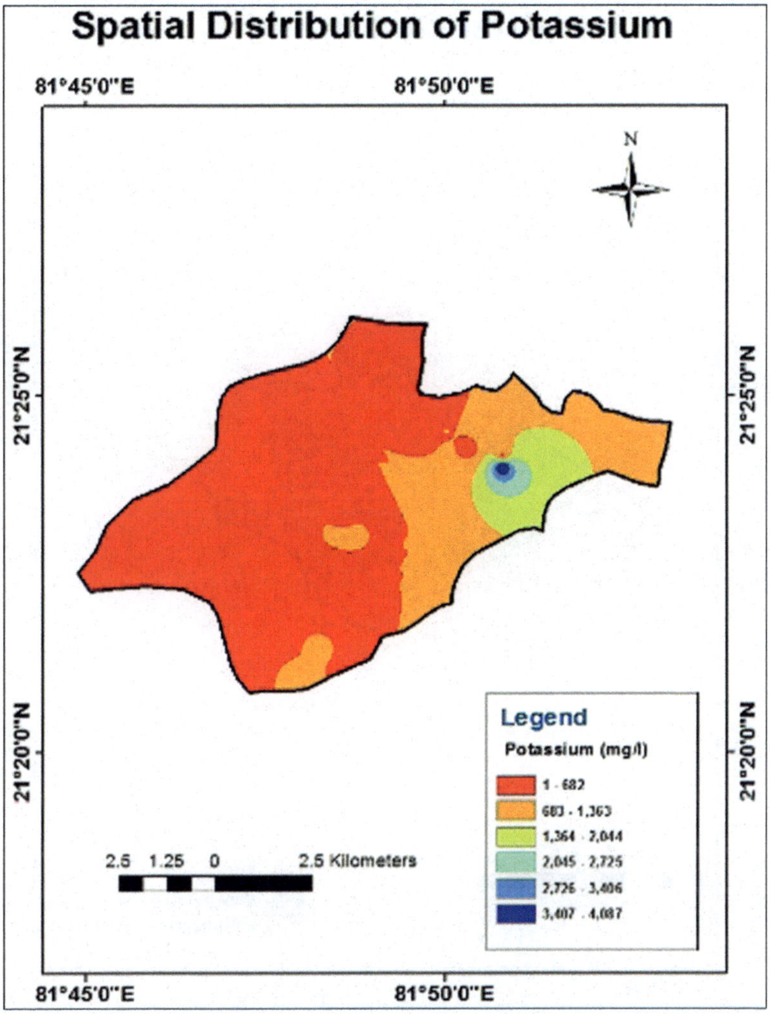

Fig. 6.14 Spatial distribution map of Potassium

can be utilised after proper treatment. Suitable remedial measures and groundwater augmentation structures are to be constructed in the study area to improve the quality of water. This study specially focused on the seasonal assessment of water quality in the study area and analysis suggests that water in the study area in general needs specific treatment before supplying to the general public as potable water. Developed WQI map can be used as a ready reckoner for suggesting suitable and economic water treatment measures in the study area. It can also for suitable crop selection and for planning conjunctive use along with available meagre surface water sources.

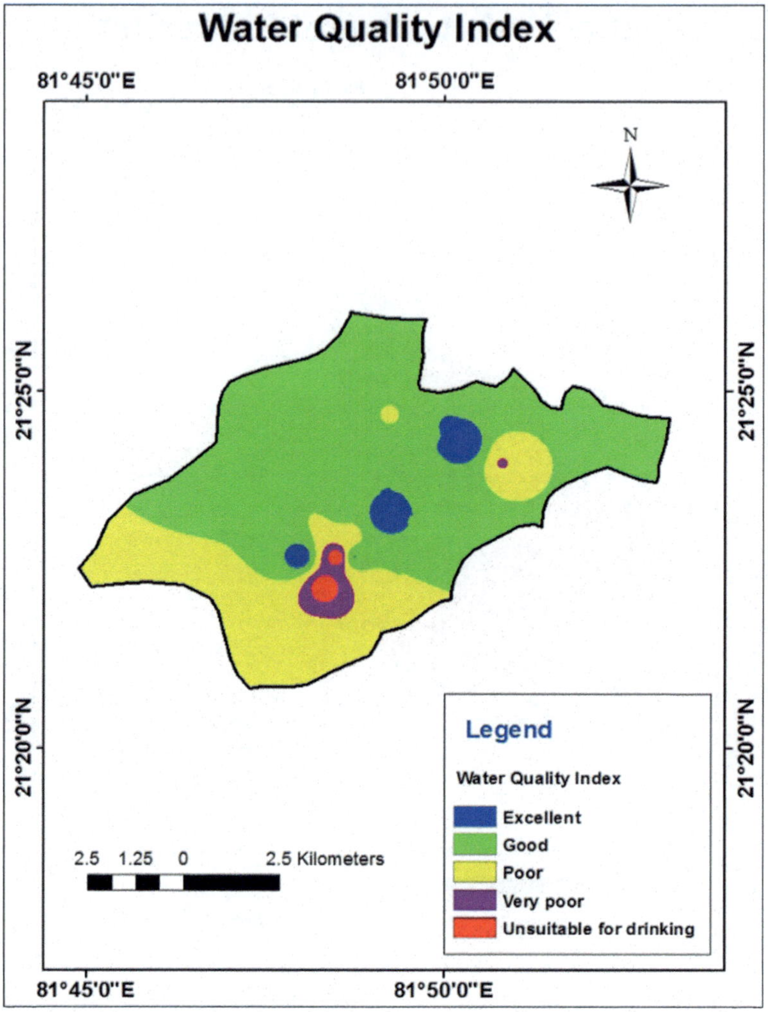

Fig. 6.15 Water quality index map

Table 6.2 Water quality index (WQI) rating

Water quality index value	Water quality status
0–25	Excellent
26–50	Good
51–75	Poor
76–100	Very poor
>100	Unsuitable for drinking

Table 6.3 Computed values for water quality index (WQI)

No. of samples	WQI	Status of water quality
1	65.3	Poor
2	58.7	Poor
3	19.5	Excellent
4	68.4	Poor
5	60.3	Poor
6	76.1	Very poor
7	63.8	Poor
8	64.0	Poor
9	65.6	Poor
10	52.0	Poor
11	66.8	Poor
12	66.6	Poor
13	36.3	Good
14	107.0	Unsuitable for drinking
15	103.0	Unsuitable for drinking
16	45.7	Good
17	19.6	Excellent
18	45.6	Good
19	57.0	Poor
20	58.8	Poor
21	47.2	Good
22	32.3	Good
23	20.0	Excellent
24	46.9	Good
25	35.8	Good

References

Chatterjee C, Raziuddin M (2002) Determination of water quality index (WQI) of a degraded river in asansol Industrial area, Raniganj, Burdwan, West Bengal. Nat Environ Pollut Technol 1(2):181–189

Naik S, Puohit KM (1996) Physico-chemical analysis of some community ponds of Rourkela. Indian J Environ Prot 16(9):679–684

Nas B, Berktay A (2010) Groundwater quality mapping in urban groundwater using GIS. Environ Monitor Assess 160(1–4):215–227

Rupal M, Bhattacharya T, Chakrabarty S (2012) Quality characterisation of groundwater using water quality index (WQI). India Int Res J Environ Sci 1(4):14–23

Selvam S, Manimaran G, Sivasubramanian P, Balasubramanian N, Seshunarayana T (2013) GIS-based evaluation of water quality index of groundwater resources around Tuticorin coastal city, south India. Environ Earth Sci 71:2847–2867

Shomar B, Fakher SA, Yahyal A (2010) Assessment of groundwater quality in the Gaza strip, Palestine using GIS mapping. J Water Resour Prot 2:93–104
Singh AA, Singh PK, Dhakate R, Singh NP (2013) Groundwater quality appraisal and its hydrochemical characterization in Ghaziabad (a region of Indo-Gangetic plain), Uttar Pradesh, India. Appl Water Sci 3(1):13–137
Srikanthan L, Jayaseelan J, Narendran D, Manikandan I, Singh SH (2013) Determination of groundwater quality using WQI in parts of Chennai city, Tamil Nadu. Indian J Sci 3(8):81–84
Srinivasamoorthy, Nanthakumar C, Vasanthavigar M (2009) Groundwater quality assessment from a hard rock terrain, Selam district of Tamil Nadu. Arab J Geosci, 112–115
Tiwari TN, Mishra M (1985) A preliminary assignment of water quality index of major Indian rivers. Indian J Environ Prot 5(4):276–279

Chapter 7
Improvement of Ground Water Quality Index Using *Citrus Limetta* Peel Powder

M. Rupas Kumar, M. Saravanakumar, S. Amarendra Kumar, V. Likhita Komal, and M. Sree Deepthi

Abstract In the recent times, the rapid rise in urbanisation complemented with population growth is necessitating much attention for potable water across the world. Due to the unavailability of adequate fresh water resources, Groundwater became the only potable source to majority of places across the world. The dependency on ground water due to scarce fresh water resources is commonly observed in densely populated regions such as India. Due to the persistence of extravagant anthropogenic activities, the quality of the groundwater is observed to be at critical levels in regions such as India. The present study concentrates on a water quality indicator called Ground Water Quality Index (WQI) representing overall water quality based on physico-chemical characteristics of a water sample. As the WQI is quantified based on the weighted average of the physical-chemical concentrations of water quality characteristics, the study considered sixteen parameters to ascertain accurate results. The ground water samples were collected from Kadapa City in Andhra Pradesh, India and physicochemical tests were performed to evaluate WQI. It is observed in the present study that higher values of WQI indicating poor water quality has been found due to excessive concentrations of Hardness, Total dissolved solids, Electrical conductivity, Turbidity, Alkalinity, and Fluorides in the collected water samples. Furthermore, the study proposes Citrus limetta (Sweet lemon) peel powder as a coagulant and the variability of WQI Values with coagulant concentrations are investigated. In addition, sensitivity analysis to examine the impact of coagulant dosage, mixing speed and stirring time on WQI is executed and optimal levels of these conditions resulting in minimum WQI values are found. A comparative study is also accomplished to examine the effect of different proportions of Alum and Citrus limetta combination on WQI. The present study recommends Citrus limetta peel powder as a natural, environmental friendly, locally available, cheap, and effective coagulant to treat Groundwater and to improve water quality status.

M. Rupas Kumar (✉) · M. Saravanakumar · S. Amarendra Kumar · V. Likhita Komal · M. Sree Deepthi
Department of Civil Engineering, Rajiv Gandhi University of Knowledge Technologies, RK Valley, Kadapa, Andhra Pradesh, India
e-mail: rupaskumar@rguktrkv.ac.in

A. Pandey et al. (eds.), *Climate Impacts on Water Resources in India*, Water Science and Technology Library 95, https://doi.org/10.1007/978-3-030-51427-3_7

7.1 Introduction

Water is a limited renewable natural resource essential for the existence of life. As the surface water sources are confined to few regions, ground water became the major dependable source of drinking water to majority parts of urban and rural regions in the world. The demand for ground water has been aggravated in the recent times due to rapid industrialization, urbanization and population growth which made ground water as the world's most extracted raw material. The increased population and industrial growth have not only escalated the exploitation of groundwater but have also ensued in declination of water quality which made the preservation and improvement of ground water qualities to be vital for existence of life.

The present study concentrates on expression of overall quality of ground water in terms of a number called Water Quality Index (WQI). Horton (1965) coined WQI as a contemplation of composite influence of individual water quality parameters on the overall quality of water. The principal objective of WQI is to transform a large quantity of complex water quality data into a simple value that can be perceivable by the public (Bharti and Katyal 2011). The present study considers sixteen physico-chemical parameters in evaluating WQI as the number of parameters considered in weighing influences accuracy of the index.

7.2 Study Area

The study concentrates on groundwater quality of Kadapa City, the district head-quarters of YSR district of Andhra Pradesh state in India. The city covers an area of 164.08 km^2 catering the population of 343,054 (Sunitha et al. 2016; Census of India 2011). The annual rainfall is about 710 mm and the area is found to experience minimum and maximum rainfall levels in the months of January an October respectively (Kumar 2013).

7.3 Objectives and Scope of the Study

Coagulation is a conventional water treatment process that involves destabilizing and aggregating colloidal water contaminants into large flocs by neutralizing their charge, then absorbing and enmeshing them. The present study proposes citrus limetta peel powder as a natural coagulant which is often discarded as a waste material. The study accentuates in estimating the influence of the proposed coagulant on WQI under different conditions such as mixing speed, stirring time and dosage of the ground water samples in Kadapa city, Andhra Pradesh, India. The optimal levels of coagulant dosage, mixing speed, and stirring time resulting in best WQI values are evaluated.

7.3.1 Evaluation of Water Quality Index

In the present study, the water quality samples were evaluated using the weighted arithmetic index method proposed by Brown et al. (1970), where the unit weights, measured values, ideal values and their respective standard permissible values of water quality parameters are driving factors in computation. The water quality index has been calculated by using the standard values of drinking water quality as per the recommendations of Bureau of Indian Standards (BIS), and World Health Organisation (WHO). The weighted arithmetic index method evaluates WQI using the following expression.

$$WQI = \frac{\sum W_n Q_n}{\sum W_n}$$

where "Q_n" and "W_n" are quality rating and unit weight of the nth water quality parameter respectively.

The weighted arithmetic index method expresses the quality rating "Q_n" by the equation

$$Q_n = 100\left(\frac{V_n - V_i}{S_n - V_i}\right)$$

where "V_n" and "S_n" are the measured and standard permissible values of nth parameter respectively. The ideal value of the water quality parameter is denoted by "V_i", which is taken as 7 for pH, 1 for Flouride, and 0 for all other water quality characteristics.

Unit weight of a parameter is expressed as an inverse proportion to its respective standard value "S_n" and can be evaluated using the equation.

$$W_n = \frac{K}{S_n}$$

where "K" is a proportionality constant computed by using the equation

$$K = \frac{1}{\sum \frac{1}{S_n}}$$

The standard values of the 16 water quality parameters considered in the present study are adopted from the recommendations of Bureau of Indian Standards (BIS) and World Health Organization (WHO) and their respective unit weight factors are computed as shown in Table 7.1. Brown et al. (1972) and Chatterji and Raziuddin (2002) indicated the status of the water quality and recommended the utility of the water samples with reference to the values of water quality index as given in Table

Table 7.1 Relative weights (W_n) of the parameters used for WQI determination

Water quality parameter	BIS/WHO standard (S_n)	Units	Unit weight (W_n)	Water quality parameter	BIS/WHO standard (S_n)	Units	Unit weight (W_n)
Colour	5	Hazen	0.012976	Magnesium	30	mg/l	0.002163
pH	8.5	–	0.007633	Sodium	200	mg/l	0.000324
Electrical conductivity	1000	μS/cm	0.000064	Chloride	250	mg/l	0.000260
Total dissolved solids	500	mg/l	0.000130	Sulphate	200	mg/l	0.000324
Turbidity	1	NTU	0.064878	Nitrate	45	mg/l	0.001442
Total hardness	200	mg/l	0.000324	Fluoride	1.5	mg/l	0.043252
Alkalinity	200	mg/l	0.000324	Iron	0.3	mg/l	0.216260
Calcium	75	mg/l	0.000865	Manganese	0.1	mg/l	0.648780

Table 7.2 Status of water quality

Water quality index	Water quality status	Possible usage
0–25	Excellent	All purposes (Drinking, Irrigation and Industrial)
25–50	Good	All purposes (Drinking, Irrigation and Industrial)
50–75	Poor	Irrigation and Industrial purposes
75–100	Very poor	Industrial purposes
>100	Unfit for drinking	Treatment required before usage to any purpose

7.2. The present study considers these indications to adjudicate the collected ground water samples under different conditions such as dosage, mixing speed and stirring time of the coagulant.

7.3.2 Influence of Coagulant on WQI

Coagulation is process used in treatment of both surface and ground water as it results in the removal of particles and their attached microorganisms by the mechanism of adsorption (Belayneh and Batu 2014). The process of coagulation could be achieved by the usage of natural and artificial materials called coagulants (Choy et al. 2014). Nowadays, the attention is being focused on the use of natural coagulants which are locally available recommending environmental friendly operations in water purification to minify the water treatment costs. Several attempts have been made in the past on the application of Moringa Oleifera as a coagulant for water treatment due to their efficiency in water purification and easy availability (Muyibi and Evison 1995; Egbuikwem and Sangodoyin 2013; Martin et al. 2012; Hendrawati et al. 2016; Dehghani and Alizadeh 2016). Furthermore, few attempts commended for the utilization of other natural materials as potential coagulants such as water melon seed (Muhammad et al. 2015), Chitosan (Bina et al. 2009), Rice husk (Asif and Chen 2017; Syuhadah and Rohasliney 2012), Peanut Seeds (Birima et al. 2013), and Plantago ovate (Ramavandi 2014). However, the studies are limited to the removal of suspended matter and microorganisms attached to the particles.

The present study concentrates on use of Citrus limetta peel powder as a coagulant material to ameliorate WQI, which is an indication of several physico-chemical characteristics of water. Citrus limetta peel powder collected from local market is washed several times with water to remove the adhering dirt particles. Furthermore, the sample was dried under sunlight for 7 days and chopped by manual cutters into fine pieces. Then, the sample was crushed to obtain small particle size powder and sieved with different fine mesh sizes. The powder is again washed with distilled water to remove any acidity or alkalinity, and oven dried at 80 °C for 24 h. The obtained sample powder was used as a coagulant in the present study.

7.4 Results and Discussion

The arithmetic mean values of physico-chemical water quality parameters of the ground water samples collected from 24 locations in Kadapa city are used for the measurement of WQI as shown in Table 7.3. The WQI evaluated by this procedure is found to be 62.92 indicating poor water quality status of the study area. It is observed that the higher concentrations of parameters such as Hardness, Total Dissolved Solids, Electrical Conductivity, Turbidity, Alkalinity, and Fluorides than their permissible levels ensued in poor water quality of the study area.

7.4.1 Effect of Coagulant Dosage on WQI

To evaluate the effect of Citrus limetta peel powder on WQI, a site located on the banks of Pennar River called "Water gandi" is selected among the 24 locations in the case study as the location is principal source of drinking water to Kadapa city. The concentrations of water quality parameters and their corresponding WQI under the dosage of different coagulant concentrations analyzed and tabulated in Table 7.4. From the obtained results, it is observed that coagulant dosage of 100 mg/l resulted in best improvement in WQI compared to other concentrations. The water quality parameters such as Electrical conductivity, Total dissolved solids, Alkalinity, and Fluorides have increased significantly when the coagulant concentrations are above 100 mg/l resulting in increase of WQI.

Table 7.3 Observed water quality parameters of Kadapa City

Water quality parameter	Units	Maximum value	Minimum value	Arithmetic mean (V_n)
Colour	Hazen	1.1	0.95	1
pH	–	8.8	7.6	8.4
Electrical conductivity	μS/cm	2400	1678	1790
Total dissolved solids	mg/l	1430	860	1200
Turbidity	NTU	3.2	2.9	3.1
Total hardness	mg/l	245	190	230
Alkalinity	mg/l	270	206	235
Calcium	mg/l	62	50	55
Magnesium	mg/l	25	20	23
Sodium	mg/l	210	183	196
Chlorides	mg/l	180	171	176.5
Sulphates	mg/l	156	141	150
Nitrates	mg/l	13	11.5	11.8
Fluorides	mg/l	1.4	1	1.2
Iron	mg/l	0.15	0.08	0.1
Manganese	mg/l	0.05	0.05	0.05
			WQI	62.92

Table 7.4 Values of water quality parameters (V_n) under different coagulant concentrations

Water quality parameter	Units	0 mg/l	50 mg/l	100 mg/l	150 mg/l	200 mg/l
Colour	Hazen	1	1	1	0.9	0.9
pH	–	8.2	7.4	7.3	7.6	8
Electrical conductivity	μS/cm	1786	1684	1634	1740	1840
Total dissolved solids	mg/l	1020	983	860	965	1100
Turbidity	NTU	3	2.2	1.5	1.8	2.1
Total hardness	mg/l	225	110	102	175	190
Alkalinity	mg/l	240	185	170	210	230
Calcium	mg/l	54	47	46	46	40
Magnesium	mg/l	22	20	20	21	20
Sodium	mg/l	190	180	175	176	175
Chlorides	mg/l	174.5	140	138	136	135
Sulphates	mg/l	145	130	132	131	130
Nitrates	mg/l	12	11	11	11.2	11
Fluorides	mg/l	1.3	1.1	0.9	1.05	1
Iron	mg/l	0.1	0.1	0.08	0.09	0.09
Manganese	mg/l	0.05	0.05	0.046	0.05	0.048
	WQI	63.0216	55.63167	45.26329	51.97927	52.3937

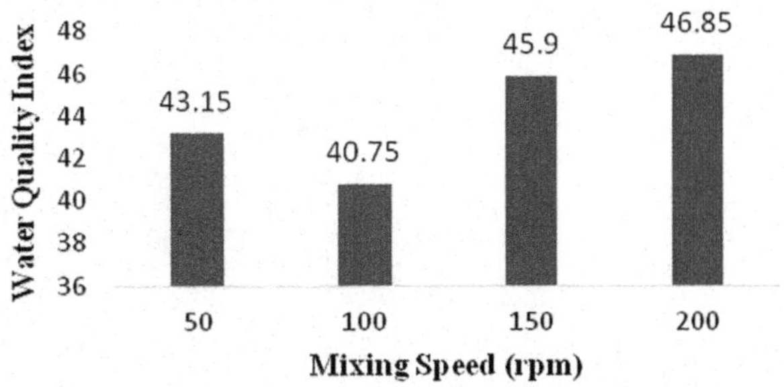

Fig. 7.1 Variation of water quality index with mixing speed (rpm)

7.4.2 *Effect of Coagulant Mixing Speed on WQI at Constant Dosage*

Coagulant mixing speed is a vital parameter which influences the activity of flocculation as it affects the dispersion of the coagulant and formation of flocs. While the low mixing speeds results in ineffective floc formation, higher mixing speeds ensues the breakage of flocs as the flocs are highly fragile in general (Martin et al. 2012). Therefore, the mixing speed should be adjusted at appropriate levels to produce good flocs achieving higher particle removal efficiency during the coagulation process (Othman et al. 2008).

To evaluate the effect of coagulant mixing speed on WQI, an interval of 50 rpm is chosen between 50 and 200 rpm at the obtained optimal coagulant dosage i.e., 100 mg/l. The variations of WQI with coagulant mixing speed are as shown in Fig. 7.1. From the obtained results, the best WQI indicating good water quality is observed at a mixing speed of 100 rpm.

7.4.3 *Effect of Stirring Time on WQI at Constant Dosage and Mixing Speed*

Stirring is a process of agitation exercised to enhance the agglomeration of particles for better settling. The variation of coagulant stirring time on WQI for the observed optimal coagulant constant dosage and mixing speed conditions are analyzed and presented in Fig. 7.2. The results show that the optimal stirring time is 20 min to achieve the best value of WQI when operated at constant mixing speed of 100 rpm.

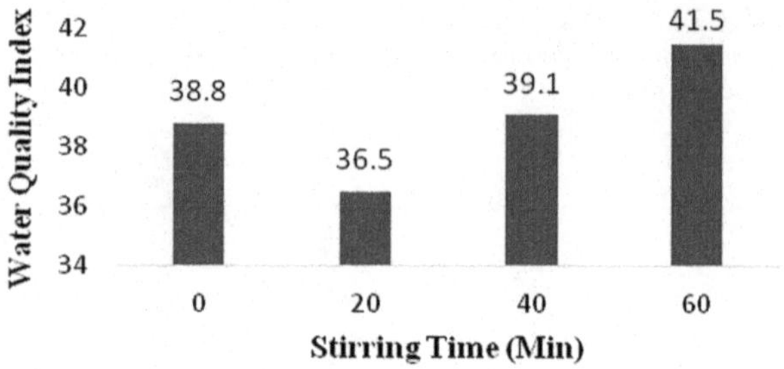

Fig. 7.2 Variation of stirring time with water quality index

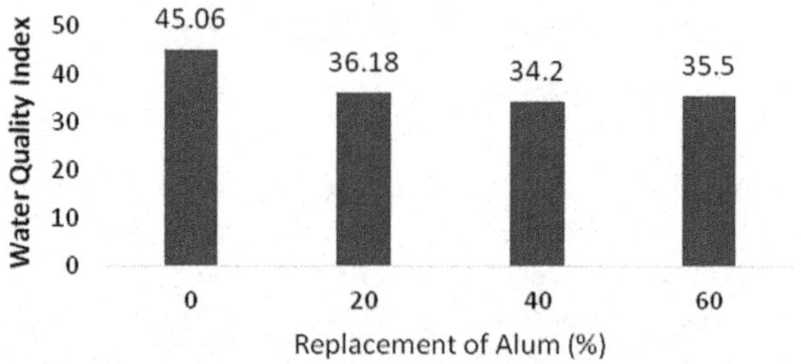

Fig. 7.3 Variation of water quality index with replacement of Alum

7.4.4 Effect of Citrus Limetta Peel Powder and Alum Combination on WQI

In order to evaluate the effect of coagulant and coagulant aid combination on WQI, Alum was used as Coagulant aid in addition to the natural coagulant citrus limetta. The combination is commingled in different proportions and the variations of WQI are recorded as shown in Fig. 7.3. The results exhibit that best value of WQI indicating "good" water quality status is obtained when natural coagulant is replaced with Alum by 40%. In other words, the ratio of citrus limetta and alum mixture is observed to be 3:2 to minify the magnitude of WQI indicating better water quality status.

7.5 Conclusion

In the present study, Citrus limetta (Sweet lemon) peel powder is proposed as a natural coagulant to improve water quality status of the ground water samples collected. WQI is used as a measure to indicate the status of water quality and therefore, the variations in WQI under different concentrations of the coagulant are recorded. The key factors influencing the effectiveness of coagulation process such as mixing speed, and stirring time are considered in the present study and optimal levels of these factors ensuing in best WQI values of the collected samples are found. The study determined the optimal levels of the conditions viz. coagulant dosage as 100 ppm, mixing speed as 100 rpm, and stirring time is 20 min. Furthermore, the study proposed Alum as a coagulant aid and revealed the proportion of citrus limetta and alum mixture as 3:2 to result in best WQI value. From the obtained results, the present study recommends Citrus limetta peel powder as a natural, environmental friendly, locally available, cheap, and effective coagulant to treat Groundwater and to improve WQI.

References

Asif Z, Chen Z (2017) Removal of arsenic from drinking water using rice husk. Appl Water Sci 7:1449–1458

Belayneh A, Batu W (2014) Application of biosorbent derived from cactus peel for removal of colorful manganese ions from ground water. J Water Resour Ocean Sci 4(1):18–23. https://doi.org/10.11648/j.wros.20150401.13

Bharti N, Katyal D (2011) Water quality indices used for surface water vulnerability assessment. Int J Environ Sci 2(1):154–173

Bina B, Mehdinejad MH, Nikaeen M, Movahedian Attar H (2009) Effectiveness of Chitosan as natural coagulant aid in treating turbid waters. Iran J Environ Health Sci Eng 6(4):247–252

Birima AH, Hammad HA, Desa MNM, Muda ZC (2013) Extraction of natural coagulant from peanut seeds for treatment of turbid water. IOP Conf Ser Earth Environ Sci 16:012065

Brown RM, McClelland NI, Deininger RA, O'Connor MF (1972) A water quality index-crashing the physiological barrier. Indic Environ Qual 1:173–182

Brown RM, McClelland NI, Deininger RA, Tozer RG (1970) Water Quality Index-do we care? Water Sewage Works 117(10):339–343

Census of India (2011) Office of the Registrar General & Census Commissioner, New Delhi

Chatterjee C, Raziuddin M (2002) Determination of water quality index (WQI) of a degraded river in Asansol Industrial area, P.O. Raniganj, District Burdwan, West Bengal. Nat Environ Pollut Technol 1(2):181–189

Choy SY, Prasad KMN, Wu TY, Raghunandan ME, Ramanan RN (2014) Utilization of plant-based natural coagulants as future alternatives towards sustainable water clarification. J Environ Sci 26:2178–2189

Dehghani M, Alizadeh MH (2016) The effects of the natural coagulant Moringa Oleifera and Alum in wastewater treatment at the Bandar Abbas Oil Refinery. Environ Health Eng Manag J 3(4):225–230

Egbuikwem PN, Sangodoyin AY (2013) Coagulation efficacy of Moringa Oleifera seed extract compared to Alum for removal of turbidity and *E. coli* in three different water sources. Eur Int J Sci Technol 2(7):13–20

Hendrawati, Yuliastri IR, Nurhasni, Rohaeti, E, Effendi H, Darusman LK (2016) The use of Moringa Oleifera seed powder as coagulant to improve the quality of wastewater and ground water. Earth Environ Sci 31

Horton RK (1965) An index number system for rating water quality. J Water Pollut Cont Fed 37:300–305

Kumar JS (2013) Ground Water Brochure—Y S R Kadapa District, Andhra Pradesh, Central Ground Water Board, Ministry of Water Resources, Government of India

Martin JS, Heredia JB, Peres JA (2012) Improvement of the flocculation process in water treatment by using Moringa Oleifera seeds extract. Braz J Chem Eng 29(03):495–501

Muhammad IM, Abdulsalam S, Abdulkarim A, Bello AA (2015) Water melon seed as a potential coagulant for water treatment. Global J Res Eng Chem Eng 15(1)

Muyibi SA, Evison LM (1995) Optimizing physical parameters affecting coagulation of turbid water with Moringa Oleifera seeds. Water Res 29(12):2689–2695

Othman Z, Bhatia S, Ahmad AL (2008) Influence of the settleability parameters for palm oil mill effluent (pome) pretreatment by using Moringa Oleifera seeds as an environmental friendly coagulant'. International Conference on Environment, Malaysia, pp 1–9
Ramavandi B (2014) Treatment of water turbidity and bacteria by using a coagulant extracted from Plantago ovata. Water Resour Ind 6:36–50
Sunitha V, Muralidhara Reddy B, Sumithra S (2016) Assessment of groundwater quality index in the Kadapa Municipal City, Y.S.R. District, Andhra Pradesh. Indian J Appl Res 6(2):545–548
Syuhadah NS, Rohasliney H (2012) Rice husk as biosorbent: a review. Health Environ J 3(1):89–95

Chapter 8
Paper Mill Effluents: Identification of Emerging Pollutants in Taranga Beel of Assam, India

Khanindra Sharma, Neelotpal Sen Sarma, and Arundhuti Devi

Abstract The present research work deals with the study of some of the essential physicochemical parameters of water samples of "Taranga beel" (Beel is a local term which defines wetland) of Jagiroad, Assam, India. This study indicates a decline in the water quality of Taranga beel due to the paper industry's influence in its vicinity. The collection of samples and the determination of various physicochemical parameters and heavy metal concentration was done according to the standard protocol. The physicochemical parameters such as Total Solids (TS), Total Dissolved Solids (TDS), Turbidity, pH, conductivity, Chemical Oxygen Demand (COD), Sulphate, Dissolved Oxygen (DO), Fluoride, and heavy metals such as Cd, Cr, Cu, Fe, Mg, Mn, Na, Ni and Pb were analyzed thoroughly. Among all the physicochemical parameters and heavy metals, the concentration of TS, TDS, Turbidity, DO, COD, Cr, Pb, Mn, and Fe were above the permissible limits prescribed by the Bureau of Indian Standards (BIS).

8.1 Introduction

Wetlands are the most important ecosystems offering shelter to a wide diversity of plants and animals and deliver many ecological, climatic, and societal functions. In addition to this, they play a vital role in provisioning and maintaining water quality for endless living organisms. Wetlands are the most fertile and productive

K. Sharma · A. Devi (✉)
Environmental Chemistry Laboratory, Life Sciences DivisionResource Management and Environment Section, Institute of Advanced Study in Science and Technology, Guwahati, Assam 781035, India
e-mail: deviarundh2@yahoo.co.in

K. Sharma
e-mail: khanindrasharma72@gmail.com

N. S. Sarma
Advanced Materials Laboratory, Physical Sciences Division, Institute of Advanced Study in Science and Technology, Guwahati, Assam 781035, India

A. Pandey et al. (eds.), *Climate Impacts on Water Resources in India*, Water Science and Technology Library 95, https://doi.org/10.1007/978-3-030-51427-3_8

sites (Hazarika 2013). However, due to the breakneck growth of human civilization, unplanned urbanization, and industrialization, aquatic resources are being used continuously as dumping grounds for industrial and other wastes (Hynes 1980). The contamination of natural water bodies by industrial effluent has appeared as a most burning and challenging issue in various developing countries (Ijeoma and Achi 2011). Industrial effluent contains enormous loads of organic and inorganic pollutants, and it is hazardous if discharged untreated to any natural or artificial water body.

Due to pollution and other anthropogenic activities, the water and sediment quality of various wetlands declines abruptly, and eventually, it affects humans, animals and plant life (Ijeoma and Achi 2011). A study reveals that about 70% of industrial waste in developing countries is dumped untreated into water bodies (UN-water 2009). Among all the industries, pulp and paper manufacturing industry is the sixth largest waste-producing industry and which eventually responsible for the degradation of its native environment (Sharma et al. 2019). This study was undertaken to explore the effect of paper mill effluent on the physicochemical properties of water of Taranga beel (a wetland) of Jagiroad, Assam, India.

8.2 Materials and Methods

8.2.1 Chemicals

All the chemicals used in the experiment were of analytical grade and purchased from Merck Chemicals. Chemicals were used without further purifications.

8.2.2 Study Area

Taranga beel is located between 92° 3′ 51″ E–92° 17′ 7″ longitude & 26° 8′ 2″–26° 11′ 41″ N latitude, in the Morigaon district of Assam, India (Saikia and Lohar 2012). Some parts of the wetland are covered with some floating pits and the exposed areas having a depth of approximately 1.0–1.5 m in winter and 6.5–7.5 m in rainy session.

8.2.3 Collection and Analysis of Water Samples

Water samples were collected from seven different sampling sites of Taranga beel by maintaining approximately a uniform distance between two sampling sites. Locations, from where samples were collected are shown in Fig. 8.1. All the Physico-chemical parameters were measured by using the standard protocols given by APHA (2005) and Trivedy and Goel (1986).

Again concentrations of various heavy metals were determined by analyzing the samples using Atomic Absorption Spectrophotometer (Shimadzu AA-7000). For AAS analysis, water samples were digested with 2N HNO_3 and few drops of concentrated $HClO_4$ at a temperature of 60–80 °C (Calace et al. 2002).

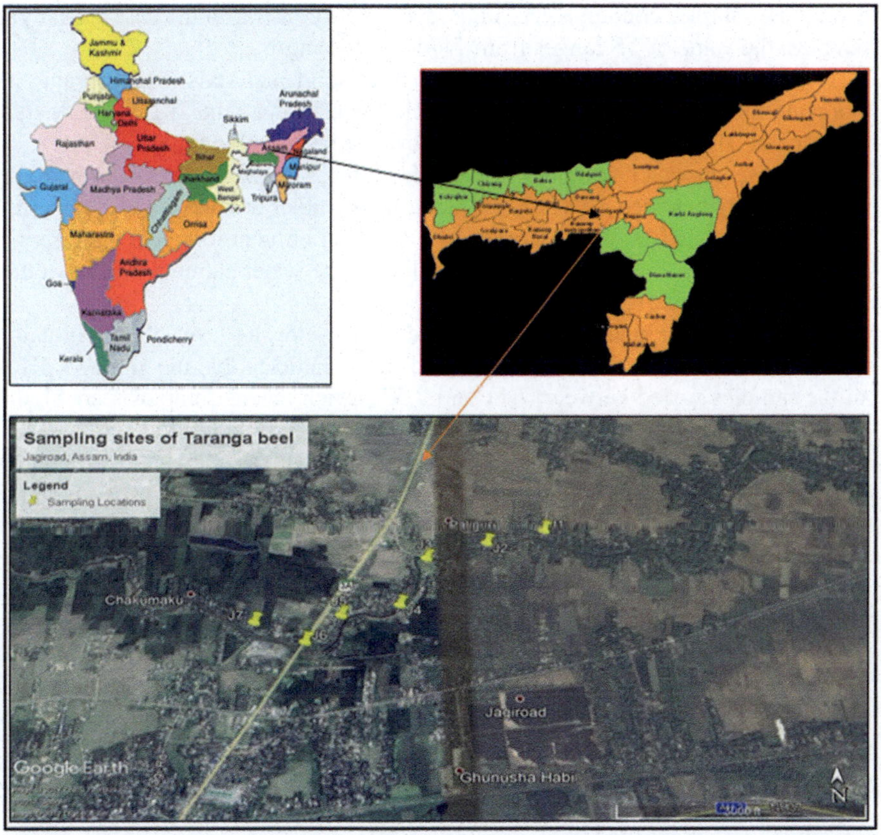

Fig. 8.1 Sampling locations of Taranga beel (*Source* Internet and Google Earth)

8.3 Results and Discussion

8.3.1 Physicochemical Characterization of Water Samples

All the collected water samples were characterized regarding their Physico-chemical parameters like TS, TDS, Turbidity, pH, Electrical conductivity, DO, COD, Sulphate, and Fluoride. Because these are the parameters, which help elucidate the level of contamination in water samples.

Experimental results of all the Physico-chemical parameters were compared with the values provided by BIS 2012 and are depicted in Table 8.1.

TS is a measurement of the combination of both dissolved and undissolved materials present in the water samples. From the obtained results, it can be observed that all the samples contain a very high level of TS content. In the case of analyzed samples, the value of TS ranges from 1121 to 2101 mg/L.

TDS is a quantitative measurement of total solid materials (both organic and inorganic substances) dissolved in the sample (Brraich and Saini 2015). In this study, the value of TDS is found to be present above the permissible limit.

All the seven water samples collected from the wetland are found to be highly turbid, ranging between 10.8–16 NTU, and these values are above the BIS permissible limit (BIS 2012). The discharge of paper mill effluent to the wetland may be responsible for such a high level of turbidity of the water samples of the wetland (Saikia and Lohar 2012).

pH is also a significant parameter, used to measure the level of contamination in an aqueous solution (Bhatnagar 2015). Results indicate that the value of pH for all the samples varied between 7.11 and 7.55, which means samples were slightly alkaline. However, pH values are within the BIS permissible limit.

The electrical conductivity (EC) is a parameter which gives an idea about the concentration of ions present in the sample. The electrical conductivity of any sample is directly proportional to its salinity. Higher the value of conductivity means the sample is highly saline (Nirgude et al. 2013). Here it is found that electrical conductivity is above the permissible limit, the maximum value of EC recorded among all the samples was 2.14 mS/cm.

DO content is below the desired limit in some parts of the wetland and indicates an improper environment for aquatic life. Discharge of paper mill effluent might be responsible for such a low DO level, as it contains a low level of dissolved oxygen (Kumar et al. 2011). Papermill effluent contains different organic and inorganic nutrients and various microorganisms used to degrade these nutrients using the oxygen dissolved in the water body and this eventually leads to the decreases in the level of DO (Sharma et al. 2019).

Table 8.1 Physicochemical parameters of water samples

Parameters	BIS permissible limits	Sample 1	Sample 2	Sample 3	Sample 4	Sample 5	Sample 6	Sample 7
TS (mg/L)	–	2101 ± 5.8	2022 ± 4.51	1225 ± 4.91	1409 ± 4.73	1206 ± 4.4	1121 ± 4.5	1220 ± 4.51
TDS (mg/L)	500	1321 ± 5.1	1300 ± 3.12	1237 ± 4.5	1010 ± 4.08	938 ± 4.03	706 ± 4.05	625 ± 4.03
Turbidity (NTU)	5	16 ± 1.51	10.8 ± 0.25	11.2 ± 0.25	13.2 ± 0.31	14 ± 0.45	12.2 ± 0.23	12 ± 0.21
pH	6.5–8.5	7.55 ± 0.05	7.89 ± 0.05	7.11 ± 0.04	7.56 ± 0.05	7.33 ± 0.04	7.19 ± 0.04	7.20 ± 0.04
EC (mS/cm)	–	2.14 ± 0.02	2.08 ± 0.03	0.0801 ± 0.00	0.1742 ± 0.00	0.1742 ± 0.00	0.186 ± 0.00	0.190 ± 0.00
DO (mg/L)	5.0–6.0	3.00 ± 0.03	3.113 ± 0.02	4.774 ± 0.01	5.273 ± 0.01	5.154 ± 0.01	6.289 ± 0.01	5.14 ± 0.01
COD (mg/L)	–	315 ± 2.1	317 ± 2.3	300 ± 3.01	312 ± 2.52	212 ± 3.51	220 ± 2.12	215 ± 2.41
Fluoride (mg/L)	0.6–1.2	1.91 ± 0.03	1.809 ± 0.03	1.689 ± 0.02	1.659 ± 0.02	1.9 ± 0.03	1.426 ± 0.02	1.52 ± 0.02

Now among all the measured parameters, COD is one of the most prominent parameters, usually used for the determination of the level of contamination in any water sample [3]. COD is a method used to measure the amount of oxygen required to break down of organic matters present in the sample. In any sample, a higher value of COD indicates a higher level of contamination. Here the value of COD is found to be very high, which ranges between 215 and 315 mg/L, indicating a higher level of contamination.

Fluoride content is also remarkably high (i.e., above the permissible limit) which ranges between 1.426–1.91 mg/L.

8.3.2 Metal Analysis

Metal content was analyzed by following the standard protocol given by APHA [7], and all the findings are presented in Table 8.2. From Table 8.2, it can be observed that the concentration of various heavy metals like Pb, Ni, Cr, Fe and Mn is found to be present, above their permissible limits. Individually all these heavy metals are very toxic and have an adverse effect on the ecosystem. Although Fe and Mn are considered as micronutrients and required for various metabolic and biological processes in living organisms, their elevated level is toxic to the environment.

8.4 Conclusion

Water pollution is a burning issue and the effects of water pollution are not only limited to human beings but also has destructive effects on the entire ecosystem of wetlands and other natural environments. In this study among all the Physico-chemical parameters, COD, DO, Fluoride, Turbidity, TS, TDS and the heavy metals Pb, Cr, Fe and Mn are found above their permissible limits. Thus, this study was carried out to bring awareness among the people about the water quality of the concerned wetland (Taranga beel).

Table 8.2 Different metal concentrations in collected water samples

Parameters	BIS permissible limits	Sample 1	Sample 2	Sample 3	Sample 4	Sample 5	Sample 6	Sample 7
Cd (mg/L)	0.003	0.0021 ± 0.00	0.0014 ± 0.00	0.0012 ± 0.00	0.0002 ± 0.00	0	0.001 ± 0.00	0
Pb (mg/L)	0.01	0.023 ± 0.00	0.018 ± 0.00	0.01 ± 0.00	0.0109 ± 0.00	0.0109 ± 0.00	0.01 ± 0.00	0.011 ± 0.00
Ni (mg/L)	0.02	0.026 ± 0.00	0.018 ± 0.00	0.006 ± 0.00	0.026 ± 0.00	0.016 ± 0.00	0.028 ± 0.00	0.025 ± 0.00
Cu (mg/L)	0.05	0.012 ± 0.00	0.008 ± 0.00	0.002 ± 0.00	0.007 ± 0.00	0.004 ± 0.00	0.007 ± 0.00	0.005 ± 0.00
Cr (mg/L)	0.05	0.123 ± 0.00	0.124 ± 0.00	0.06 ± 0.00	0.103 ± 0.00	0.107 ± 0.00	0.127 ± 0.00	0.1 ± 0.00
Na (mg/L)	–	2.814 ± 0.01	2.819 ± 0.01	3.04 ± 0.01	3.25 ± 0.01	2.87 ± 0.01	3.01 ± 0.01	3.19 ± 0.01
Mg (mg/L)	30	0.7 ± 0.01	0.726 ± 0.01	0.72 ± 0.01	0.728 ± 0.01	0.727 ± 0.01	0.725 ± 0.01	0.8 ± 0.01
Fe (mg/L)	0.3	8.41 ± 0.09	8.506 ± 0.08	5.447 ± 0.06	8.278 ± 0.08	8.5 ± 0.09	17.19 ± 0.08	18.59 ± 0.1
Mn (mg/L)	0.1	2.22 ± 0.05	2.192 ± 0.03	1.336 ± 0.01	1.625 ± 0.01	1.134 ± 0.01	3.532 ± 0.04	2.19 ± 0.03

Acknowledgements Authors are greatful to the Department of Science and Technology, Govt. of India and the Director of the Institute of Advanced Study in Science and Technology (IASST), Guwahati for financial support to carry out this work.

References

APHA (2005) Standard methods for the examination of water and wastewater, 22nd edition Prepared and published jointly by: American Public Health Association (APHA), American Water Works Association (AWWA), Water Pollution Control Federation

Bhatnagar A (2015) Assessment of physico-chemical characteristics of paper industry effluents. Rasayan J Chem 8:143–145

BIS (2012) Indian Standard drinking water specification. Prepared by Bureau of Indian Standards

Brraich OS, Saini SK (2015) Water quality index of Ranjit Sagar wetland situated on the Ravi river of Indus river system. Int J Adv Res 3(12):1498–1509

Calace N, Di AM, Nardi E, Petronio BM, Pietroletti M (2002) Adsorption isotherms for describing heavy-metal retention in paper mill sludges. Ind Eng Chem Res 41:5491–5497

Hazarika LP (2013) A study of certain physico-chemical characteristics of Satajan wetland with special reference to fish diversity indices, Assam, India. Eur J Exp Biol 3:173–180

Hynes HBN (1980) The biology of polluted waters. With 22 Fig. Liverpool University Press, Liverpool, p 202

Ijeoma K, Achi OK (2011) Industrial effluents and their impact on water quality of receiving rivers in Nigeria. J Appl Technol Environ Sanitation 1(1):75–86

Kumar DS, Satheeshkumar P, Gopalakrishnan P (2011) Ground water quality assessment in paper mill effluent irrigated area—using multivariate statistical analysis. World Appl Sci J 13(4):829–836

Nirgude NT, Shukla S, Venkatachalam A (2013) Physico-chemical analysis of some industrial effluents from Vapi industrial area, Gujarat, India. Rasayan J Chem 6:68–72

Saikia MK, Lohar P (2012) Structural and physico-chemical correlation of algal community of a wetland affected by pulp and paper mill effluents. Glob J Sci Front Res 12(5)

Sharma K, Pathak M, Kalita S, Bhattacharyya KG, Sarma NS, Devi A (2019) Sequential treatment of paper mill effluent with modified Fenton oxidation and bioflocculation. Environ Dev Sustain, 1–18

Trivedy RK, Goel PK (1986) Chemical and biological methods for water pollution studies. Environmental Publications

UN-Water (2009) World Water Day brochure. http//www.unwater.org/worldwaterday/downloads/WD09brochureLOW

Chapter 9
Transport of Nano-plant Nutrients in Lateritic Soils

Maheshwar Durgam and Damodhara Rao Mailapalli

Abstract Agriculture is the leading source of non-point source of water pollution mainly due to excess application of the fertilizers by the farmers to obtain more yield. The major portion of the applied fertilizer is being lost to the environment causing degradation of the land and water resources. One of the possible ways to deal with this problem in recent years comes out to be the application of nano-fertilizers in agriculture. Controlled release of nutrients is possible through the application of nano-fertilizers. Moreover, application of nutrients in nano form (nano-fertilizers) makes it easier for the plants to utilize the nutrients in a better way. But, the fate, mobility and transport of nano-fertilizers is still a question that is not been answered in recent times. An attempt was made in this study to know the mobility and transport of the nano-fertilizers such as nano-hydroxyapatite (nHAP) and urea doped nano-hydroxyapatite (Ur@nHAP) in the soil. Results from the soil column studies indicated that leaching of NH_4-N and NO_3-N were 55% less in case of Ur@nHAP treatment compared to Urea treatment and Ortho-P leaching was 128% less in case of both nano-fertilizer (nHAP and Ur@nHAP) treatments compared to Superphosphate treatments. The rate of transport of Phosphorus was found to be about half that of Nitrogen. Hence the application of nano-fertilizers to the soil may be environmentally safe and effective for crop production. However, more extensive studies with modelling tools may be needed to understand the mobility and transport of nano-fertilizers in different soil types.

9.1 Introduction

Water pollution is of foremost concern around the globe. Out of point and non-point sources, non-point source pollution is the main cause of water quality degradation. Agriculture is the principal source of the non-point source pollution. This is primarily due to the excess application of the fertilizers, plant protection chemicals and lack of

M. Durgam · D. R. Mailapalli (✉)
Agricultural and Food Engineering Department, IIT Kharagpur, Kharagpur, West Bengal 721302, India
e-mail: mailapalli@agfe.iitkgp.ac.in

A. Pandey et al. (eds.), *Climate Impacts on Water Resources in India*, Water Science and Technology Library 95, https://doi.org/10.1007/978-3-030-51427-3_9

nutrient management techniques. The ratio of macro-nutrients (Nitrogen: Phosphorous: Potassium) for optimal plant growth is 4:2:1 whereas, in India, it is practised as 10:2.7:1 (Subramanian and Tarafdar 2011), which results in two-fold supplementary nitrogen application to soil and environmental pollution. Approximately 40–75% of the applied fertilizer is being lost from the cropland, thus contaminates land and water resources (Celsia and Mala 2014). Excess application of the fertilizer will also decrease the producing capacity of the soil. Experience in developed countries has shown that as point sources are brought under control, the proportion of total loads from non-point sources became significantly higher as a percentage of total load. This is inevitable, especially as agricultural sources of non-point source effluence have proven to be particularly problematic to control (Ongley et al. 2010).

The loss of fertilizers (nitrogen and phosphorous) into the water bodies will result in eutrophication and thus pose a severe problem for the aquatic life. Eutrophication was acknowledged as the serious problem in surface waters, with agriculture affecting 60% of the river contamination, 50% of the lake contamination, and 34% of estuary contamination (U.S. EPA 1996; Carpenter et al. 1998). Nutrient use efficiencies of the N, P and K fertilizers remained constant as 30–35%, 18–20% and 35–40%, respectively leaving a major portion of the fertilizer to stay in the soil or enter into the aquatic system causing eutrophication (Subramanian and Tarafdar 2011). These problems, in turn, adversely affect the commercial, recreational and aesthetic values associated with the water resources (Sharpley and Withers 1994; Carpenter et al. 1998; Daniel et al. 1998).

One of the possible ways to deal with the above-mentioned problem is through the application of Nanotechnology in agriculture i.e. using nano-fertilizers. Nanotechnology is the interdisciplinary science which deals with the particles in the range of 1–100 nm. Nanotechnology can revolutionize the agriculture, particularly issues related to fertilizers such as nano-fertilizers are concerned (Benzone et al. 2015). As plants can only take nutrients in nano form, nutrients will be efficiently used if they are applied in nano range. It is possible through the application of nanofabricated fertilizers. Nano-fertilizers are the new products which contain rapidly available nutrients in the nanoscale range and are preferred largely due to their efficiency (Janmohammadi et al. 2016). Fertilizer nutrient use efficiency in crop production can be enhanced with the effective use of nano-fertilizers. Nitrogen use efficiency of the nano-fertilizer was found to be 82% whereas, the N-use efficiency of the conventional fertilizer was limited to 42% (Subramanian and Tarafdar 2011) Nano-fertilizers mainly delay the release rate of nutrients and extends the fertilizer effective period. The purity of elements is very high in nano-fertilizers. One of the advantages of nano-fertilizers is that they can be used in fewer amounts than ordinary fertilizers, means it saves cost by less expenditure on fertilizer which is economical for farmers point of view (Adhikari et al. 2014; Rameshaiah et al. 2015). Nanofabricated materials containing plant nutrients can be used in aqueous suspension and hydrogel forms, so as to enable hazard-free application, easy storage, and a convenient delivery system (Mukhopadhyay 2014).

However, new environmental and human hazards may emerge from nano-enhanced applications. This raises concerns for the agricultural workers who may become exposed primarily to such xenobiotics during their job tasks. Therefore, it is necessary to study the fate and transport of nanomaterials before their application in the field of agriculture (Iavicoli et al. 2017). In order to better understand the interaction of soil-landed ENMs (engineered nanomaterials) with plants, it is necessary to understand their fate, mobility, and transformations in soil. Firstly, the ageing of nanomaterials in the soil can have a dramatic impact on their toxic potential. It has been reported that toxic potential of various nanomaterials changes with their residence time in the soil (McShane et al. 2014). Because of their impact on crop nutritional quality and stress tolerance in plants, the application of nano-fertilizers is increasing. However, there are virtually no studies on the potential environmental impact of nanomaterials when used in agriculture (Morales-Diaz et al. 2017). So an attempt was made in this study to investigate the fate and transport of the nano-fertilizers (synthesized at Non-point Source Pollution Laboratory of Agricultural and Food Engineering Dept., IIT Kharagpur) in the lateritic soil. The specific objectives of this study were to conduct laboratory soil column experiments using nano-fertilizers and to measure plant nutrient analytes in leachate and to compare the nutrient transport among bulk and nano-fertilizers and to investigate their fate and transport in lateritic soil.

9.2 Materials and Methods

Soil column experiments were conducted at Non-point Source Pollution Laboratory of Agricultural and Food Engineering Department, IIT Kharagpur, to investigate the fate and transport of the nano-fertilizers in lateritic soils.

9.2.1 Soil Collection

The laterite soil was collected from the top 10–35 cm soil depth from the Research farm of Agricultural and Food Engineering (AgFE) Department, IIT Kharagpur. Large soil aggregates were sun-dried, broken up manually, and sieved to 2 mm. The physicochemical properties such as pH, EC, total N and total P were determined, are presented in Table 9.1. pH was measured using calibrated pH meter (Mettler

Table 9.1 Selected properties of field soil used for soil column studies

S. No.	Properties	Value
1	pH	7.45
2	EC(μS/cm)	38.97
3	Total P (kg/ha)	450.00
4	Total N (kg/ha)	280.00

Fig. 9.1 Soil column setup used for studying the transport of nano-nutrients in the laboratory

Toledo), EC by using calibrated EC meter (Mettler Toledo) and total N and total P were measured by using Easychem 300 (SYSTEA) machine.

9.2.2 Soil Column Design

Experiments were carried out using 9 columns fabricated using polyvinyl chloride with an inner diameter of 5.8 cm and a height of 33 cm and were housed in a chamber where the temperature was maintained constant (room temperature). Each column had two layers of cheesecloth glued to bottom of the column. Each column was plugged into a PVC cap with a drain hole at the bottom that allowed leachate to drain into collection bottles of 300 mL. The columns were placed on a wooden stand to facilitate free drainage into the collection bottles. The soil column setup is shown in Fig. 9.1.

9.2.3 Packing of Soil Columns

Soil packing inside the PVC column was performed to achieve a soil bulk density (BD) of 1.5 g/cm^3 in order to represent the field condition. A 3 cm layer of pea gravel of 161.6 g was added to the bottom and above the cheesecloth to retain soil in columns and filter soil particles from leachate. Each column received 990.78 g of the sun-dried and sieved (2 mm) soil. The soil for each column was divided into five lifts and each lift was used to pack 5 cm of column height to achieve uniform bulk density throughout the soil column length. The soil was manually packed using a PVC disc with a diameter just smaller than the inner diameter of the column. The top 5 cm of each column was kept free to facilitate irrigation. Therefore, the effective soil column height was 25 cm for all columns.

After filling, the columns were placed in a container and allowed to saturate overnight with tap water and then allowed to drain to achieve field capacity. When the wetting was completed, the soils were weighed to determine the water content at saturation and were placed on a wooden stand with a funnel placed below each column to direct leachate into 300 ml bottles. Then, distilled water was added to the top of each column. The soil columns were allowed to drain overnight and weighed to measure pot holding capacity (PHC), which is similar to field capacity (FC). The PHC was measured as the volume of water left in the soil after 24 h of free drainage (PHC, cm^3/cm^3). The columns were closed with plastic lids to avoid evaporation and volatile losses if any.

9.2.4 Synthesis of Nano-fertilizers

Nano-hydroxyapatite was synthesized using the sol-gel method. Known solutions of Calcium nitrate (200 ml) and Ammonium dihydrogen phosphate (154 ml) are mixed and to the mixture 25 ml of Triethanolamine added and the resulting mixture was neutralized to pH 11 using NaOH solution. Then, the mixture was stirred using magnetic stirrer with preheating at 75 °C for 24 h. After 24 h the mixture is kept under refrigeration for 24–48 h. After refrigeration, the mixture was centrifuged at 8500 rpm for 10 min in order to remove water with the help of Ethanol. Then, the remaining material after removal of water was kept in the oven for drying (48 h) to remove moisture at 90 °C. After drying synthesized nano-fertilizer (nHAP) was preserved carefully. Urea doped nano-hydroxyapatite (Ur@nHAP) was synthesized using the mixture of known solutions of nHAP and Urea. The resulting solution was stirred with a magnetic stirrer for 24 h and then kept for refrigeration for 48 h. After refrigeration, water was removed carefully without disturbing the bottom of the flask and then kept in the oven for drying for 60–72 h. After drying nano-fertilizer was crushed and stored carefully in the conical centrifuge tubes.

9.2.5 Nutrient Treatments

The treatments considered in soil column experiments were (1) Control, (2) Urea, (3) Superphosphate, (4) nHAP and (5) Ur@nHAP. In all the treatments (except control) same concentration (500 ppm) of the fertilizers were added. All treatments were replicated three times in all experiments.

9.2.6 Experimentation

Fertilizer was applied after washing the soil columns with DI water. The soil columns were irrigated once in every three days, using 100 ml of DI-water. The DI water was irrigated to each column using a syringe without disturbing the soil surface. The weight of soil columns was taken daily. The experiments were conducted for 30 days, from 10-10-2017 to 08-11-2017. Leachate samples were collected, one in every three days. Leachate samples were subjected to Easychem300 (Italy) machine for nitrate, ammonia, total N, total P and Ortho-P analysis. Soil samples also extracted from each column at three different depths (1) at the surface 0–1 cm, (2) 9–10 cm and (3) 24–25 cm after end of experiments. Extracted soil samples were oven dried to zero moisture and crushed manually, later sieved and digested for total N and total P analysis.

9.2.7 Statistical Analysis

The data from the soil column experiment were statistically analyzed by ANOVA single factor test. Nitrate, ammonia and ortho-P of leachate samples of different treatments were statistically analyzed by performing ANOVA single factor test at 5% level of significance.

9.3 Results and Discussion

Leachate analysis for three plant nutrients namely NO_3-N, NH_4-N and Ortho-P were presented below. Reason behind selecting only these three nutrients include, these are the major contaminants of groundwater and plants will take Nitrogen and Phosphorus in the above-mentioned forms.

9.3.1 NO_3-N Concentrations in Leachate

It was observed that nutrient front does not move as fast as the water front in case of soil (Fig. 9.2). In Fig. 9.2 Day 0 indicates the concentrations of NO_3-N in leachate before application of the fertilizer. Peak of NO_3-N leaching was observed on the 14th day (after 5th irrigation) in all treatments (except control) indicating rate of transport of NO_3-N in the soil as 7.44 mm/h (Fig. 9.2). On an average NO_3-N concentration was high in case of Urea treatment (Fig. 9.2). Leaching of NO_3-N was 55% (Table 9.3) higher in case of Urea treatment than the Ur@nHAP treatment (Table 9.3). However, their means were not significantly different among the treatments (Table

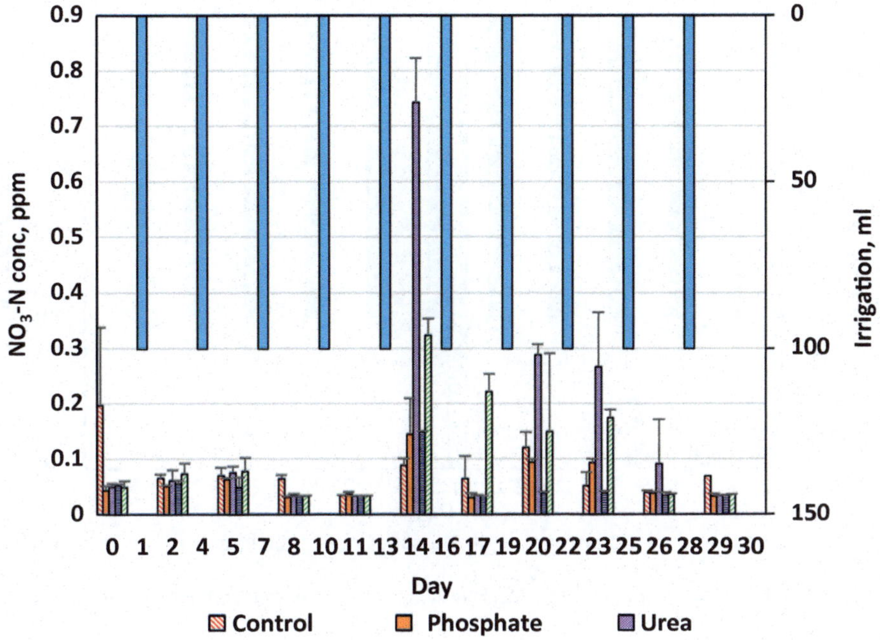

Fig. 9.2 NO_3-N concentrations in the leachate samples collected from different treatment columns

Table 9.2 ANOVA single factor test results of different treatments at 5% level of significance

Nutrients	Treatments				
	Control	Urea	Superphosphate	nHAP	Ur@nHAP
Nitrate	NS	NS	–	–	NS
Ammonia	NS	NS	–	–	NS
Ortho-P	NS	–	NS	NS	NS

NS is not significantly different

9.2). NO_3-N concentrations in the leachate were found to be less than that of USEPA targeted level (10 mg/l) in all treatments. Leaching of NO_3-N did not follow any trend indicative of the presence of some cations in the soil which are influencing the leaching.

9.3.2 *NH_4-N Concentrations in Leachate*

Leaching of NH_4-N was stopped after the 14th day (Fig. 9.3). This could be due to NH_4-N fixation in the soil and conversion of NH_4-N into other forms of Nitrogen. Peak of NH_4-N was observed on 14th day (Fig. 9.3) in all treatments indicating rate

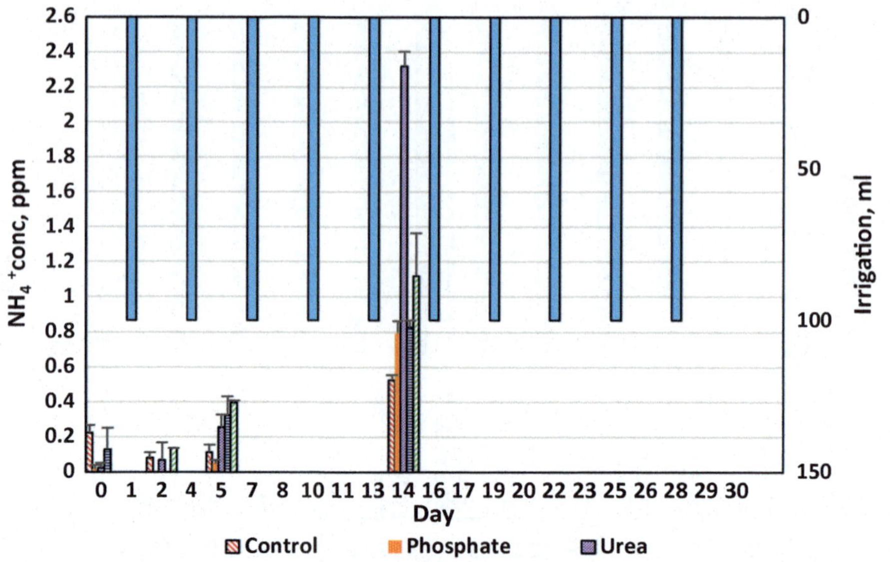

Fig. 9.3 NH_4-N concentrations in the leachate samples collected from different treatment columns

of transport of NH_4-N in the soil as 7.44 mm/h same as that of NO3-N leaching rate. Almost in every treatment leaching of NH_4-N was higher than the USEPA targeted level (0.1 mg/l) for groundwater. In Fig. 9.3, Day 0 indicates the concentration of NH_4-N in the leachate during the initial washing of the soil column after saturation. From the Fig. 9.3, it can be observed that leaching of NH_4-N was higher in case of Urea compared to the nano-fertilizer (Ur@nHAP). However, the means of the data were significantly not different from each other (Table 9.2). The higher leaching of NH_4-N in Urea was due to the fast release of the nutrients. Discontinuous leaching of NH_4-N under different treatments was due to the presence of clay particles and organic matter. Leaching of NH_4-Nwas found to be about 55% (Table 9.3) less in case of Ur@nHAP treatment in comparison to Urea treatment.

Table 9.3 Total load and mean concentration of different analytes in leachate from lateritic soil

Treatment	Nitrate		Ammonia		Ortho-P	
	Load (mg)	Concentration (mg/L)	Load (mg)	Concentration (mg/L)	Load (mg)	Concentration (mg/L)
Control	0.06	0.07	0.01	0.02	0.03	0.04
Superphosphate	0.045	0.06	0.06	0.08	0.11	0.16
Urea	0.12	0.17	0.18	0.24	0	0
nHAP	0.05	0.06	0.08	0.1	0.01	0.05
Ur@nHAP	0.08	0.11	0.12	0.15	0.02	0.07

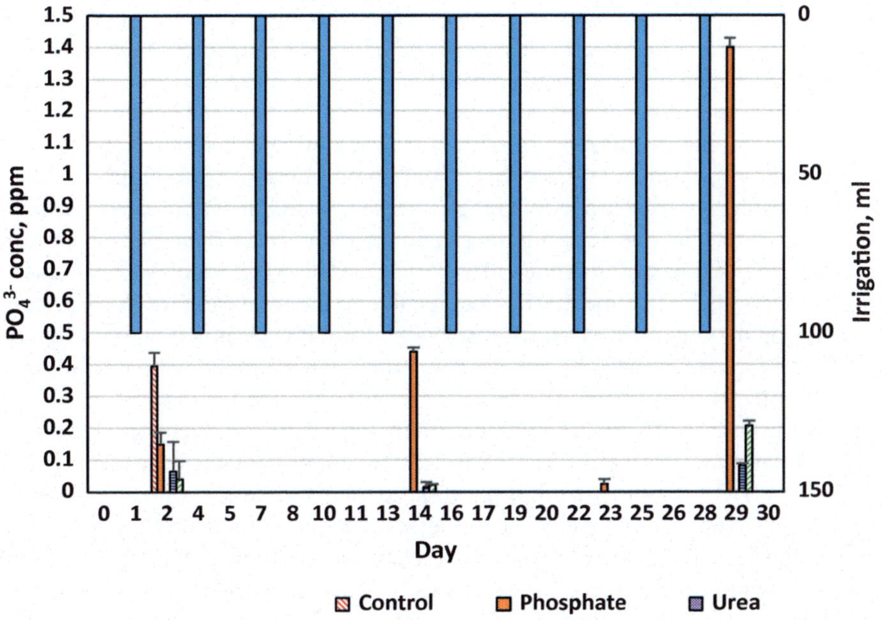

Fig. 9.4 Ortho-P concentrations in the leachate samples collected from different treatment columns

9.3.3 Ortho-P Concentrations in Leachate

Initially, there were no Ortho-P concentrations in leachate of all treatments (day 0). From Fig. 9.4, it can be observed that leaching of Ortho-P was not continuous in all treatments. Peak of Ortho-P leaching was observed on 29th day of the experiment indicating rate of transport of Ortho-P as 3.6 mm/h (Fig. 9.4). Leaching of Ortho-P was higher in case of Superphosphate treatment compared to nHAP and Ur@nHAP treatments. Out of the two nano-fertilizer treatments (nHAP and Ur@nHAP) leaching was higher in case of Ur@nHAP may be due to the surface modifications during doping with the Urea. However, there was no significant difference between the means of the three treatments (Superphosphate, nHAP and Ur@nHAP). Leaching of Ortho-P was found to be 128% higher in case of superphosphate treatment compared to two nano-fertilizer treatments (Table 9.3). Moreover, rate of transport of Phosphorus was found to be about half that of Nitrogen (Figs. 9.2, 9.3 and 9.4).

9.4 Conclusions

Soil column experiments with different fertilizer treatments suggested that plant nutrients does not move as fast as water through the soil. NH_4-N leaching was observed to be discontinuous and was influenced by the presence of clay particles and organic matter in the soil. There was no NH_4-N leaching after 5th irrigation due to NH4-N fixation in the soil and conversion into other forms of Nitrogen. Leaching of NO_3-N seems to be influenced by the presence of some metal cations in the soil. However, the rate of transport of both Nitrogen compounds (NH_4-N and NO_3-N) were found to be same (7.44 mm/h). The leaching of ortho-P was higher in case of Superphosphate fertilizer due to its high release rate and in case of nano-fertilizers, the ortho-P release was very minimal in comparison to the Superphosphate fertilizer. Out of two nano-fertilizers (nHAP and Ur@nHAP), leaching was higher in case of Ur@nHAP treatment. This may be due to the weakening of the PO_4 bonds during doping with urea. From the results obtained, it is clear that nutrient leaching of selected nutrients (NO_3-N, NH_4-N and Ortho-P) were found to be low in case of nano-fertilizers. The leaching of NO_3-N and NH_4-N were found to be 55% (Table 9.3) less in case of Ur@nHAP treatment compared to bulk fertilizer (Urea) treatment. Ortho-P leaching was 128% (Table 9.3) less in case of both nano-fertilizers (nHAP and Ur@nHAP). Mobility of Phosphorus was found to be half of that of Nitrogen (Table 9.3). However, more extensive studies are required for nano-fertilizer adaptation in the agriculture so as to minimize the non-point source pollution (particularly due to agriculture). Furthermore, experiments with the repeated application of nano-fertilizers is needed to conclude the potential risks or benefits of the nano-fertilizers in view of the increasing non-point source pollution due to commercial fertilizers used in agriculture.

Acknowledgements This work was generously supported by a major grant from the Food security-MHRD, Government of India (Grant No: 4-25/2013-TS-1).

References

Adhikari T, Kundu S, Meena V, Rao AS (2014) Utilization of nano rock phosphate by Maize (Zea mays L.) crop in a Vertisol of central India. J Agric Sci Technol A 4:384–394

Benzone HRL, Rubenecia MRU, Ultra VU, Lee SC (2015) Nanoparticles fertilizer affects the growth, development, and chemical properties of rice. Int J Agron Agric Res 7:105–117

Carpenter SR, Caraco NF, Correll DL, Howarth RW, Sharpley A, Smith VH (1998) Nonpoint pollution of surface waters with Phosphorus and Nitrogen. Ecol Appl 8(3):559–568

Celsia AR, Mala R (2014) Fabrication of nanostructured slow release fertilizer system and its influence on germination and biochemical characteristics of vigna raidata. Recent Adv Chem Eng 6:4497–4503

Daniel TC, Sharpley AN, Lemunyon JL (1998) Agricultural phosphorus and eutrophication: a symposium overview. J Environ Qual 27:251–257

Iavicoli I, Leso V, Beezhold DH, Shvedova AA (2017) Nanotechnology in agriculture: opportunities, toxicological implications, and occupational risks. Toxicol Appl Pharmacol 329:96

Janmohammadi M, Sabaghnia N, Dashti S, Nouraein M (2016) Investigation of foliar application of nano-micronutrient fertilizers and nano-titanium dioxide on some traits of barley. Biologija 62(2)

McShane HV, Sunahara GI, Whalen JK, Hendershot WH (2014) Differences in soil solution chemistry between soils amended with nanosized CuO or Cu reference materials: implications for nanotoxicity tests. Environ Sci Technol 48(14):8135–8142

Morales-Díaz AB, Ortega-Ortíz H, Juárez-Maldonado A, Cadenas-Pliego G, González-Morales S, Benavides-Mendoza A (2017) Application of nanoelements in plant nutrition and its impact in ecosystems. Adv Nat Sci Nanosci Nanotechnol 8(1):013001

Mukhopadhyay SS (2014) Nanotechnology in agriculture: prospects and constraints. Nanotechnol Sci Appl 7:63

Ongley ED, Xiaolan Z, Tao Y (2010) Current status of agricultural and rural non-point source pollution assessment in China. Environ Pollut 158(5):1159–1168

Rameshaiah GN, Pallavi J, Shabnam S (2015) Nano fertilizers and nanosensors—an attempt for developing smart agriculture. Int J Eng Res Gen Sci 3(1):2091–2730

Sharpley AN, Withers PJA (1994) Environmentally sound management of agricultural Phosphorus. Fertil Res 39:133–146

Subramanian KS, Tarafdar JC (2011) Prospects of nanotechnology in Indian farming. Indian J Agric Sci 81(10):887–893

US Environmental Protection Agency (1996) Environmental indicators of water quality in the United States

Chapter 10
Production and Characterization of Bio-surfactants from Various Lactobacillus Species: A Bioremediation Technique

T. Ashwin, Surinder Deswal, and Baljeet Singh Saharan

Abstract A promising method that can improve the bioavailability and biodegradability of polyaromatic hydrocarbons (PAH) is the use of biosurfactants. In this work, Biosurfactants produced from lactic acid bacteria were utilized for bioremediation of Benzo(a)Pyrene. Besides, lignocellulosic waste (*Luffa acutangula*) was also used as a carbon source for production of the cost effective biosurfactants. Initially, acid hydrolysis of lignocellulosic waste was done. Following, *Lactobacillus plantarum, L. casei, L. acidophilus* were employed for fermentation after nutrient supplementation separately. Then the Lactobacillus cells were submitted to extraction process with phosphate buffer saline (PBS) to evaluate the intercellular biosurfactants. Biosurfactants were measured based on surface tension reduction. The results showed that the surface tension and CMC of biosurfactants from *L. plantarum, L. casei, L. acidophilus* were 58, 51, 60 mN/m and 8, 12, 6 g/L respectively. The FTIR characterization showed that the produced biosurfactants was Glycoprotein/Glycolipid. The ^{1}H NMR results confirmed that the produced biosurfactant was Glycolipid only. The produced biosurfactants employed in industrial sludge, which was contaminated with 500 mg/kg of Benzo(a)pyrene. After mobilization process, sample was analyzed in GC resulting that 60% (approx.) degradability of Benzo(a)Pyrene. With the help of this technique, the biodegradability of PAH contaminated sites can be improved.

T. Ashwin (✉) · S. Deswal
Department of Civil Engineering (Environmental Engineering), National Institute of Technology, Kurukshetra, Haryana 136119, India
e-mail: ashwin.t12@gmail.com

S. Deswal
e-mail: sdeswal@nitkkr.ac.in

B. S. Saharan
Department of Microbiology, Kurukshetra University, Kurukshetra, Haryana, India
e-mail: Baljeet.kuk@gmail.com

A. Pandey et al. (eds.), *Climate Impacts on Water Resources in India*, Water Science and Technology Library 95, https://doi.org/10.1007/978-3-030-51427-3_10

10.1 Introduction

As per Central Pollution Control Board, India is generating more than 33,000 million litres of wastewater every day. The sewage treatment plant treats those generated wastewater through various physical and chemical process. Now, the biggest challenge to the environmentalist is the disposal and treatment of sludge (Vecino et al. 2015). Over 300 organic compounds have been identified in sewage sludge (Duarte-Davidson and Jones 1996). Recycling and reuse of wastes are the best choice for sludge disposal instead of incineration and landfilling (J. Hall 1999). Many industrialised countries like European Union member states using 30% of sewage sludge as fertilizer to their agricultural lands (Wild 1990). However, many member states regulates the use of sewage sludge utilization because it tends to concentrate with pollutants (PAH_S, PCB_S) during the physical and chemical process in waste water treatment (Bontoux et al. 1998; Harms 1996). In order to improve sewage sludge management, the 3rd draft of European Union directive regulates the limits for polyaromatic hydrocarbons, poly chlorinated biphenyls and other organic pollutants entering into the sewage sludge (3rd draft 2000). As per that 3rd draft, the total concentration of PAH (sum of 11 compounds) in sewage sludge for agriculture use should be less than 6 mg/kg dry mass.

The interest in Poly aromatic hydrocarbons (PAH) in sewage sludge is due to its low biodegradability and high lipophilicity (Alcock and Jones 1993). PAH_S are the persistent organic compounds due to incomplete combustion of organic matter (Vecino et al. 2015) and received much attention from researchers only because they are well known carcinogens and mutagens (Samanta et al. 2002). Due to high adsorption capacity on solid particles, PAH_s are yielding higher concentrations to a larger extend upto 90% in sewage sludge during waste water treatment (Pena et al. 2010). So Unites states Environmental Protection Agency indicates 16 PAH_S as a priority pollutant in the sewage sludge (Vecino et al. 2015). For the utilization of sewage sludge, these priority pollutants should be removed without any harmful effects to the environment.

The best approach to decontaminate these pollutants from sewage sludge is "**Bioremediation**". The success of bioremediation is depends upon the availability of microbes, accessibility of contaminants and environmental factors (Cameotra and Makkar 2010). Among the various bioremediation methods, the best eco-friendly remediation technology is the use of biosurfactants (BS) which mobilize and biodegrade the PAH in the sewage sludge (Vecino et al. 2015; Sponza and Gok 2012; Pacwa-Plociniczak et al. 2011). However, the utilization of biosurfactants for the bioremediation of contaminated soil is not yet well established due to its high production cost (Moldes et al. 2011). The major drawback of biosurfactants when compared to chemically synthesized surfactants is their high production cost and low production rates (Vecino et al. 2015). Among the total operational cost of biosurfactants, 10–80% cost is for the purchase of raw materials (Petrides 2003). So the researchers focused on using low cost agricultural residues as a carbon source during the production of biosurfactants (Moldes et al. 2007). The commercial success of biosurfactants is feasible by using cheap renewable agricultural substrates and low cost feedstock as a carbon source during its production (Moldes et al. 2011, 2013). In this work, *Luffa acutangula*, an agricultural residue used as a carbon source for the Lactobacillus derived biosurfactants. The main objectives of this study is

- To produce cost competitive biosurfactants using *Lactobacillus species.*
- To remove Benzo(a)pyrene (priority pollutant) from the sewage/Industrial sludge using biosurfactant.
- To convert lignocellulosic wastes into a useful carbon source for biosurfactant production.

10.2 Materials and Methods

10.2.1 Chemicals and Reagents

All the chemicals used in the present study were of analytical grade/GR grade from Hi-media Pvt. Ltd. India and Titan biotech Ltd. India.

10.2.2 Raw Materials

After harvesting, the agricultural residues of Ridge gourd plant (scientifically *Luffa acutangula*), will become as waste. So it was used in this work as a carbon source for biosurfactant production. By the use of this low cost substrates, the cost of biosurfactant production can be reduced. Its local name is peerkangai in Tamil. It was collected from the agricultural fields in Bahadurpur village, Muzaffarnagar, India.

10.2.3 Bacterial Cultures

Lactobacillus casei ssp. *casei* (NCDC accession number-357), *Lactobacillus acidophilus* (NCDC accession number-291) and *Lactobacillus plantarum* (NCDC accession number-25) were collected from National Dairy Research Institute (NCDC), Karnal, India. The collected strains were in freeze dried form. After collection, these strains were stored in refrigerator at 2–5 °C.

10.2.4 Sludge Sample Collection

The sludge coming after from the secondary treatment was used in this study for remediation process. It was collected from Sewerage treatment plant (capacity 39 MLD), Panchkula, India. The collected sludge was dried to a room temperature prior to its utilization for its stabilisation and reduction of water content.

10.2.5 Contamination of Sludge

For the bioremediation experiments, 100 g of sludge was taken in conical flask and contaminated with 50 mg of Benzo(a)pyrene. Therefore, the sludge was contaminated with 500 mg/kg of BaP. 100 ml of acetone also added as a solvent. The whole mixture was stirred vigorously for 30 min for the homogenous distribution of BaP. Before starting bioremediation process, the sludge was left to rest for 48 h at 30 °C in a shaker at 150 rpm.

10.2.6 Acid Hydrolysis of Raw Materials

Agricultural residues of vine shoots (*Luffa acutangula*) was collected dried, sieved to a particle size less than 1 mm and stored in for usage. The stored agricultural residue (*Luffa acutangula*) was subjected to acid hydrolysis using H_2SO_4 (2–4%). It was carried out in a vertical autoclave for 30 min at 129 °C. Hydrolysates was formed at the end of hydrolysis and it was used during media preparation.

10.2.7 Bacterial Growth on Media

Skim milk was prepared by dissolving skim milk powder at 12.5% in distilled water and MRS media was prepared using the complete media containing (per litre) Dextrose 20 g, proteose peptone 10 g, Beef extract 10 g, Yeast extract 6 g, Sodium acetone 6 g, ammonium citrate 6 g, Disodium phosphate 2 g, Tween 80: 1 g, Magnesium sulphate 0.1 g, Manganese sulphate 0.06 g. Three Lactobacillus strains were grown on MRS plates separately at 23 °C for 24 h. Suspension of cells from plates into MRS media, Skim Milk, hemicellulosic hydrolysates was done to prepare Inocula. The bacterial cells per ml (biomass) was determined by using optical density at 600 nm.

10.2.8 Fermentation

Fermentation was carried out in a 2 L working volume flask separately for the three Lactobacillus in a medium (at controlled pH 6.5) containing hemicellulosic hydrolysates (neutralized to pH 6.7) and 10 g of yeast extract. The media was inoculated with 1% (v/v) pre culture and the fermentation was carried out for 48 h @ 31 °C. Samples for determination of biomass and surface tension were collected after fermentation.

10.2.9 Extraction of Biosurfactants

Initially, phosphate buffer saline (PBS 1X) was prepared using sodium chloride (8 g), potassium chloride (0.2 g), disodium hydrogen phosphate (1.42 g), potassium dihydrogen phosphate (0.24 g) and its pH was adjusted to 7.4. The samples withdrawn during fermentation of each Lactobacillus were centrifuged and re-suspended in 50 ml of PBS buffer at room temperature for upto 2 to 3 h. The bacterial cells were removed by centrifugation and the remaining supertenant was collected in aliquots and sent for analysis (Sharma D et al. 2014).

10.2.10 Surface Tension Measurement

Using stalagmometer, the surface tension of the produced biosurfactants were measured. The supertenants (PBS extract + biosurfactants) from *L. plantarum*, *L. casei*, and *L. acidophilus* were tested for surface tension at temperature 28 ± 2 °C. The surface tension was calculated using the following formula.

$$\gamma_o = \gamma(n/n_0) \tag{10.1}$$

γ_o, γ are the surface tension of water (72 mN/m) and biosurfactant respectively.
n_0, n are the drop numbers of water and biosurfactant respectively.

10.2.11 Biochemical Analysis of Surfactant

Phenol sulphuric acid method was conducted to estimate carbohydrate content in the biosurfactant. The supertenant sample collected after centrifugation was placed in a test-tube, then phenol and sulfuric acid were added. The solution turned into yellow-orange colour as a result of the interaction between the carbohydrates and the phenol. The absorbance at 420 nm was calculated using spectrophotometer. Using the calibration curve, the carbohydrate quantity was estimated. The protein content was also measured using Biuret method. The Biuret reagent was added in the supertenant solution and allowed to stand for 15–30 min. Then the absorbance was read at 540 nm using UV–visible spectrometer. The lipid content of the sample was done by following the procedure proposed by Folch et al. (1957).

10.2.12 FTIR Analysis

Fourier Transform Infra-Red (FTIR) Spectroscopy is mostly done to identify the functional groups and therefore chemical structure of unknown sample can be identified. The produced crude biosurfactant was freeze dried using lyophilizer. Pellets were made by grinding freeze dried biosurfactant with KBr. The pellets obtained were analyzed by FTIR spectrometry (Resolution: 4 cm^{-1}; Range 450–4000 cm^{-1}).

10.2.13 NMR Structural Elucidation

Using NMR spectroscopy, further structural characterization was done. The biosurfactant produced from *L. plantarum, L. casei,L. acidophilus* were dissolved in deuterated chloroform ($CDCl_3$) separately and their respective proton NMR (^1H NMR) spectra were obtained after analysis.

10.2.14 Mobilization Process

Sewage sludge (10 g) contaminated with 500 mg/kg of BaP were placed in 250 mL Erlenmeyer flasks, and 100 mL of an aqueous solution containing the biosurfactant, at 2.5 times its critical micelle concentration (CMC), were added. The experiments were conducted in a shaker at 150 rpm, for 21 days at 30 °C.

10.2.15 Gas Chromatography Analysis

The sample was collected after the treatment of 21 days and analysed in gas chromatography. Before analysis, the sample was filtered by allowing it to pass through celite and cotton plug. The filtered sample was injected into the GC. The oven temperature was increased from 60 to 170 °C at 6 °C/min, further increased at a rate of 3 °C/min until 240 °C and finally held at 300 °C for 7 min. Injector and transfer line temperature was set at 280 and 300 °C respectively. The peaks observed in the graph was compared with calibration curves.

10.3 Results and Discussion

10.3.1 Surface Tension Results

The surface tension of the biosurfactants produced from *L. plantarum*, *L. casei*, *L. acidophilus* were plotted in graphs Figs. 10.1, 10.2 and 10.3 respectively.

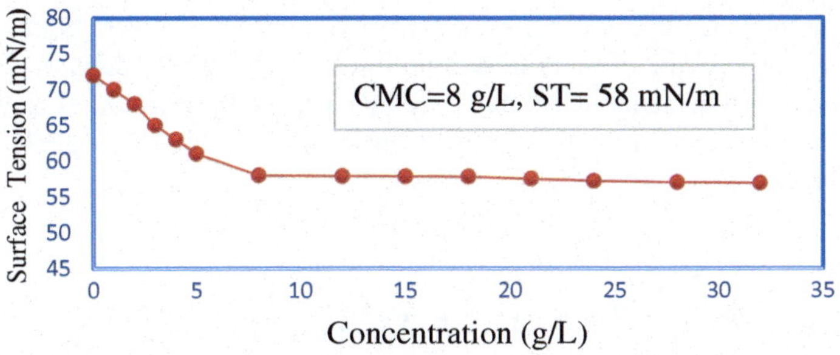

Fig. 10.1 ST of the biosurfactant from *L. plantarum*

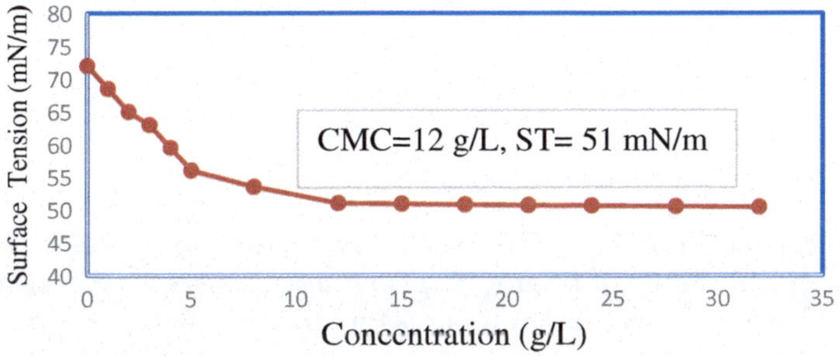

Fig. 10.2 ST of the biosurfactant from *L. casei*

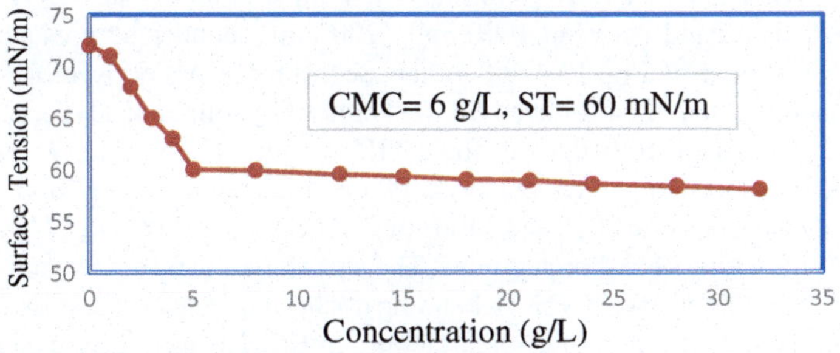

Fig. 10.3 ST of the biosurfactant from *L. acidophilus*

Table 10.1 Biochemical analysis (*A, B, C are the samples of *L. plantarum*, *L. casei*, *L. acidophilus* respectively)

Sample	Carbohydrate (μg/ml)	Lipid (mg/g)	Protein (μg/ml)
A*	415.5	0.45	0.57
B	356.5	0.41	0.45
C	397.9	0.39	0.52

From the graphical results plotted below, it was concluded that the surface tension of biosurfactants from *L. plantarum*, *L. casei*, *L. acidophilus* were 58, 51, 60 mN/m respectively. The CMC of that same sample were 8, 12, 6 g/L respectively. Since all the produced biosurfactants can able to reduce the ST by 8 mN/m, it is effective for application. And also biosurfactants were more effective for bioremediation due to lower CMC values.

10.3.2 Biochemical Analysis Results

Biochemical analysis confirms that the biosurfactant was mainly composed of carbohydrates and lipids (small fraction of protein also). Results were shown in Table 10.1.

10.3.3 FTIR Results

The spectrum obtained from FTIR were shown in the figures below. Figure 10.4 depicts the FTIR spectra of BS from *L. plantarum*. The presence of peptide linkage (O–H and N–H) was confirmed by the peaks at 3100 and 3400 cm^{-1}. The absorbance band between 2900 and 2750 cm^{-1} suggested the confirmation of C–H stretching in BS. The peak at 1800 cm^{-1} specified C=O stretching of carbonyl groups. The absorbance band at 1500 cm^{-1} showing protein related weak bonds (N–H bending). The peaks between 1450 and 1350 cm^{-1} demonstrates C–H stretching. The absorbance band between 1000 and 1200 cm^{-1} confirmed presence of sugar moiety. On comparing Fig. 10.4 with other spectrum of LAB derived biosurfactants, it was concluded that the produced BS was either Glycolipid or glycoprotein.

Figures 10.5 and 10.6 depicts the FTIR spectra of BS from *L. casei* and*L. acidophilus* respectively. The presence of peptide linkage (O–H and N–H) was confirmed by the peaks at 3050 and 3400 cm^{-1}. The peak at 1800 and 1747 cm^{-1} specified C=O stretching of carbonyl groups. The absorbance band at 1650 cm^{-1} showing protein related weak bonds (N–H bending). The absorbance band at 2800 cm^{-1} suggested the confirmation of C–H stretching in BS. The absorbance band between 1037 and 1200 cm^{-1} confirmed presence of sugar moiety. On comparing Figs. 10.5

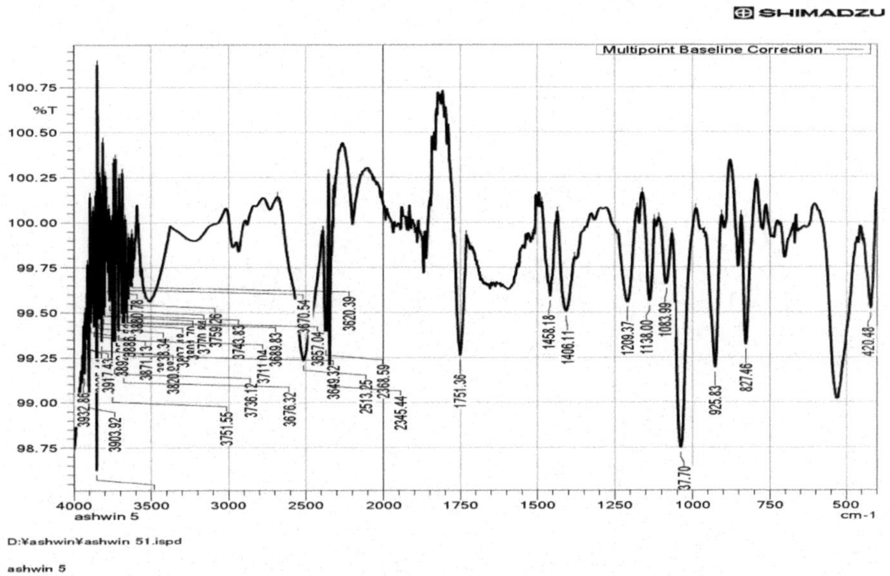

Fig. 10.4 FTIR spectrum of biosurfactant produced by *Lactobacillus plantarum*

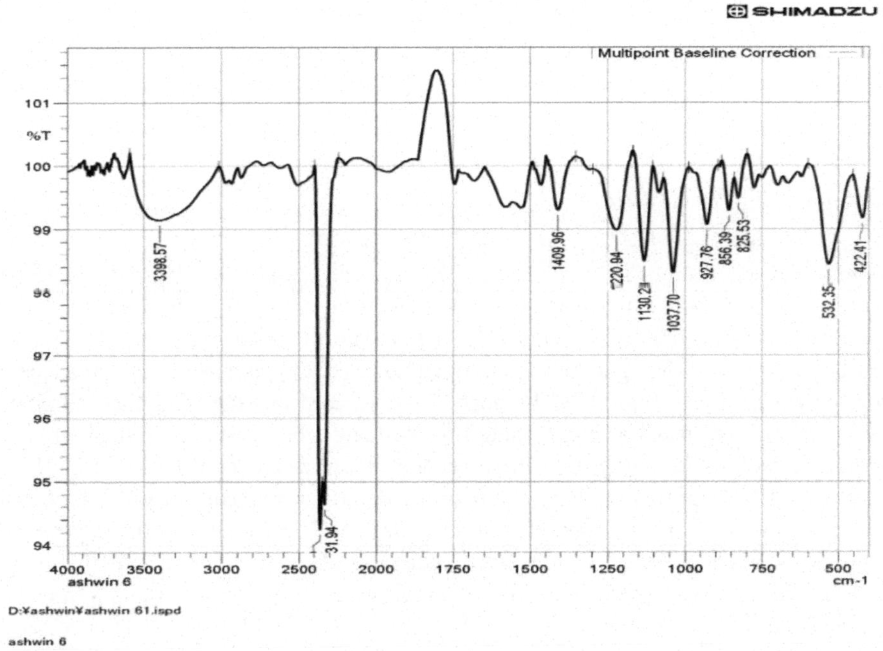

Fig. 10.5 FTIR spectrum of biosurfactant produced by *L. casei* ssp. *casei*

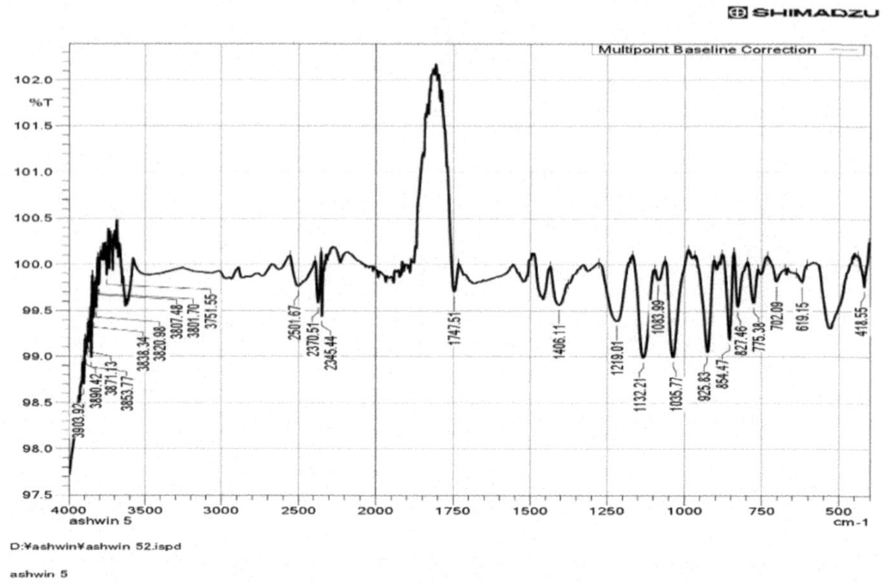

Fig. 10.6 FTIR spectrum of biosurfactant produced by *L. acidophilus*

and 10.6 with other spectrum of LAB derived biosurfactants, it was concluded that the produced BS was either Glycolipid or glycoprotein.

10.3.4 NMR Results

The NMR spectrum of BS from by *L. plantarum, L. casei, L. acidophilus* are represented in Figs. 10.7, 10.8 and 10.9 respectively. In Fig. 10.7, the peak at δ 1.68 confirms the long chain aliphatic fatty acids. The peaks at δ 4.81 and 4.77 indicates the presence of sugar moiety. Another peak at 7.35 suggesting the presence of carboxylic acids. In Fig. 10.8, the peak at δ 1.62 confirms the long chain aliphatic fatty acids. The peaks at δ 4.74 and 4.58 indicates the presence of sugar moiety. Another peak at 7.2 suggesting the presence of carboxylic acids. In Fig. 10.9, the peak at δ 1.66 confirms the long chain aliphatic fatty acids. The peak at δ 4.79 indicates the presence of sugar moiety. Another peak at 7.33 suggesting the presence of carboxylic acids. From all the observed peaks from Figs. 10.7, 10.8 and 10.9, it was concluded that the biosurfactant from *L. plantarum, L. casei* and *L. acidophilus* was Glycolipid.

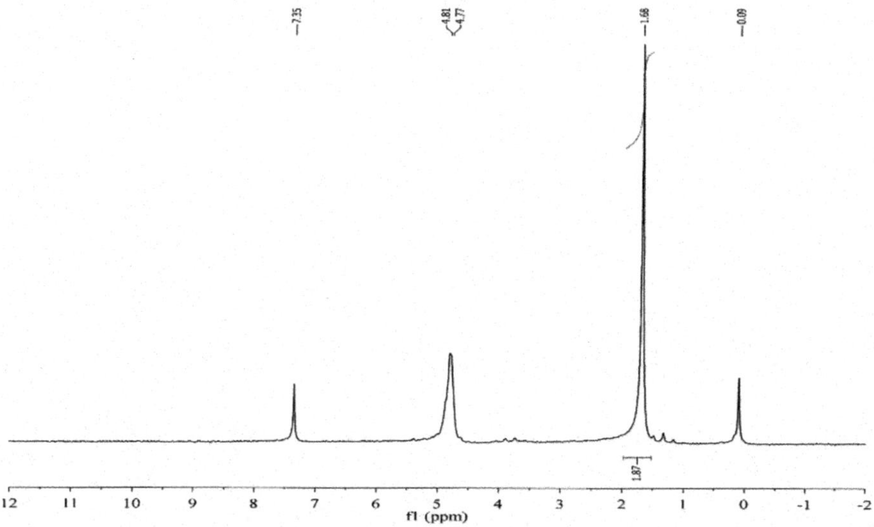

Fig. 10.7 NMR spectrum of BS from *L. plantarum*

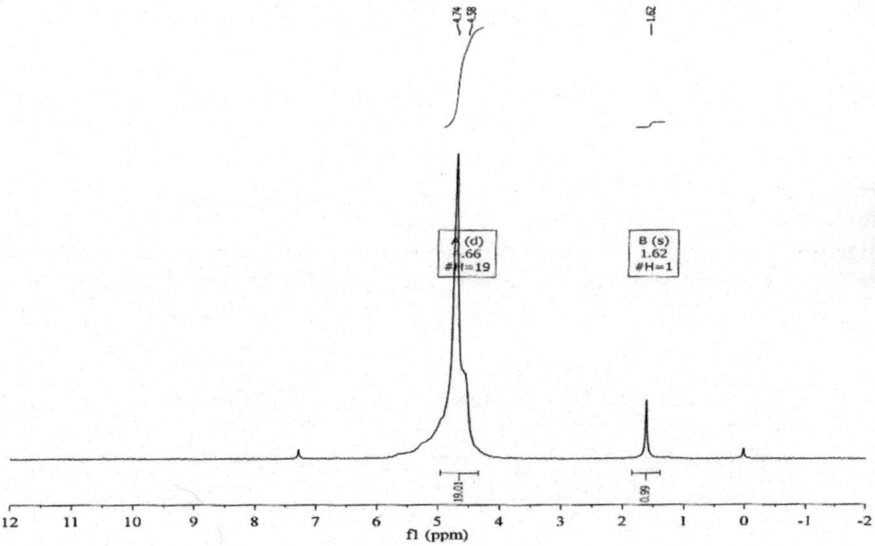

Fig. 10.8 NMR spectrum of BS from *L. casei*

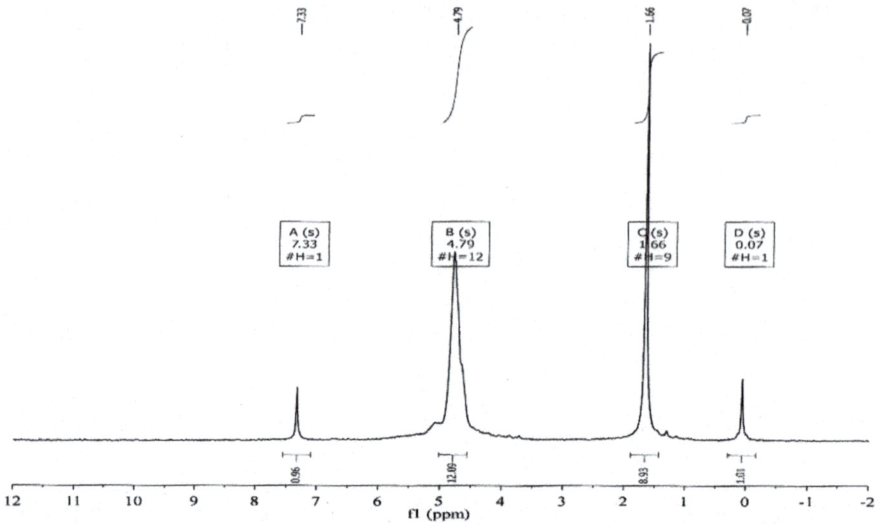

Fig. 10.9 NMR spectrum of BS from *L. acidophilus*

10.3.5 Results of Biosurfactant Enhanced BaP Remediation

The bioremediation experiment was carried out in gas chromatography for 21 days. The variation of BaP (polyaromatic hydrocarbon) and the percentage of biodegradation in the presence of both acetonitrile (Control) and biosurfactants were shown in the following graphs. Figure 10.10 depicts the bioremediation of BaP in the sludge in the presence of control solvent Acetonitrile (ACN). The initial concentration of BaP was 500 mg/kg of sludge. During the days of bioremediation, the concentration of BaP was decreasing due to mobilization and increase in bioavailability to the microorganisms. It was observed that 22% of BaP was biodegraded after 10 days

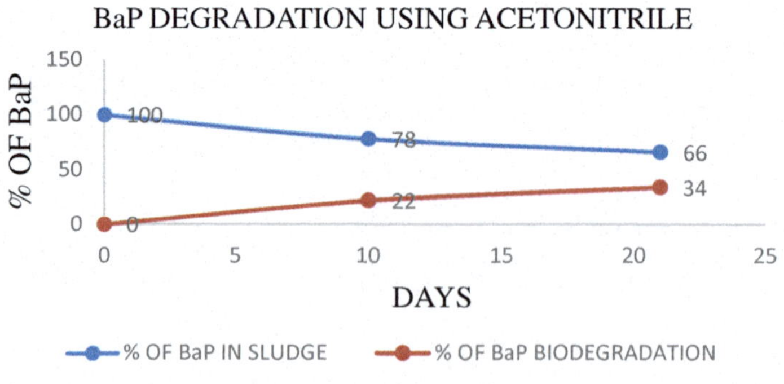

Fig. 10.10 BaP remediation in sludge in the presence of ACN

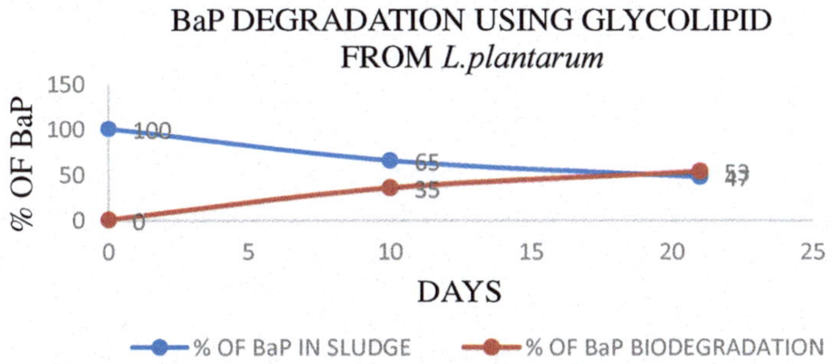

Fig. 10.11 BaP remediation in sludge using glycolipid from *L. plantarum*

of treatment with acetonitrile. After 21 days of treatment with ACN, 34% of BaP biodegradation was achieved.

Figure 10.11 depicts the bioremediation of BaP in the sludge in the presence of biosurfactant (glycolipid) derived from *L. plantarum.* It was observed that BaP concentration in sludge was reducing from 500 to 325 mg/kg at the end of 10 days. It indicates only 65% of BaP remaining in the sludge and 35% BaP was biodegraded. At the end of 21 days of treatment, the BaP concentration was observed as 235 mg/kg. This result indicates that the 53% of BaP was biodegraded and only 47% BaP present in the sludge. Figure 10.12 depicts the bioremediation of BaP in the sludge in the presence of biosurfactant (glycolipid) derived from *L. casei.* It was observed that BaP concentration in sludge was reducing from 500 to 305 mg/kg at the end of 10 days. It indicates only 61% of BaP remaining in the sludge and 39% BaP was biodegraded. At the end of 21 days of treatment, the BaP concentration was observed as 215 mg/kg. This result indicates that the 57% of BaP was biodegraded and only 43% BaP present in the sludge.

Figure 10.13 depicts the bioremediation of BaP in the sludge in the presence of biosurfactant (glycolipid) derived from *L. acidophilus.* It was observed that BaP concentration in sludge was reducing from 500 to 290 mg/kg at the end of 10 days.

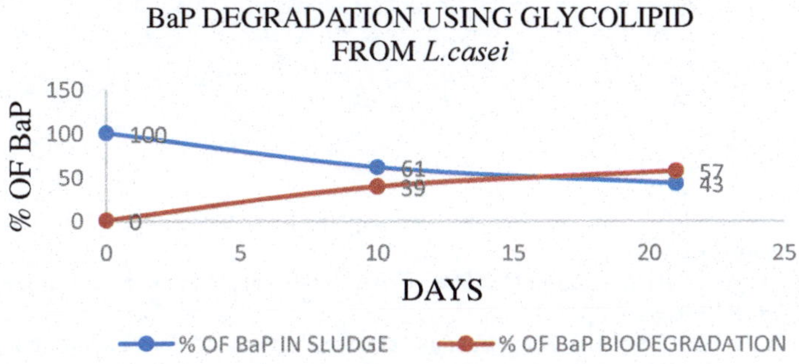

Fig. 10.12 BaP remediation in sludge using glycolipid from *L. casei*

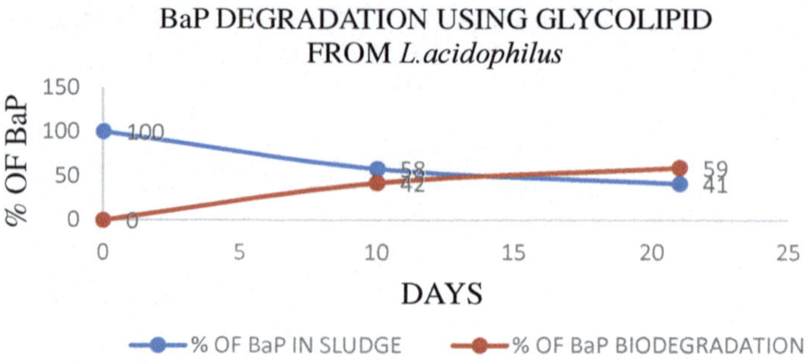

Fig. 10.13 BaP remediation in sludge using glycolipid from *L. acidophilus*

It indicates only 58% of BaP remaining in the sludge and 42% BaP was biodegraded. At the end of 21 days of treatment, the BaP concentration was observed as 205 mg/kg. This result indicates that the 59% of BaP was biodegraded and only 41% BaP present in the sludge. For the utilization of sludge in agriculture, the BaP concentration should be 6 mg/kg of sludge. But the test results exceeded the limits. If this bioremediation process continues for two to three months, 99% of BaP biodegradation can be achieved. And then the treated sludge can be utilized for agriculture as a manure since it has high organic manic matter and free from pollutants.

10.4 Conclusion

The above study was undertaken to devise the approach of simultaneous production of biosurfactants using probiotic lactic acid bacteria with a low cost agricultural residues as a production medium, and further their application in bioremediation of polyaromatic hydrocarbon (BaP) contaminated sludge. As a result, Biosurfactants were produced in a cost effective manner with the help of utilization of low cost substrates. And also, 59% of BaP degradation was achieved during bioremediation process.

References

Alcock ER, Jones KC (1993) Polychlorinated biphenyls in the digested UK sewage sludges. Chemosphere 26:2199–2207

Bontoux L, Vega M, Papameletiou D (1998) Urban wastewater treatment in Europe: what about the sludge? IPTS Report, pp 5–14

Cameotra SS, Makkar RS (2010) Biosurfactant enhanced bioremediation of hydrophobic pollutants. Pure Appl Chem 82:97–116

Duarte-Davidson R, Jones KC (1996) Screening the environmental fate of organic contaminants in the sewage sludge applied to agricultural soils: II. The potential for transfers to plants and grazing animals. Sci Total Environ 185:771–779

Folch et al (1957) A simple method for the isolation and purificaion of total lipides from animal tissues. J Biol Chem 226(1):497–509

Hall J (1999) Ecological and economical balance for sludge managememt options. In: Proceedings of the workshop on Problems around sludge, pp 155–172

Harms HH (1996) Bioaccumulation and metabolic fate of sewage sludge derived from organic xenobiotics in plants. Sci Total Environ 185:83–92

Moldes AB et al (2007) Evaluation of biosurfactant production from various agricultural residues by *Lactobacillus pentosus*. J Agric Food Chem 55:4481–4486

Moldes AB et al (2011) Ex situ treatment of hydrocarbon-contaminated soil using biosurfactants from *Lactobacillus pentosus*. J Agric Food Chem 59:9443–9447

Moldes AB et al (2013) Partial characterization of biosurfactant from *Lactobacillus pentosus* and comparison with sodium dodecyl sulphate for the bioremediation of hydrocarbon contaminated soil. BioMed Res Int, Article number 961842

Pacwa-Plociniczak M et al (2011) Environmental applications of biosurfactants: recent advances. Int J Mol Sci 12:633–654

Pena MT et al (2010) Development of a sample preparation procedure of sewage sludge samples for the determination of polycyclic aromatic hydrocarbons based on selective pressurized liquid extraction. J Chromatogr A 1217:425–435

Petrides D (2003) Bioprocess design and economics. In: Bioseparations science and engineering. Oxford University Press, Oxford

Samanta SK et al (2002) Polycyclic aromatic hydrocarbons: environmental pollution and bioremediation. Trends Biotechnol 20:243–248

Sharma D, Singh Saharan B (2014) Simultaneous production of biosurfactants and bacteriocins by probiotic *Lactobacillus casei* MRTL3. Int J Microbiol, Article ID 698713

Sponza DT, Gok O (2012) Aerobic biodegradation and inhibition kinetics of poly-aromatic hydrocarbons (PAHs) in a petrochemical industry wastewater in the presence of biosurfactants. J Chem Technol Biotechnol 87:7658–7672

Vecino X et al (2015) Bioremediation of sewage sludge contaminated with fluorine using a lipopeptide biosurfactant. Int J Chem Mol Nucl Mater Metall Eng 9:751–754

Wild SR (1990) Organic contaminants in an agricultural soil with a known history of sewage sludge amendments: polynuclear aromatic hydrocarbons. Environ Sci Technol 24:1706–1711

Chapter 11
Settling Velocity of Suspended Sediments in Muthupet Estuary, India and Bouregreg Estuary, Morocco

K. L. Priya, S. Haddout, and S. Adarsh

Abstract Studies were conducted during the wet and dry seasons to understand the settling velocity of cohesive suspended sediments in the Muthupet estuary, a Ramsar site in Tamil Nadu, India. Settling velocity was identified to be a function of suspended sediment concentration, turbulence and salinity gradient. An empirical model was developed for settling velocity with its influencing parameters for two scenarios based on the stratification parameter $\delta S/\langle S \rangle$, which relates the difference in salinity between the bottom and the surface (δS) with the average salinity of the water column ($\langle S \rangle$). The model was extended to another estuary with similar estuarine characteristics in terms of hydrodynamic classification, tidal range and freshwater flow regimes in the northwestern part of Morocco, the Bouregreg estuary. Settling velocity was observed to be an inverse function of suspended sediment concentration at the meandering zone. The relationship between settling velocity and suspended sediment concentration was difficult to predict due to dredging activities in the location near the mouth. The settling velocity exhibited a direct relationship with turbulence shear at low values of turbulence, while an inverse relationship was observed at higher turbulence shear. Thus, the model predicted settling velocity fairly well at the upstream location, but failed to prove its applicability at the downstream stations due to natural and anthropogenic interventions such as meandering and dredging respectively.

K. L. Priya (✉) · S. Adarsh
Department of Civil Engineering, TKM College of Engineering, Kollam, Kerala, India
e-mail: klpriyaram@gmail.com

S. Adarsh
e-mail: adarsh_lce@yahoo.co.in

S. Haddout
Department of Physics, Faculty of Science, Ibn Tofail University, Kenitra, Morocco
e-mail: soufian.haddout@gmail.com

A. Pandey et al. (eds.), *Climate Impacts on Water Resources in India*, Water Science and Technology Library 95, https://doi.org/10.1007/978-3-030-51427-3_11

11.1 Introduction

The dynamics of cohesive sediments in estuaries is highly complex. The water column is expected to have a greater suspended sediment concentration when estuarine bed consists of cohesive sediments ie. fine silt and clay. The cohesive suspended sediments are subjected to flocculation and settling. The estimation of settling velocity of suspended sediments is vital for modelling the behavior of cohesive sediments (Lumborg and Pejrup 2005; Manning et al. 2007). The settling velocity of cohesive sediments is affected by salinity (Portela et al. 2013; Krone 1962), suspended sediment concentration (Manning et al. 2006; Johansen and Larsen 1998; Shi et al. 2003), turbulence (Maa and Kwon 2007; Gratiot and Manning 2007; Wolanski et al. 1992; Pejrup and Mikkelsen 2010). Recent studies highlighted the influence of vertical salinity gradient on the settling velocity of suspended sediments (Priya et al. 2015a, b). In estuaries, salinity varies over a horizontal and vertical scale. The vertical salinity gradient is experienced due to freshwater flow and tidal flow. The vertical salinity gradient is the difference of bottom salinity and surface salinity divided by the water depth. Higher the freshwater flow, higher will be the salinity gradient. This higher gradient causes a denser water at the bottom, thereby imparting a density difference in the vertical direction. The density difference is said to influence the settling velocity of suspended sediments. This phenomenon is first proposed for the settling velocity of cohesive suspended sediment in a shallow estuary (Muthupet estuary) on the south-east coast of India. However, this phenomenon has to be established in other estuaries having similar and different hydrodynamic characteristics to come out with a universal relationship.

Numerous models are available relating settling velocity (w_s) with suspended sediment concentration (SSC) and turbulence shear (G). The influence of salinity gradient (s) on settling velocity has been studied by Priya et al. 2015a and a new model for settling velocity incorporating salinity gradient of the form $w_s = a\ SSC^x\ G^y\ s^{-z}$ has been developed and tested in the Muthupet estuary, India (Priya et al. 2015a, b). This paper is an extension of the work done by Priya et al. 2015a to examine the applicability of the model for Bouregreg estuary, Morocco. This estuary has been selected for two reasons: (i) The basic estuarine characteristics of the estuary is similar to that of Muthupet estuary; (ii) Very limited studies are available on the sediment dynamics of Moroccan estuaries and no studies have reported the behavior of settling velocity of suspended sediments and this study is the first of its kind.

11.2 Theoretical Concept

The vertical distribution of suspended sediment concentration is given by Rouse equation (Rouse 1938) as

$$\frac{C}{C_a} = \left[\left(\frac{h-z}{z}\right).\left(\frac{z_a}{h-z_a}\right)\right]^{w_s/ku_*} \tag{11.1}$$

where C is the SSC, C_a the reference concentration at distance a above the estuarine bed, z the vertical co-ordinate above the bed, h the depth of water, w_s/ku_* defined as Rouse number of suspended sediments, where u_* the shear velocity, k the von Karman's constant with a value of 0.4 for unstratified flow.

The shear velocity u_* is obtained from the horizontal flow velocity u (Thorn and Parsons 1980) as

$$\frac{u}{u_*} = \frac{1}{k}.\ln\left(\frac{z}{z_0}\right) \tag{11.2}$$

where z_0 is the bed roughness length, which is related to the median sediment size d_{50} as $d_{50}/12$ (Masselink et al. 2009).

Taking logarithm on both sides of Eq. (11.1),

$$ln\left(\frac{C}{C_a}\right) = \frac{w_s}{ku_*} ln\left[\left(\frac{h-z}{z}\right).\left(\frac{z_a}{h-z_a}\right)\right] \tag{11.3}$$

Discretizing the water column over a vertical scale with space step of dz with n number of steps, the settling velocity is given by,

$$w_{s,t} = \frac{k}{n}\int_{z=0}^{h} u_*(z).\frac{\ln\left(\frac{C(z)}{C_a}\right)}{ln\left[\left(\frac{h-z}{z}\right).\left(\frac{z_a}{h-z_a}\right)\right]}dz \tag{11.4}$$

Turbulence intensity is generally expressed as the root mean square velocity gradient G, which is given by (Camp and Stein 1943).

$$G = \sqrt{\frac{\varepsilon}{\nu}} \tag{11.5}$$

where ε is turbulence energy dissipation and ν the kinematic viscosity of the fluid (0.9×10^{-6} m^2/s at 25 °C). Considering steady and uniform flow, ε is equal to P, the turbulent energy production (van Leussen 1994).

$$\varepsilon = P = \frac{\tau}{\rho}\frac{dU}{dz} \tag{11.6}$$

The velocity is assumed to have a logarithm variation,

$$\frac{dU}{dz} = \frac{u_*}{kz} \tag{11.7}$$

Further considering a linear variation of shear with maximum at bed and 0 at the water surface (Nezu and Nakagawa 1993),

$$\frac{\tau(z)}{\rho} = {u_*}^2\left[1 - \frac{z}{h}\right] \tag{11.8}$$

Thus,

$$G = \sqrt{\frac{u_*^3\left(1 - \frac{z}{h}\right)}{\nu k z}} \tag{11.9}$$

11.3 Application Sites

11.3.1 Muthupet Estuary, Tamil Nadu, India

Point Calimere is a Ramsar site since 2002. It consists of a forest, a wetland and the Muthupet estuary. It is located on the delta of Cauvery river (Fig. 11.1a). The estuary is surrounded by mangroves dominated by Avicennia marina. The estuary is quite shallow having a mean depth of 1 m. The tidal range at the mouth is only 0.5 m, thereby coming under the microtidal category. The estuary is diverged at the mouth in the east and west branches, but the tides predominantly discharge in the eastern branch due to the peculiar geomorphology of the estuary (Priya et al. 2012). The bed sediments of the estuary is made of clay and fine silty sediments.

The major freshwater flow occurs from the Korayar river and Paminiyar river, the majority of flow is from the Korayar. The river discharge into the estuary varies from 0–80 $m^3\ s^{-1}$. The river discharge is available only for a period of 3–4 months. During the dry season, there is no freshwater input into the estuary. This has resulted in the formation of salt plug at intermediate locations inside the estuarine lagoon part (Priya et al. 2015b). The formation of zone of turbidity maximum is also a peculiar feature of the Muthupet estuary and it occurs at similar locations as that of the salt plug (Priya et al. 2015b). The estuary has been found to trap metal such as iron within the turbidity maxima zone (Priya et al. 2014). The concentration of Fe (II) in the water column is observed to be influenced by the salt transport through the estuary (Priya et al. 2016).

11.3.2 Bouregreg Estuary, Morocco

Located in the northwestern part of Morocco, the Bouregreg river stretches for 240 km from the Middle Atlas mountain and discharges to Atlantic Ocean (Fig. 11.1b). The

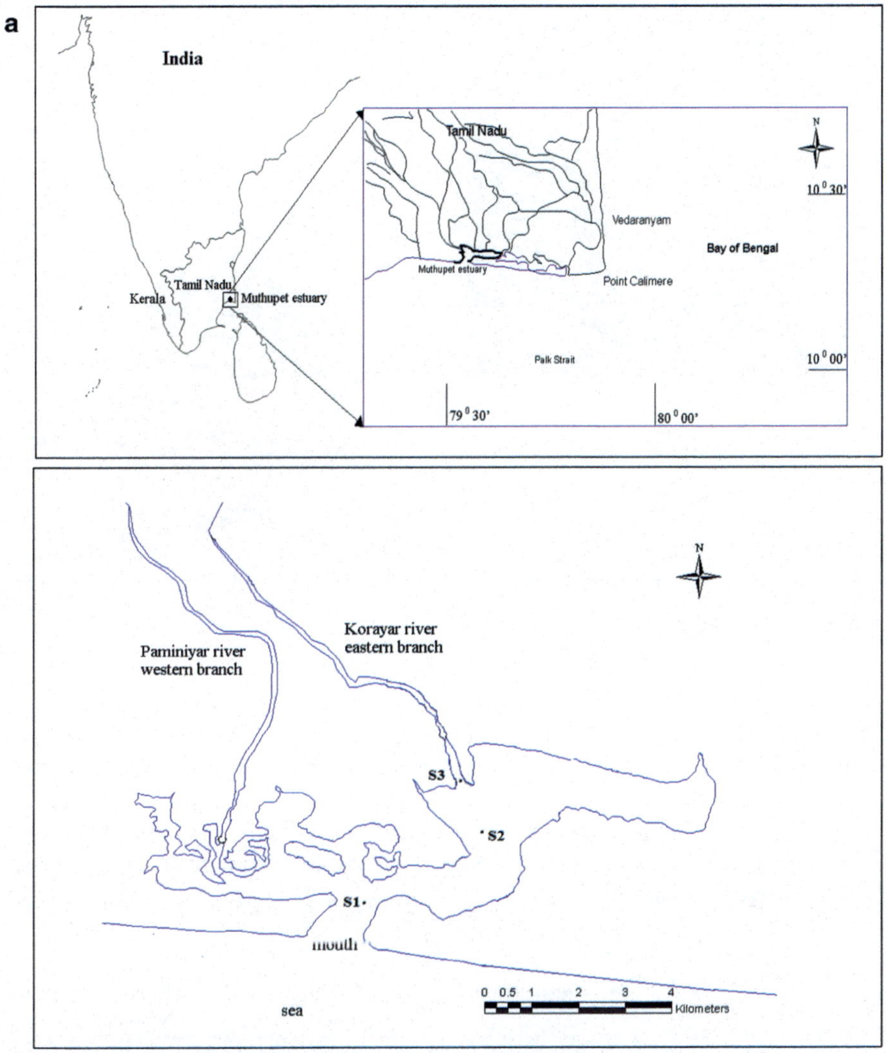

Fig. 11.1 **a** Location map of Muthupet estuary showing sampling locations. **b** Location map of Bouregreg estuary showing sampling locations

watershed of the river covers an area of ca. 9800 km^2 (Zellou and Rahali 2017). The average annual rainfall is 450 mm and average annual temperature is 18 °C. The length of the estuary is 23 km, with Sidi Mohammed Ben Abdellah dam on the upstream side and the average width being 150 m. The freshwater discharge varies between 3 and 84 $m^3 s^{-1}$. The tide at the mouth of the estuary is of semi-diurnal type having a tidal cycle of 44,100 s and a tidal range of 1.4 m, coming under micro-tidal category. The average depth of the estuary is ca. 3 m. The estuary is tidal dominated with minimal river discharge for most of the year. The land use near the estuary mouth

b

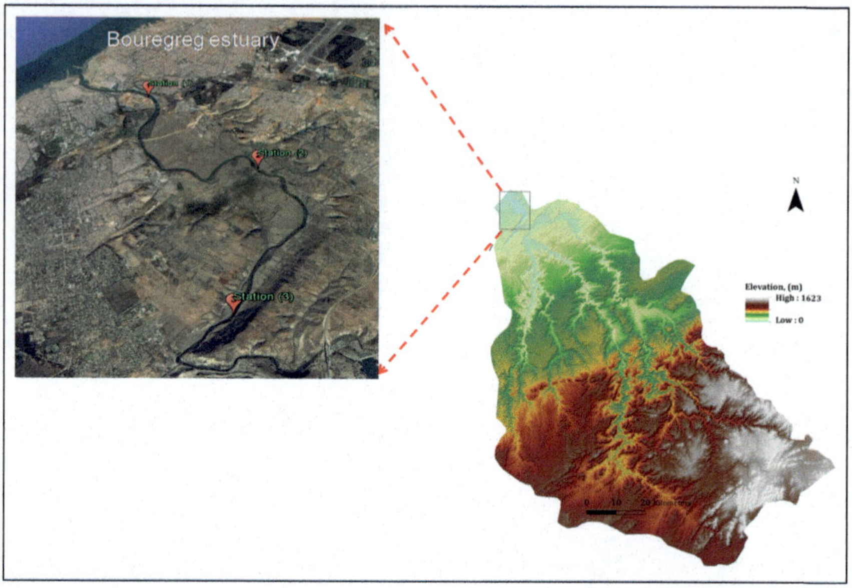

Fig. 11.1 (continued)

is of agricultural and industrial regions. Domestic sewage enters the estuary from the near-by cities. The Bouregreg estuary is categorized by an increase in suspended organic solids and nutrients and an the enhancement in salinity. The salinity intrusion into the estuary was observed to increase after the commission of Sidi Mohammed Ben Abdellah dam. The dissolved oxygen level of the estuary is satisfactory and is well oxygenated. Table 11.1 gives a comparison of the different features of Muthupet and Bouregreg estuary.

Table 11.1 Basic features of Muthuper and Buregreg estuary

Basic feature	Muthupet estuary, India	Bouregreg estuary, Morocco
Climatic condition	Semi-arid type	Semi-arid type
Salinity intrusion (km)	10–30	8–23
Av. depth (m)	1	3
Tidal range	Micro-tidal	Micro-tidal
Tidal cycle	Semi-diurnal tide	Semi-diurnal tide
Freshwater discharge	0–80 m^3/s	3–84 m^3/s
Mixing type	Well mixed/partially stratified type	Well mixed/partially stratified type
Oxygenation	Well oxygenated	Well oxygenated

11.4 Field Experiments

The data set utilized for the investigations has been described by Priya et al. 2015a for the Muthupet estuary. Three sampling stations namely S1, S2 and S3 were selected for the study. The station S1 is situated near the estuarine mouth, S2 at an intermediate location and S3 at the river end. Station S2 is reported to be located at the turbidity maxima. The sampling stations in the Muthupet estuary are denoted as M-S1, M-S2 and M-S3 to avoid confusion. The field experiments were carried out during the spring tide and neap tide of dry (March) and wet (January) seasons in 2012, where the freshwater flow varied between 0 and 64 $m^3 s^{-1}$.

The sampling stations adopted for the present study in the Bouregreg estuary include S1, S2 and S3 where S1 is located near the mouth where dredging activities prevail, S2 at an intermediate location where the river meanders, and S3 at the river end. These stations are denoted as B-S1, B-S2 and B-S3. The same sampling protocol was carried out in the Bouregreg estuary during the spring and neap tides of wet (November) and dry (May) seasons in 2017. The river discharge varied between 9 and 40 $m^3 s^{-1}$ during the sampling period. All the field experiments were executed for duration of one tidal cycle during which the water level was measured every one hour. Current was measured at 0.3 D and 0.6 D (where D represents the water depth) every one hour using OTT current meter. Samples were collected from 0.3D and 0.6D every one hour and analysed for salinity and suspended sediment concentration. Salinity was measured using YSI Pro 30 Conductivity meter and suspended sediment concentration using gravimetric method.

11.5 Results and Discussion

11.5.1 Water Level, Current, Salinity, SSC and Settling Velocity

The variation of water level at the sampling stations of the Muthupet estuary and Bouregreg estuary are shown in Fig. 11.2a and b respectively. The maximum, minimum and average values of tidal averaged water level are furnished. The water level variation and hence the tidal range was high during the spring tides compared to that during the neap tides in both the estuaries. A higher variation near the mouth is evident and it progressively decreases towards the upstream side, as the tide dampens due to the geomorphology. Figure 11.3a and b shows the tidal averaged depth averaged current in the Muthupet and Bouregreg estuary. The current was at its highest near the mouth station in the Muthupet estuary and lowest at the intermediate station M-S2, while the current decreased gradually from the mouth station (B-S1) to the upstream station (B-S3) in the Bouregreg estuary. On an average, the current was highest during the spring tides than during the neap tides.

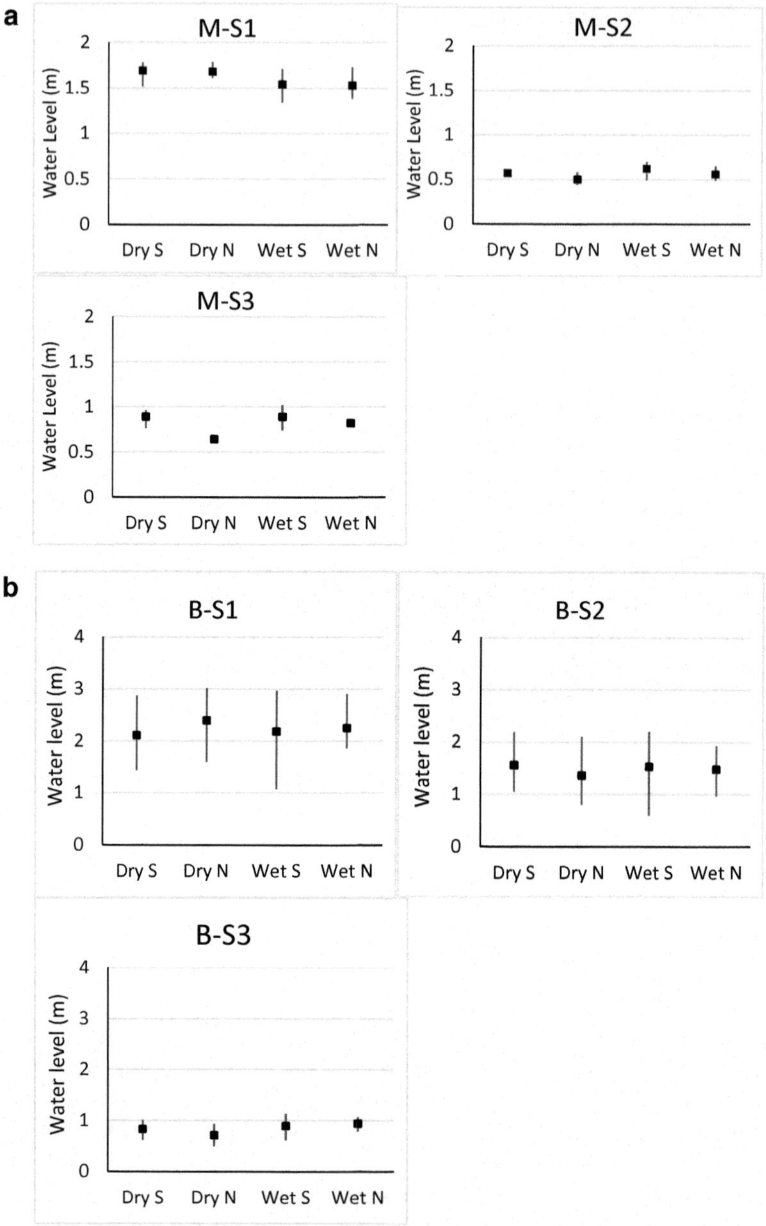

Fig. 11.2 **a** Tidal averaged water level at different stations during different seasons in the Muthupet estuary. *Note* Dry S: Dry season and Spring tidal condition; Dry N: Dry season and Neap tidal condition; Wet S: Wet season and spring tidal condition; Wet N: Wet season and neap tidal condition. **b** Tidal averaged water level at different stations during different seasons in the Bouregreg estuary. *Note* Dry S: Dry season and Spring tidal condition; Dry N: Dry season and Neap tidal condition; Wet S: Wet season and spring tidal condition; Wet N: Wet season and neap tidal condition

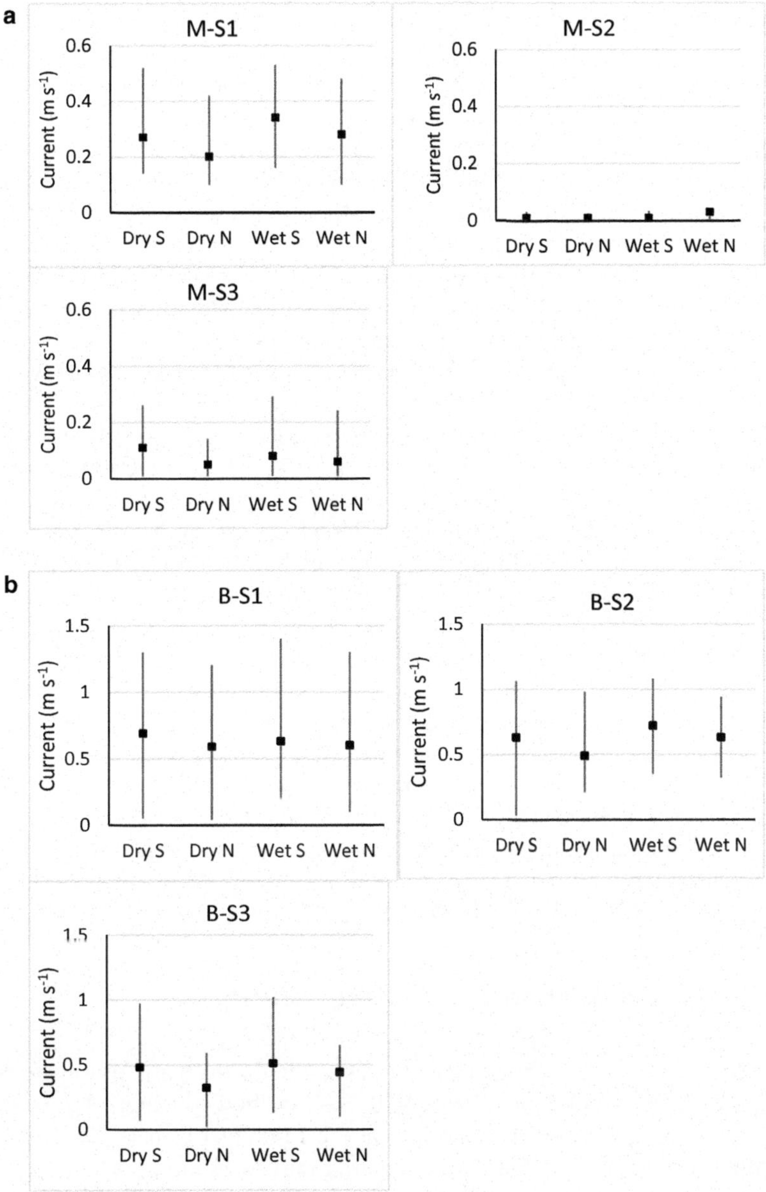

Fig. 11.3 **a** Tidal averaged depth averaged current at different stations during different seasons in the Muthupet estuary. *Note* Dry S: Dry season and Spring tidal condition; Dry N: Dry season and Neap tidal condition; Wet S: Wet season and Spring tidal condition; Wet N: Wet season and Neap tidal condition. **b** Tidal averaged depth averaged current at different stations during different seasons in the Bouregreg estuary. *Note* Dry S: Dry season and Spring tidal condition; Dry N: Dry season and Neap tidal condition; Wet S: Wet season and Spring tidal condition; Wet N: Wet season and Neap tidal condition

There was a prominent variation in the tidal averaged depth averaged values of salinity over a seasonal scale in both the estuaries (Fig. 11.4a and b). The freshwater flow is a vital controlling parameter of salinity in the estuary. The higher the freshwater discharge, the lower is the salinity and was in agreement in both the estuaries. The averaged salinity values ranged between 22 and 29 ppt at M-S1 and between 9 and 31 ppt at M-S3 in the Muthupet estuary, while the range was 24 and 32 ppt at B-S1 and 2 and 7.5 ppt at B-S3 in the Bouregreg estuary. M-S3 is located at a distance of 7 km upstream of sea mouth in the Muthupet estuary, while in the Bouregreg estuary, B-S3 is located at a distance of 17 km upstream the sea mouth. Salinity values at a distance of 17 km from the mouth in the Muthupet estuary was in the similar range as that of the Bouregreg estuary (Priya et al. 2015b).

The tide averaged depth averaged SSC in the estuaries during dry and wet seasons are shown in Fig. 11.5a and b. SSC was at its highest at M-S2 in the Muthupet estuary and lowest at M-S3. In the Bouregreg estuary, SSC was at its highest at B-S3 and lowest at B-S1. SSC ranged between 1.47 and 0.15 g L^{-1} in the Muthupet estuary, while the range was 0.174 and 0.02 g L^{-1} in the Bouregreg estuary. The tidal averaged depth averaged settling velocity during different seasons in the estuaries are shown in Fig. 11.6a and b. The settling velocity ranged between 0.1 and 2.5 mm s^{-1} with lowest at M-S2 and highest at M-S3 in the Muthupet estuary. Its range was from 0.61 to 0.01 mm s^{-1} with minimum at B-S2 and maximum at B-S3 in the Bouregreg estuary. The settling velocity of suspended sediments reported in literatures are of similar range (Dyer et al. 1996; Manning et al. 2006; Winterwerp 2002). A tidal stage variation in the settling velocity was evident in the Muthupet estuary with higher settling velocity during the neap tides than during spring tides. Thus, the settling velocity variation occurred over a spring-neap tidal cycle and spatial scale. However, only the spatial variation in settling velocity was significant in the Bouregreg estuary. Table 11.2 summarizes the significant characteristics with regard to Muthupet and Bouregreg estuary.

11.5.2 Influence of SSC on Settling Velocity

The influence of SSC on settling velocity was studied with regard to both the estuaries and the variation of settling velocity with SSC is shown in Fig. 11.7a and b. The variation of settling velocity with SSC in the Muthupet estuary has been clearly described in Priya et al. 2015a, b. To summarize the observations: the settling velocity was a positive power function of SSC at stations M-S1 and M-S3 with a relationship of the form $w_s = a\ SSC^x$. Nevertheless, a negative variation of settling velocity with SSC was observed at M-S2, which shows a spatial variability in its relation. The observations in the previous study were that M-S2 was located at the turbidity maximum zone and that hindered settling occurs at this zone leading to a negative relationship. In the case of Bouregreg estuary, even though a positive relation was observed between settling velocity and SSC at stations B-S1 and B-S2, the influence

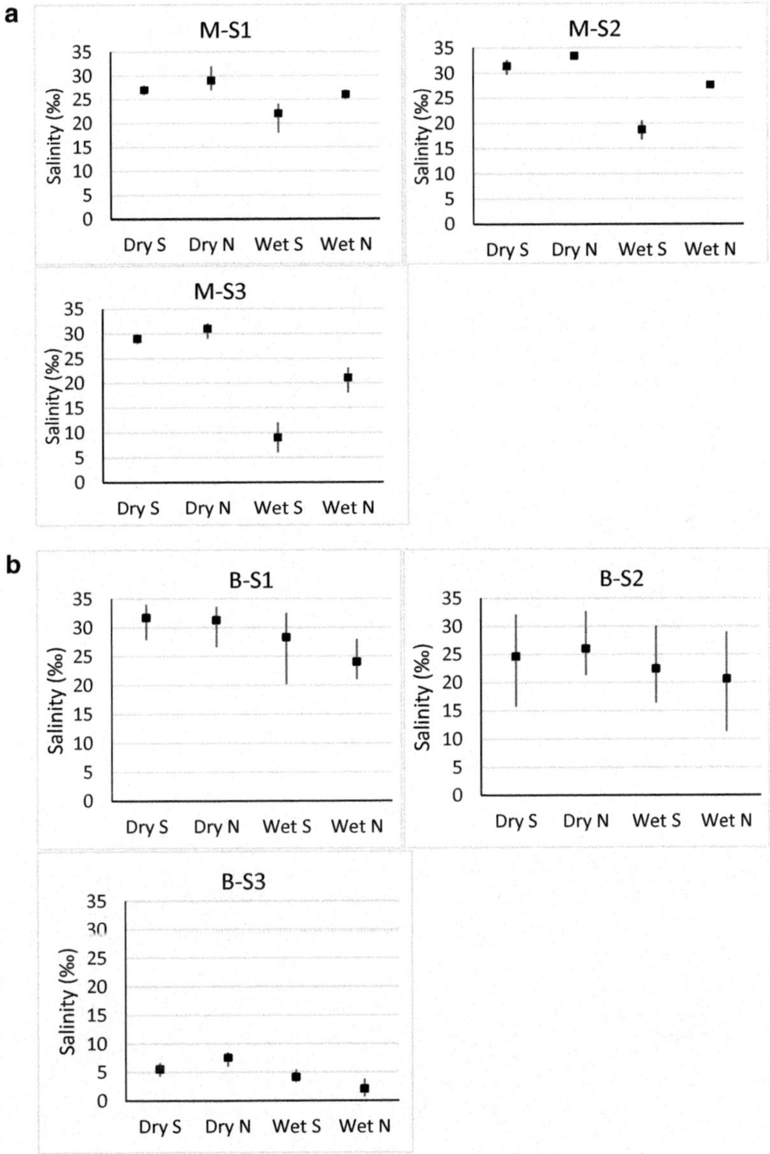

Fig. 11.4 **a** Tidal averaged depth averaged salinity at different stations during different seasons in the Muthupet estuary. *Note* Dry S: Dry season and Spring tidal condition; Dry N: Dry season and Neap tidal condition; Wet S: Wet season and Spring tidal condition; Wet N: Wet season and Neap tidal condition. **b** Tidal averaged depth averaged salinity at different stations during different seasons in the Bouregreg estuary. *Note* Dry S: Dry season and Spring tidal condition; Dry N: Dry season and Neap tidal condition; Wet S: Wet season and Spring tidal condition; Wet N: Wet season and Neap tidal condition

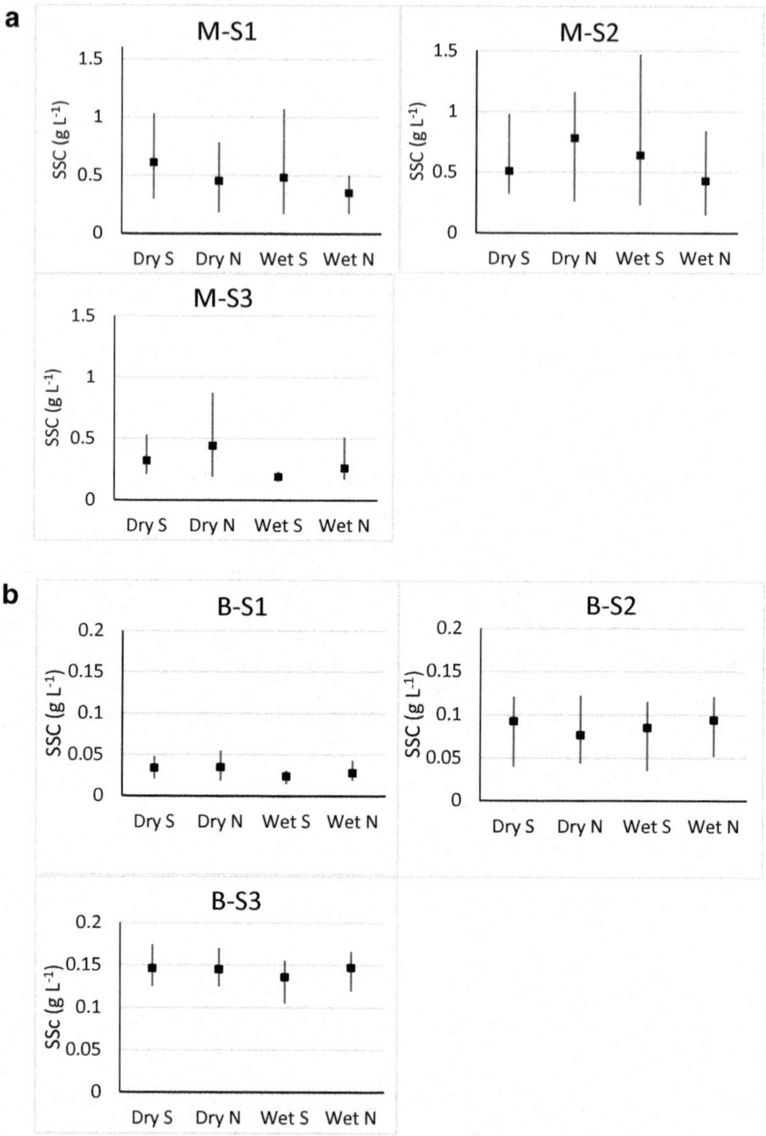

Fig. 11.5 **a** Tidal averaged depth averaged SSC at different stations during different seasons in the Muthupet estuary. *Note* Dry S: Dry season and Spring tidal condition; Dry N: Dry season and Neap tidal condition; Wet S: Wet season and Spring tidal condition; Wet N: Wet season and Neap tidal condition. **b** Tidal averaged depth averaged SSC at different stations during different seasons in the Bouregreg estuary. *Note* Dry S: Dry season and Spring tidal condition; Dry N: Dry season and Neap tidal condition; Wet S: Wet season and Spring tidal condition; Wet N: Wet season and Neap tidal condition

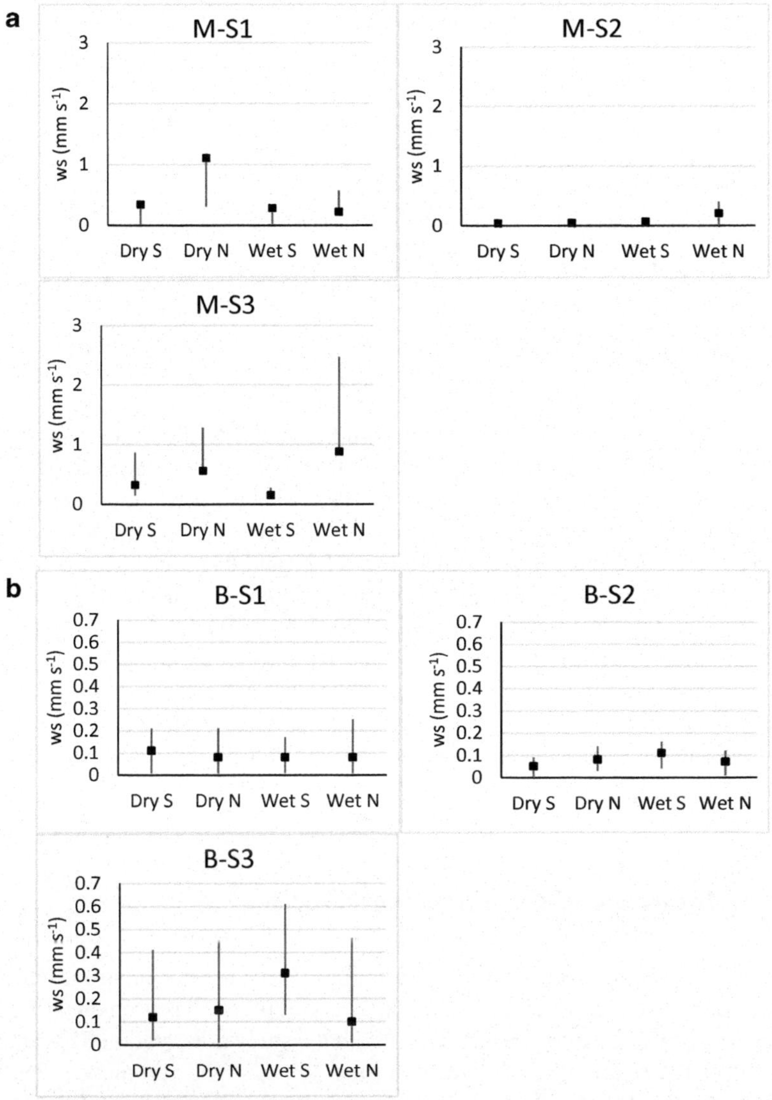

Fig. 11.6 **a** Tidal averaged depth averaged settling velocity at different stations during different seasons in the Muthupet estuary. *Note* Dry S: Dry season and Spring tidal condition; Dry N: Dry season and Neap tidal condition; Wet S: Wet season and Spring tidal condition; Wet N: Wet season and Neap tidal condition. **b** Tidal averaged depth averaged settling velocity at different stations during different seasons in the Bouregreg estuary. *Note* Dry S: Dry season and Spring tidal condition; Dry N: Dry season and Neap tidal condition; Wet S: Wet season and Spring tidal condition; Wet N: Wet season and Neap tidal condition

Table 11.2 Some significant observations between Muthupet estuary and Bouregreg estuary

SI. No	Significant characteristics	Muthupet estuary, India	Bouregreg estuary, Morocco
1	Salt plug	Present during dry season (Priya et al. 2015b)	Absent
2	Turbidity maxima	Present (Priya et al. 2015b)	Absent
3	Transport of salt and SSC	Pronounced seasonal variation (Priya et al. 2016)	Always upstream transport (unpublished data)
4	Key parameter of salt and SSC transport	Fluvial advection (Priya et al. 2016)	Fluvial advection (unpublished data)
5	Settling velocity	High at upstream	High at upstream

of SSC on settling velocity was not much significant at B-S1. Nevertheless, station B-S2 exhibited a negative relationship between settling velocity and SSC. The settling velocity at B-S3 was much influenced by SSC where a sharp increase in settling velocity with SSC was observed. The relationship between w_s and SSC was of the form $w_s = a\ SSC^x$ where a varied between 0.025 and 647 and x varied between −0.367 and 4.414 (Table 11.3). The station B-S1 is located at the dredging zone. The geometry is not constant and is influenced by the dredging activities. Hence, it is difficult to establish a relationship between settling velocity and SSC. Station B-S2 is located at the meandering zone. At this zone, three effects are liable to occur: (i) reversal of vertical profile (ii) turbulence effect and (iii) erosion of banks. There is more possibilities of a resuspension than settling in this zone and hence a negative relation was observed between settling velocity and SSC.

11.5.3 Influence of G on Settling Velocity

Turbulence is a significant parameter which influences the flocculation, settling and resuspension of sediments in the water column. Turbulence has been quantified using root-mean-square velocity gradient G. At both the study area, station S2 was omitted to establish the relationship between settling velocity and turbulence. The variation of settling velocity with turbulence is depicted in Fig. 11.8a and b. Generally, settling velocity was directly varying with turbulence in both the estuaries. Increase in turbulence in the water column causes more aggregation of particles enhancing the floc formation. The flocs are formed due to the influence of particulate organic matter, which are rich in both the estuaries. The flocs settle down faster than individual particles, thereby enhancing settling. However, at higher turbulence, the shear is high enough to cause breakage of the flocs formed, resulting in a reduction in settling velocity. In the case of Muthupet estuary, the G values were low enough to prevent the breakage of flocs ($G < 20\ s^{-1}$). In the Bouregreg estuary, G ranges between 0.35 and 89 s^{-1}. At higher value of G, a decreasing trend in settling velocity is noted (Fig. 11.8b). Thus, the analysis was made separately for lower and higher values of

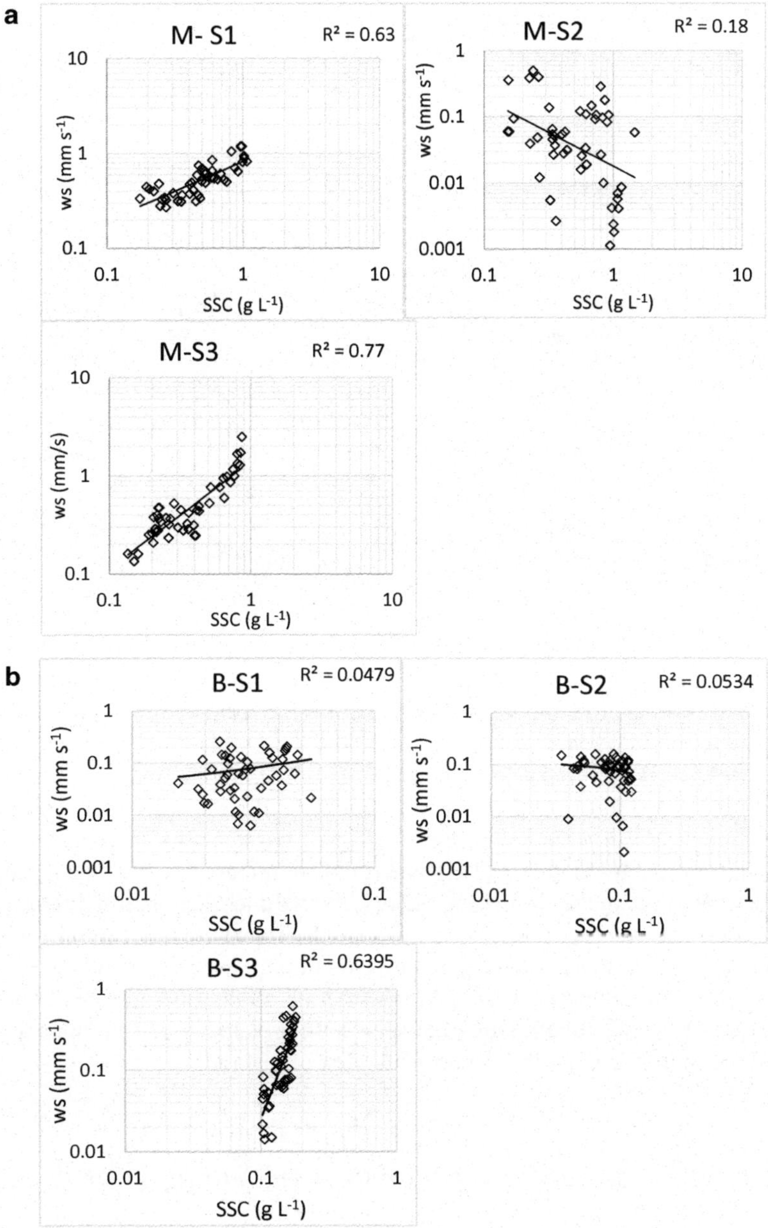

Fig. 11.7 **a** Variation of settling velocity with SSC in Muthupet estuary at different stations (*Source* Priya et al. 2015a). **b** Variation of settling velocity with SSC in Bouregreg estuary at different stations

Table 11.3 Results of sensitivity tests

Approach		Stations considered	Muthupet estuary, India			Bouregreg estuary, Morocco		
			Coefficient	Exponents	R^2	Coefficient	Exponents	R^2
$w_s = a\,SSC^x$		M-S1 and B-S1	0.816	0.619	0.63	0.618	0.698	0.04
		M-S2 and B-S2	0.018	−1.02	0.18	0.025	−0.367	0.02
		M-S3 and B-S3	1.39	1.09	0.77	647	4.414	0.64
$w_s = a\,G^y$ G < 20 G > 20		M-S1, M-S3 and B-S1, B-S3	0.253 –	0.322 –	0.42 –	0.015 2.29	0.648 −0.761	0.54 0.2
$w_s = a\,SSC^x$ $G^y\,s^{-z}$	S-I S-II	M-S1 and M-S3 (Muthupet) and B-S3 (Bouregreg)	2.95 1	0.624, 0.47, −0.2 1.188, 0.36, – 0.03	0.86 0.64	0.4 –	0.71, 0.32, −0.5 –	0.78 –

Note S-I Scenario I; *S-II* Scenario II

G and is shown in Fig. 11.8c and d. At lower values of G (here G < 20 s^{-1}), a positive relation of the form $w_s = a\,G^y$ was observed, while at higher values of G, settling velocity was a negative function of G. An converse relationship between w_s and G has been reported by Winterwerp (1999) and a direct relationship by Manning and Dyer (1999) and Pejrup and Mikkelsen (2010). The laboratory studies of Milligan and Hill (1998) yielded a higher rate of formation of flocs at higher concentration and low turbulence. The variation on the stability of the floc with the type of sediment has been reported by Li et al. (1993), where clay dominant floc was much stable than silt dominant floc. In the Bouregreg estuary, the sediments are predominantly clayey silt causing less stability thereby the flocs are destroyed at higher turbulence. Other researchers have also reported the negative relation between w_s and G, but all those studies are on a laboratory scale (Wolanski et al. 1992; Ha and Maa 2010). Field validation of the negative relation between settling velocity and turbulence has not been done in the past and this report is the first of its kind.

11.5.4 Influence of Salinity Gradient on Settling Velocity

Salinity gradient *s* is the difference of bottom and surface salinities divided by water depth (Priya et al. 2015a). It is controlled by the freshwater flow and tidal flow. At higher freshwater flows, a well-defined vertical salinity gradient is developed in the water column. When the turbulence is high, a vertical mixing causes the reduction in the salinity gradient. In earlier studies, it was reported that the salinity gradient influences the settling velocity for a stratification parameter greater than 0.1 in the

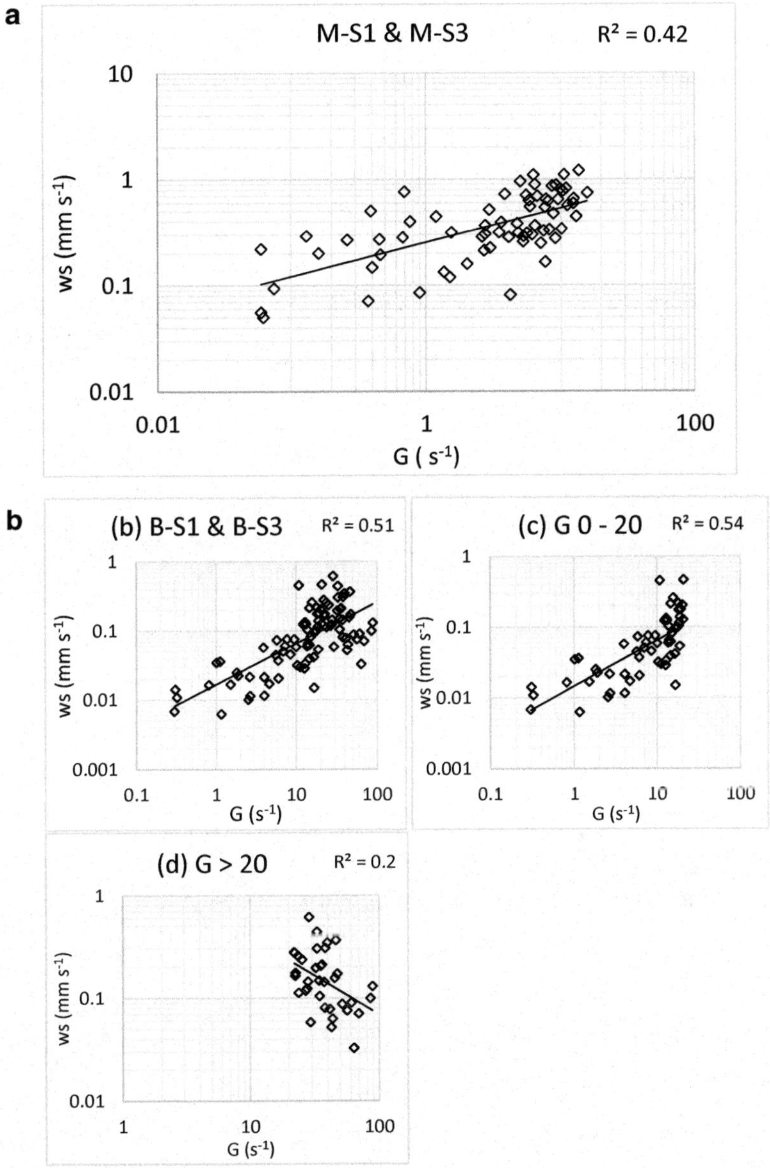

Fig. 11.8 **a** Variation of settling velocity with turbulence in Muthupet estuary and Bouregreg estuary (*Source* Priya et al. 2015a). **b** when all the samples were considered **c** for those samples whose G < 20 s^{-1} **d** for those samples whose G > 20 s^{-1}

Muthupet estuary (Priya et al. 2015a, b). Hence, for the analysis in Bouregreg estuary, those samples having a stratification parameter >0.1 are considered. The variation of settling velocity with *s* in the Muthupet estuary & Bouregreg estuary is depicted respectively in Fig. 11.9a and b. A negative relationship was observed in the case of both the estuaries, which suggest that higher salinity gradient hinders the settling of suspended sediments. The theory was put forward by Priya et al. (2015a) in which salinity gradient causes high density near the estuarine bed, resulting in enhancing the viscosity. The flocs, on its path of settling, when reaches the bottom dense waters, the density of which is greater than that of the flocs, encounters a retardation in settling. Thus, the flocs remain in suspension causing a reduction in the settling velocity. However, a detailed laboratory study comparing the density of water column and that of flocs in such conditions is called for in future.

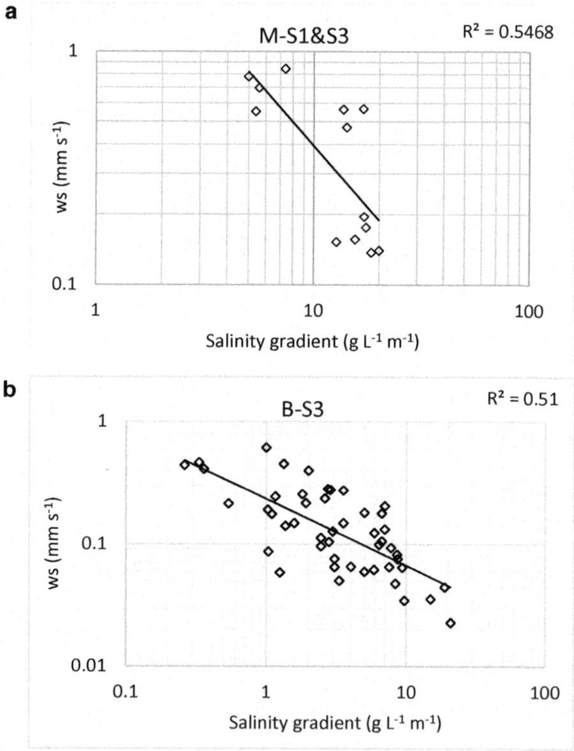

Fig. 11.9 **a** Variation of settling velocity with salinity **b** Variation of settling velocity with gradient for $\partial S/S_0 > 0.1$ in Muthupet estuary salinity gradient in Bouregreg estuary (*Source* Priya et al. 2015a)

11.5.5 Application of Empirical Model Relating w_s, SSC, G and s

An empirical model connecting ws, SSC, G and s was developed by Priya et al. (2015a) and the model was applied to the data set of Muthupet estuary for predicting the settling velocity. The model predicted fairly better than the available models. The model considers two scenarios based on the stratification parameter $\partial S/S_0$: Scenario I: when the stratification parameter varies between 0.1 and 1; Scenario II: when the stratification parameter is less than 0.1. In the case of scenario II, there was not much of an influence of *s* on settling velocity. The model is advantageous in the fact that it considers the effect of concentration of suspended sediments, fluid shear on flocculation and accounts for the resistance for settling down due to *s*. The predictableness of settling velocity using this model has enhanced by 32% in the R^2 value compared to existing models (Priya et al. 2015a). The model was applied to the dataset of Bouregreg estuary for station B-S2 alone, as the behaviour of suspended sediments at other stations were affected by dredging and meandering. The applicability of the model in predicting settling velocity was assessed by undertaking a sensitivity test. Figure 11.10a and b shows the observed settling velocity vs predicted settling velocity using the model. Table 11.3 gives the statistical parameters of the test. The 45° line shown can be used to examine the quality of fit obtained. It can be observed that the model connecting SSC, G and s significantly improves the predictability of settling velocity of suspended sediments in the Bouregreg estuaty. The R^2 value is higher for the model, thereby suggesting that it gives a better description of settling velocity.

11.6 Conclusions

The influencing factors of settling velocity of suspended sediments in the Bouregreg estuary, Morocco were studied. The settling velocity was a power function of concentration of suspended sediments at river end station. The behavior of settling velocity was difficult to understand near the mouth, as it was a dredging one. Another station located at the meandering zone was dominated by resuspension and hence a negative relationship between settling velocity and suspended sediment concentration was obtained. The study evaluated the applicability of a recently developed model for the settling velocity of suspended sediments. The model relates settling velocity with suspended sediment concentration, turbulence shear and salinity gradient. The model was originally developed for predicting the settling velocity of suspended sediments of Muthupet estuary, India. The model predicted the settling velocity of suspended sediments in the Bouregreg estuary fairly well, except at the locations of dredging and meandering. The behavior of sediments at the meandering zone needs special attention and a detailed study is called for in the future.

Fig. 11.10 **a** Observed versus predicted settling velocity in Muthupet estuary for Scenario I and Scenario II (*Source* Priya et al. 2015a). *Note* Station M-S1 & M-S3 are considered; bold line represents 45° line. **b** Observed vs predicted settling velocity in Bouregreg estuary for Scenario I. *Note* Station B-S3 alone is considered; bold line represents 45° line

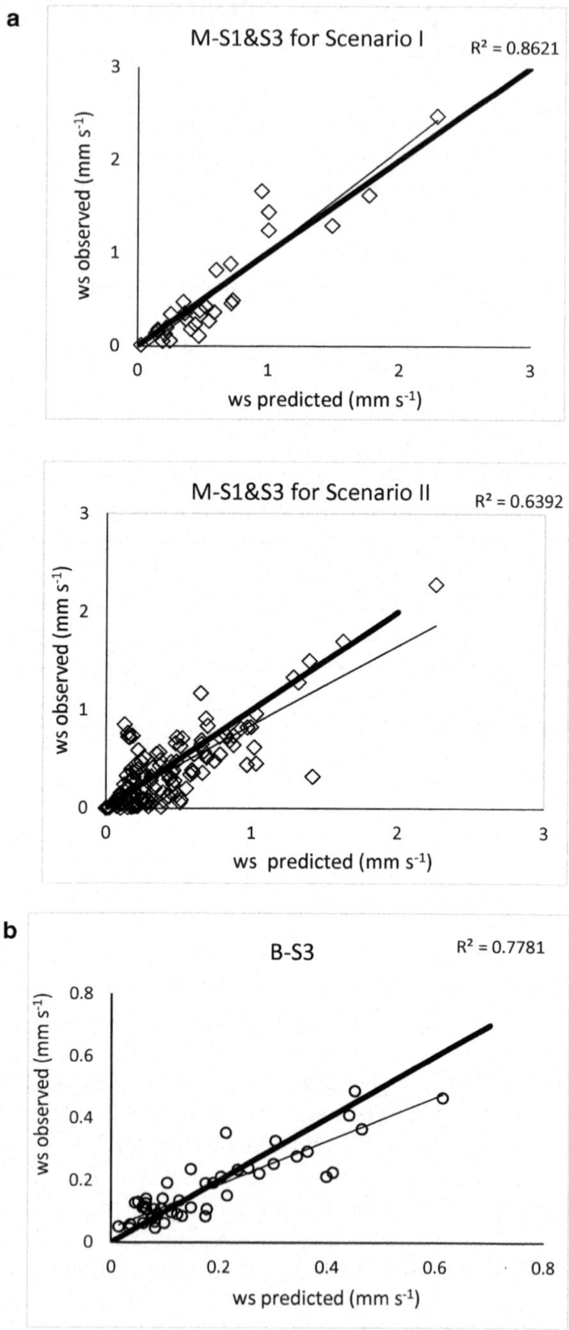

References

Camp TR, Stein PC (1943) Velocity gradients and internal work in fluid motion. J Boston Soc Civil Eng 30:219–237

Dyer KR, Cornelisse J, Dearnaley MP, Fennessy MJ, Jones SE, Kappenberg J, McCave IN, Pejrup M, Puls W, van Leussen W, Wolfstein K (1996) A comparison of in situ techniques for estuarine floc settling velocity measurements. J Sea Res 36:15-29

Gratiot N, Manning AJ (2007) A laboratory study of dilute suspension mud floc characteristics in an oscillatory diffusive turbulent flow. J Coast Res SI 50:1142–1146

Ha HK, Maa JP-Y (2010) Effects of suspended sediment concentration and turbulence on settling velocity of cohesive sediment, Goesci J 14(2). 10.1007/s12303-010-0016-2

Johansen C, Larsen T (1998) Measure of settling velocity of fine sediment using a recirculated settling column. J Coast Res 14(1):132–139

Krone RB (1962) Flume studies of the transport of sediment in estuarial shoaling processes—final report. Hydraulic Engineering Laboratory and Sanitary Engineering Research Laboratory, University of California, Berkeley, p 110

Li Y, Wolanski E, Xie Q (1993) Coagulation and settling of suspended sediment in the Jiaojiang River estuary, China. J Coast Res 9(2):390–402

Lumborg U, Pejrup M (2005) Modelling of cohesive sediment transport in a tidal lagoon—an annual budget. Marine Geol 218:1–16

Maa JP-Y, Kwon J-I (2007) Using ADV for cohesive sediment settling velocity measurements. Estuar Coast Shelf Sci 73:351–354

Manning AJ, Dyer KR (1999) A laboratory examination of floc characteristics with regard to turbulent shearing. Mar Geol 160:147–170

Manning AJ, Bass SJ, Dyer KR (2006) Floc properties in the turbidity maximum of a mesotidal estuary during neap and spring tidal conditions. Mar Geol 235:193–211

Manning AJ, Martens C, De Mulder T, Vanlede J, Winterwerp JC, Ganderton P, Graham GW (2007) Mud floc observations in the zone of maximum turbidity of the Scheldt estuary during neaptides. J Coast Res 50:832–836 (Special Issue)

Masselink G, Cointre L, Williams J, Gehrels R, Blake W (2009) Tide-driven dune migration and sediment transport on an intertidal shoal in a shallow estuary in Devon. Marine Geol UK. https://doi.org/10.1016/j.margeo.2009.03.009

Milligan TG, Hill PS (1998) A laboratory assessment of the relative importance of turbulence, particle composition, and concentration in limiting maximal floc size and settling behavior. J Sea Res 39:227–241

Nezu I, Nakagawa H (1993) Turbulence in open channel flows. International Association for Hydraulic Research Monograph Series, Rotterdam

Pejrup M, Mikkelsen OA (2010) Factors controlling the field settling velocity of cohesive sediment in estuaries. Estuar Coast Shelf Sci 87:177–185

Portela LI, Ramos S, Trigo-Teixeira A (2013) Effect of salinity on the settling velocity of fine sediments of a harbour basin. J Coast Res 65:1188–1193 (Special Issue)

Priya KL, Jegathambal P, James EJ (2012) Seasonal behavior of a shallow estuary of the lower Cauvery basin, India. Environ Res Eng Manag 3(61):6–13

Priya KL, Jegathambal P, James EJ (2014) Trace metal distribution in a shallow estuary. Toxicol Environ Chem 96(4):579–593

Priya KL, Jegathambal P, James EJ (2015a) On the factors affecting the settling velocity of fine suspended sediments of a shallow estuary. J Oceanogr 71:163–175

Priya KL, Jegathambal P, James EJ (2015b) Seasonal dynamics of Turbidity Maximum in the Muthupet estuary, India. J Ocean Univ China 14(5):765–777

Priya KL, Jegathambal P, James EJ (2016) Salinity and suspended sediment transport in a shallow estuary on the east coast of India. Reg Stud Mar Sci 7:88–99

Rouse H (1938) Experiments on the mechanics of sediment suspension. In: Proceedings of the fifth International Congress for applied mechanics, Cambridge, pp 550–554.

Shi Z, Zhou HJ, Eittreim SL, Winterwerp JC (2003) Settling velocities of fine suspended particles in the Changjiang estuary, China. J Asian Earth Sci 22:245–251.

Thorn MFC, Parsons JG (1980) Erosion and cohesive sediments in estuaries: an engineering guide. In: Proceedings of the third international symposium on dredging technology, Bordeaux, pp 349–358.

van Leussen W (1994) Estuarine macro flocs and their role in fine-grained sediment transport. Dissertation. University of Utrecht: Utrecht, The Netherlands.

Winterwerp JC (1999) On the dynamics of high-concentrated mud suspension. Dissertation, Technical University of Delft, Delft, The Netherlands.

Winterwerp JC (2002) On the flocculation and settling velocity of estuarine mud. Cont Shelf Res 22:1339–1360

Wolanski E, Gibbs RJ, Mazda Y, Mehta A, King B (1992) The role of turbulence in the settling of mud flocs. J Coastal Res 8(1):35–46

Zellou B, Rahali H (2017) Assessment of reduced-complexity landscape evolution model suitability to adequately simulate flood events in complex flow conditions. Nat Hazards 86(1):1–29

Chapter 12
Enviro–Economic Analysis and Production Cost of Distilled Water Obtained from Cooling Condensing Active Single Slope Solar Still

Poonam Joshi, G. N. Tiwari, and T. S. Bhatti

Abstract This communication primarily deals with the effect of water cooling condensing cover and helically coiled heat exchanger on single slope solar still integrated with Photovoltaic Thermal Compound Parabolic Concentrators (PVT-CPCs). The design parameters for the active distillation system have been optimized and furthermore production cost of distilled water (₹ 0.11/kg) is obtained. The computational work show that carbondioxide mitigated is 13 ton CO_2/annum and the revenue generated is $108/annum on the basis of energy. The proposed active distillation system gives higher efficiency of solar module (12.03% June, New Delhi, India) with higher distilled water (32.50 kg, daily) because of lower solar cell temperature (68 °C) and cooling condensing cover.

12.1 Introduction

Solar still (Renewable energy technology) converts brackish water (500–1000 ppm), saline water (1000–3500 ppm), hard water or impure water into potable water with the usage of solar energy. In this work we have focussed on active distillation system (solar still), which includes the integration of external thermal collectors for preheating of impure water eg. flat plate collectors (FPCs), solar concentrators, evacuated thermal collectors (ETCs), photovoltaic thermal collectors (PVT) etc. Al-harahsheh et al. (2018) studied experimentally single slope solar still with thermal energy storage (phase change material) and heat exchanger in the basin integrated to solar collectors. They obtained 4300 ml/day m^2 of distilled water from, which 40% was obtained during night time as phase change material acts as a thermal storage. Muthu Manokar et al. (2018) peformed an experiment to analyze a solar still

P. Joshi (✉) · T. S. Bhatti
Centre for Energy Studies, Indian Institute of Technology, New Delhi, India
e-mail: poonam_joshi@aol.com

G. N. Tiwari
Bag Energy Research Society (BERS), Sodha Bers Complex, Plot No. 51, Mahamana Nagar, Karaudi, Varanasi, UP 221005, India

A. Pandey et al. (eds.), *Climate Impacts on Water Resources in India*, Water Science and Technology Library 95, https://doi.org/10.1007/978-3-030-51427-3_12

performance integrated to photovoltaic panel with and without insulation. The daily efficiency was found to be 71.2%, 38.2% and 34.5% for the desalination configuration with insulation, side wall insulation and without insulation. Mahmoud et al. (2018) developed a desalination system in, which a humidification and dehumidification effects were incorporated with solar concentrator and photovoltaic modules. The daily productivity was increased at the operating conditions i.e. three concentration ratio and 0.15 m of water depth in the basin. Recently, Singh and Tiwari (2017) reported that active single slope solar still performs better at higher depth of impure water mass (0.31 m) compared to active double slope solar still.

In the present work we have proposed a helical coil heat exchanger placed inside the basin of cooling condensing single slope solar still integrated with photovoltaic thermal compound parabolic concentrators. The integration of heat exchanger increases the life of the proposed system as the impure water does not corrode the tubes of thermal collectors.

12.2 System Description

The distillation system comprise of a helical coil copper heat exchanger (Fig. 12.1a), N–photovoltaic thermal–compound parabolic concentrator (N–PVT–CPC), single slope solar still and a water storage tank. The concentration ratio $\left(\frac{A_a}{A_r}\right)$ of PVT–CPC is two. The PVT–CPCs can be arranged in series, parallel or series parallel combination to obtain the desired outlet fluid temperature (°C). In the proposed distillation system the thermal collectors are arranged in series combination to obtain increased outlet fluid temperature at the Nth PVT-CPC collector. The outlet from the Nth PVT-CPC collector is the inlet to the heat exchanger (immersed in basin water) and the outlet from the heat exchanger is the inlet to the first PVT-CPC collector. Thus, the circulation of working fluid is in the closed loop. Additionally an arrangement has been made for the cooling of condensing cover so that the water can extract the latent heat of condensation. A DC motor is connected in the insulated connecting pipes to overcome the pressure drop.

Incident global radiation on the basin liner heats the blackened surface ($\alpha = 0.8$), which transfer thermal energy to the impure/hard/brackish water kept in the solar still. Additionally, thermal energy is transferred from the heated fluid (water) inside the heat exchanger to the impure water. It leads to higher evaporation process towards inner condensing cover and its temperature increases. To maintain lower inner glass temperature $\left(T_{gi}\right)$, fluid (water) is allowed to flow over it, which extracts the latent heat of condensation. The temperature of flowing water over condensing cover is initially kept at a lower temperature (T_a − 4) °C, during the summers and (T_a − 1) °C during winters so that it does not evaporate. The whole set up is oriented towards southward to obtain maximum solar energy throughout the life span with the help of the mild steel stand (Fig. 12.1b).

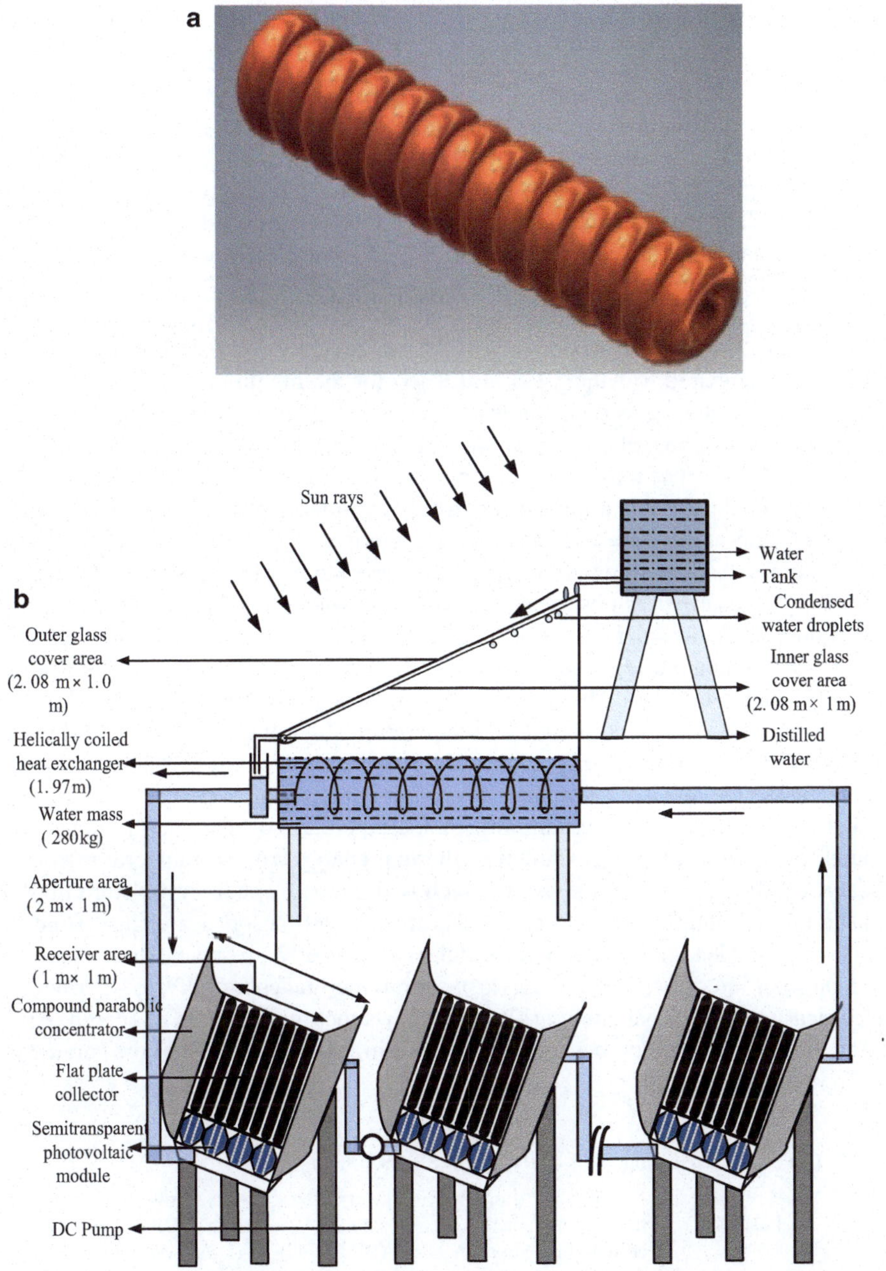

Fig. 12.1 **a** Helically coiled copper tube heat exchanger. **b** Schematic diagram of N-PVT-CPC integrated with cooling condensing single slope solar still with the help of a helical coil heat exchanger

12.3 Results and Discussion

The optimized design parameters (Table 12.1) have been numerically computed for the proposed distillation system. With the help of Table 12.1 higher outlet fluid temperature $\left(T_{foN}\right)$, temperature of basin (T_b) and water (T_w), yield $(\dot{m}_{ew})$ and lower inner glass temperature $\left(T_{gi}\right)$ is obtained (Fig. 12.2a) for two different climatic conditions, January and June of New Delhi, India. The graph shows a similar trend as solar radiation extracted from Lui and Jordan formula, Tiwari (2014).

Figure 12.2b depicts a lower cell temperature of 75.6 °C and 60.5 °C corresponding to higher solar module efficiency of 11.5 and 12% for the two months. The proposed distillation system (PVT-CPC-SS-SS) gives 37% higher solar module efficiency compared to Atheaya et al. (2015) for the month of January because of cooling condensing cover. As there is a constant flow of water (kg/s) the water extracts the latent heat of condensation, which affects the overall performance of the proposed system. The photovoltaic thermal compound parabolic concentrator gives 37% higher solar module efficiency compared to Atheaya et al. (2015) for the month of January because of the cooling condensing cover. As in silicon cells the voltage decreases by 0.4% per degree increase in temperature, which leads to decrease in efficiency. This effect is observed in Fig. 12.2b, when the solar cell temperature is maximum 75 °C the solar module efficiency becomes minimum 0.11%. The effect of water cooling condensing cover on electrical efficiency and cell temperature can be observed from Fig. 12.3a. The graph shows that on increasing the water mass flow rate (kg/s) from 0.025 to 0.2 kg/s the cell temperature slightly increases from 0.1144 to 0.1146 °C and correspondingly solar module efficiency decreases. The reason behind as the mass flow rate increases, there is a lesser contact time period between flowing water and inner glass temperature thus, the temperature of inner condensing cover increases. As a result lesser amount of thermal energy will be transferred from the heat exchanger to the impure water, which will increase water temperature flowing in the tubes of the thermal collectors and thus, it will absorb lesser heat from the solar cells and lead to lower module efficiency. Figure 12.3b depicts the variation of annual solar intensity (kWh), yield (kg), overall thermal energy (kWh) (thermal and electrical) and electrical energy (kWh). It shows that in the month of March, April, May, September and October solar radiation is highest and the rest of the parameters follows the same trend.

Table 12.1 Optimized parameters of PVT-CPC SS-SS system

S. No.	Area of basin (m^2)	No. of thermal collectors	Length of heat exchanger (m)	Mass flow rate (PVT-CPC) (kg/s)	Mass flow rate (CC) (kg/s)	Depth of water (m)
1	2	7	1.97	0.04	0.025	0.14
2	2	6	1.97	0.04	0.025	0.14

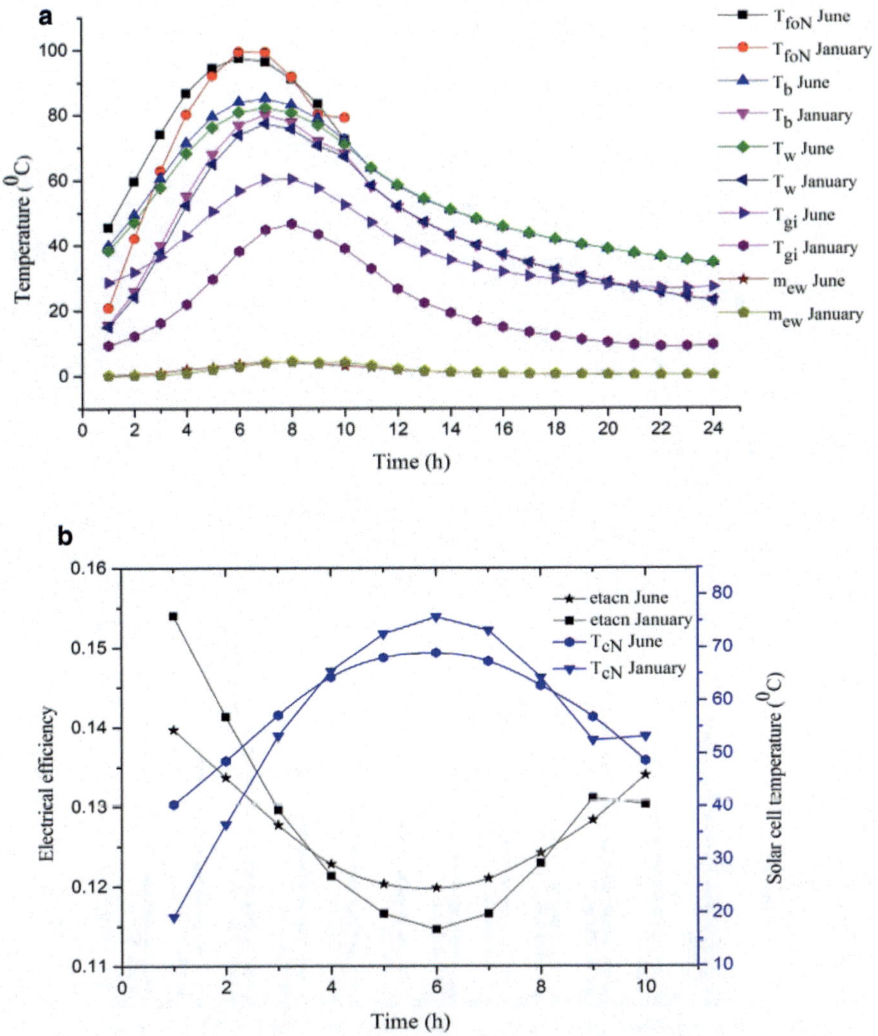

Fig. 12.2 **a** Hourly variation of temperature of different components of PVT-CPC-SS-SS. **b** Hourly variation of electrical efficiency and cell temperature for the PVT-CPC SS-SS

12.3.1 Economic Analysis

Initially monthly overall thermal energy and exergy (kWh), yield (kg), solar intensity (kW/m^2) and electrical energy (kWh) have been calculated (Fig. 12.3b). Then energy pay back time is calculated on the basis of energy and exergy with the help of Eqs. (12.1) and (12.2)

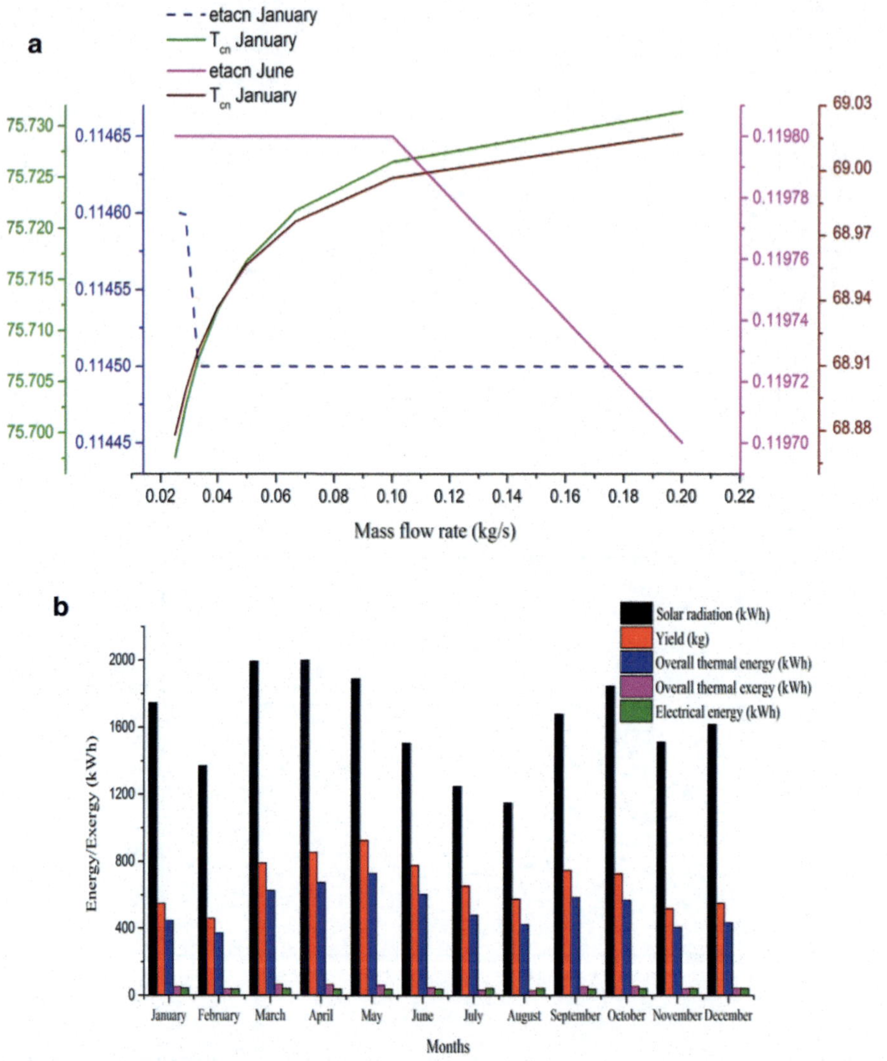

Fig. 12.3 **a** Mass flow rate. **b** Monthly variation of solar intensity, yield, overall thermal energy and exergy, and electrical energy

$$\text{EPBT\_energy} = \frac{\text{Embodied energy}\ (E_{\text{in}})}{\text{Annual overall hermal energy}\ (E_{\text{out}})} \tag{12.1}$$

$$\text{EPBT\_exergy} = \frac{\text{Embodied energy}\ (E_{\text{in}})}{\text{Annual overall exergy gain}\ \left(G_{\text{ex, annual}}\right)} \tag{12.2}$$

Numerical calculations show that energy pay back time on the basis of energy and exergy is two years and seventeen years. Then Uniform Annualized Cost (UAC) at different rate of interest (2, 5, 10%), production cost of disilled water $\left(C_{\text{wp}}\right)$ and electricity (C_e) is calculated using Eqs. (12.3) and (12.4)

$$\text{UAC} = \left(P_s \times F_{\text{CR,in}}\right) + \left(M_s \times F_{\text{CR,in}}\right) - \left(S_s \times F_{\text{CR,in}}\right) \tag{12.3}$$

$$C_{\text{wp}} = \frac{(\text{UAC} - R_e)}{M_w};\ C_e = \frac{(\text{UAC} - R_w)}{E_e} \tag{12.4}$$

Table 12.2 calculates the production cost of distilled water varying from ₹ 0.11 to 1.03/kg for different rate of interest varying from 2 to 5%. As the minimum production cost is ₹ 0.11/kg therefore, the proposed system is viable at 2% rate of interest. The production cost obtained from the proposed distillation system i.e. ₹ 0.11/kg is 336% lesser than the production cost of distilled water obtained by Singh and Tiwari (2016). Equation (12.4) also calculates production cost of electricity (C_e), which is null (Table 12.2), which means higher distilled water (kg) is obtained from the proposed system and thus, the revenue generated from the distilled water can overcome the cost of the proposed active distillation system. After obtaining production cost of distilled water $\left(C_{\text{wp}}\right)$, environmental cost is calculated (Eq. 12.5) because it tells carbondioxide mitigated year (ton CO_2/annum) and revenue earned ($/annum) due to mitigation on the basis of energy and exergy

$$\varphi_{\text{CO}_2} = \frac{2.08 \times (\text{Over all thermal energy/Over all thermal exergy})}{10^3};$$
$$Z_{\text{CO}_2} = \varphi_{\text{CO}_2} \times P_{\text{CO}_2} \tag{12.5}$$

P_{CO_2} is assumed as $14.5. On the basis of energy and exergy carbon dioxide mitigated and carbon credits earned is 13 t CO_2/annum, 1.22 t CO_2/annum and $188/annum, $17.69/annum.

12.4 Summary and Conclusions

The water flowing above condensing cover affects the overall performance of the proposed system, i.e. increased distilled water (8126.48 kg), lowers production cost of distilled water (₹ 0.11 /kg) and higher carbon dioxide is mitigated (13 ton CO_2/annum). This can fulfill small scale purposes in domestic as well as industries usage, hospitals, laboratories, pharmaceutical companies etc. without global warming. The result are encouraging in support of commercialization as energy pay–back time (EPBT) is 2 years based of energy analysis.

Table 12.2 Production cost of distilled water and electricity obtained from PVT-CPC SS-SS

S.No.	i (%)	UAC (₹)	M_w (kg)	$(SP)_w/(SP)_e$ (₹/kg)/(₹/kWh)	R_w (₹)	E_e (kWh)	R_e (₹)	UAC-R_e(₹)	C_{wp} (₹/kg)	C_e (₹/kWh)
1	2	3386	8126	5	40,632	490	2450	936	0.11	0
2	5	5960	8126	5	40,632	490	2450	3510	0.43	0
3	10	10,876	8126	5	40,632	490	2450	8426	1.03	0

12.5 Appendix A

A_a	Aperture area of concentrator
A_r	Receiver area of concentrator
UAC	Uniform annualized cost
R_e	Revenue generated from electricity
M_w	Mass of water
R_w	Revenue generated from water
E_e	Electricity generated
P_s	Present cost
$F_{\mathrm{CR,in}}$	Capital recovery factor
M_s	Mass of water
$F_{\mathrm{SR,in}}$	Sinking fund factor
S_S	Salvage value

References

Al-harahsheh M, Abu-Arabi M, Mousa H, Alzghoul Z (2018) Solar desalination using solar still enhanced by external solar collector and PCM. Appl Therm Eng 128:1030–1040

Atheaya D, Tiwari A, Tiwari GN, Al-Helal IM (2015) Analytical characteristic equation for partially covered photovoltaic thermal (PVT) compound parabolic concentrator (CPC). Sol Energy 111:176–185

Mahmoud A, Fath H, Ahmed M (2018) Enhancing the performance of a solar driven hybrid solar still/humidification-dehumidification desalination system integrated with solar concentrator and photovoltaic panel. Desalination 430:165–179

Muthu Manokar A, Prince Winston D, Kabeel AE, Sathyamurthy R (2018) Sustainable fresh water and power production by integrating PV panel in inclined solar still. J Cleaner Prod 172:2711–2719

Singh DB, Tiwari GN (2016) Effect of energy matrices on life cycle cost analysis of partially covered photovoltaic compound parabolic concentrator collector active solar distillation system. Desalination 397:75–91

Singh DB, Tiwari GN (2017) Performance analysis of basin type solar stills integrated with N identical photovoltaic thermal (PVT) compound parabolic concentrator (CPC) collectors: A comparative study. Sol Energy 142:144–158

Tiwari GN (2014) Solar energy, fundamental, design, modelling and applications. Narosa publishing house pvt. ltd. ISBN 978-81-8487-277-4, 1–40.

Chapter 13
Hydro-ecological Assessment of Environmental Flows for Satluj River

Pradeep Kumar, Jai Prakash Nayak, and Shobha Ram

Abstract The present study was envisaged to estimate environmental flows (EFs) of Satluj river with limited ecological information. For this study, Satluj river basin up to Kasol gauging site has been considered and the EF assessment has been done at three locations viz. Rampur, Suni, and Kasol. The daily discharges of Satluj river at these sites for the years 1964–2011 were used for the analysis. The Global Environmental Flow Calculator (GEFC) developed by Smakhtin and Anputhas (An assessment of environmental flow requirements of Indian river basins p. 42, 2006); Smakhtin and Erivagama (2008) was used to estimate the EFs for different Environmental Management Classes (EMCs). The assessment of ecological status of the river in terms of EMC has been carried out as prescribed by Smakhtin et al. (Developing procedures for assessment of ecological status of Indian river basins in the context of environmental water requirements p. 40, 2007). The ecological status of Satluj River is found to be representative of EMC 'C' at all the three sites. The average monthly EFs at Rampur, Suni, and Kasol sites vary from 63.06 m^3/s (February) to 397.92 m^3/s (July), 71.32 m^3/s (January) to 431.62 m^3/s (July) and 77.78 m^3/s (January) to 455.45 m^3/s (July) respectively. The EF in terms of percentage of long-term natural mean annual runoff at Rampur, Suni, and Kasol sites range from 42.34, 41.50 and 41.49% respectively at Rampur, Suni, and Kasol sites.

13.1 Introduction

An environmental flow is the water regime provided within a river to maintain ecosystems and their benefits where there are competing water uses and where flows

P. Kumar (✉)
Scientist 'D', National Institute of Hydrology, Roorkee, India
e-mail: pradeep.nihr@gov.in

J. P. Nayak
Department of Civil Engineering, PDM University, Bahadurgarh, Haryana, India

S. Ram
Department of Civil Engineering, Gautam Buddha University, Greater Noida, India

A. Pandey et al. (eds.), *Climate Impacts on Water Resources in India*, Water Science and Technology Library 95, https://doi.org/10.1007/978-3-030-51427-3_13

are regulated. Environmental flows provide critical contributions to river health, economic development, and poverty alleviation. The concept of minimum flows in rivers came into practice in the 1970s. The United Nations Conference on Environment and Development (UNCED) at Rio de Janeiro in 1992 established the idea that the health and integrity of the entire ecosystem is fundamental to sustain human well-being. Environmental flows serve to represent water allocation for ecosystems. As ecosystems, in turn, provide services to people, providing for environmental flows is not exclusively a matter of sustaining ecosystems but also a matter of supporting livelihoods of village people who make direct use of river water for a variety of purposes including religious worship. A river reach is deprived of its natural flows due to diversion at control structure. There may be critical reaches in the river where altered flows are not able to sustain the ecosystem services existing prior to implementation of these types of projects. The main objective of this research is to estimate environmental flows of Satluj River with limited ecological information.

13.2 Study Area

For this present study, the Satluj River basin up to Kasol gauging site has been considered. The basin map of the study area has been shown as Fig. 13.1. The basin area of Satluj River at Rampur, Suni, and Kaol sites are 50,800, 52,983, and 53,768 km^2 respectively. The daily discharge data of Satluj River at Rampur, Suni and Kasol for the years 1964–2011 have been used for the analysis.

13.3 Methodology

The methodology of this study divided into two parts: (i) Assessment of ecological status (in terms of EMC) and (ii) assessment of e-flows of Satluj river.

13.3.1 Assessment of Ecological Status (Environmental Management Class)

Ideally, the definition of the environmental management class (EMC) should be based on existing empirical relationships between flow changes and ecological status/conditions, which are associated with clearly identifiable thresholds. Despite some documented examples, limited evidence or knowledge is available of such thresholds (Beecher 1990; Puckridge et al. 1998). Therefore, EMC is a management concept that has been developed and used in the world because of a need to make decisions regardless of the limited lucid hydro-ecological knowledge available. In

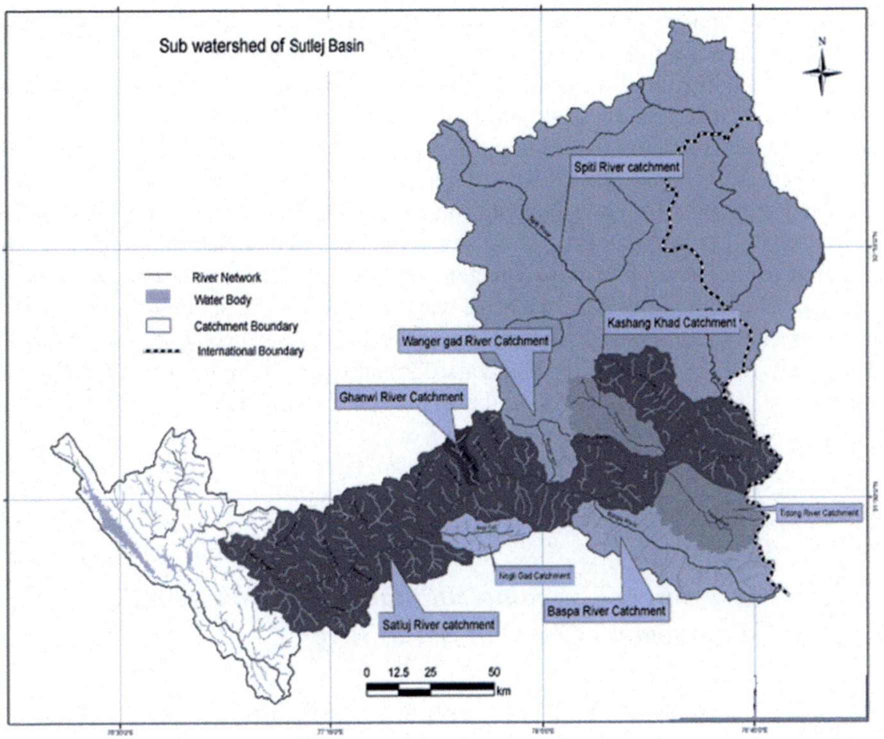

Fig. 13.1 Satluj basin showing major sub-basins and tributaries

these conditions of uncertainty with regard to which EMC is required for a particular river, the EMCs may be used as default 'scenarios' of environmental protection and associated environmental flows as 'scenarios' of environmental water demand (Smakhtin and Anputhas 2006). It is possible to estimate environmental demand corresponding to all or any of such default EMCs and then consider which one is the most feasible for a river in question, given the existing and future basin developments. Alternatively, it is also possible to use expert judgment in order to place a river into the most 'achievable' EMC. An approximation of the overall water quality in a river is indexed using Indian national water quality categorization, which has several classes, from 'A' to 'E' depending on the level of pollution expressed by ranges of several constituents.

13.3.2 Assessment of E-Flows by Using IWMI Approach

Smakhtin and Anputhas (2006) reviewed various hydrology based environmental flow assessment methodologies and their applicability in the Indian context. Based on the study, they suggested a flow duration curve based approach which links environmental flow requirement with environmental management classes. This EFA method is built around a period-of-record FDC and includes several subsequent steps. The first step is the calculation of a representative FDC for each site where the environmental water requirement (EWR) is to be calculated. Based on the study, they suggested a flow duration curve based approach which links environmental flow requirement with environmental management classes. EF aim to maintain an ecosystem in, or upgrade it to, some prescribed or negotiated condition/status also referred to as "environmental management class (EMC)" have been used in this method. The higher the EMC, the more water will need to be allocated for ecosystem maintenance or conservation and more flow variability will need to be preserved. Six EMCs are used in this approach and six corresponding default levels of EWR may be defined. The set of EMCs is similar to the one described in DWAF (1997).

13.4 Results and Discussion

13.4.1 Assessment of Ecological Status (Environmental Management Class) of Satluj River

Normally, the ecological status of a river is assessed based on the Environmental Management Class of that river. The definition of EMC should be based on existing empirical relationships between flow changes and ecological status/conditions, which are associated with clearly identifiable thresholds (Smakhtin et al. 2007). In this connection, EMC is a management concept that has been developed and used globally because of a need to make decisions regardless of the limited hydro-ecological knowledge available (Smakhtin et al. 2007). In these conditions of uncertainty with regard to which EMC is required for a particular river, the EMCs may be used as default 'scenarios' of environmental protection and associated environmental flows as 'scenarios' of environmental water demand (Smakhtin and Anputhas 2006). As per the assessment, the ecological status of Satluj River is representative of Environmental Management Class 'C' (Table 13.1).

13.4.2 Assessment of Environmental Flows

The Global Environmental Flow Calculator (GEFC) developed by International Water Management Institute (IWMI) has been used in the present study. This method is based on the flow duration curve which is further modified for different Environmental Management Classes (EMCs). The method is available in the form of freely available software and it requires average monthly discharge data of sufficient duration for assessing the environmental flows. In the present study, the daily hydrological

Table 13.1 Parameters and their scores for assessing the ecological status (EMC) of Satluj river

Indicator	Rampur	Suni	Kasol	Remarks
Rare and endangered aquatic biota	3	3	3	There are at least 51 fish species in the reach. However, among Himalayan species, moderate numbers of threatened species occur in this reach
Unique aquatic biota	4	4	4	Although this reach has endemic species, but many species are Himalayan element and have adapted to live in torrential river that too in cold water
Diversity of aquatic habitats	3	3	3	Occurrence of diversity of substratum, pools and riffles, sandy banks, slow and fast flowing reaches, rafts, lagoons, confluences of different streams, formation of islands during summer and winter offers relatively diverse habitats for fish and other wildlife
Presence of protected or pristine areas	2	2	2	Although small portions of reaches are inside the protected areas, majority of reaches are outside and are relatively disturbed due to hydro electric projects
Sensitivity of aquatic ecosystems to flow reduction	4	4	4	There are a number of species largely restricted to Himalaya that occur in this reach which have evolved to live in fast flowing water. Therefore, any change in the flow reduction will be detrimental to the populations of several Himalayan species
% of watershed remaining under natural vegetation	4	4	4	Larger portion of catchment are relatively undisturbed and with natural vegetation
% of floodplains remaining	4	4	4	There is not much reduction in the floodplains area within study area, but degradation of floodplains is observed
Degree of flow regulation	3	3	3	Because of presence of hydro electric projects, the reach has already been fragmented and water flow is highly regulated

(continued)

Table 13.1 (continued)

Indicator	Rampur	Suni	Kasol	Remarks
% of watershed closed to movement of aquatic biota by structures or degree of flow fragmentation	4	4	4	Movement of migratory species has been blocked by a number of hydropower projects. There is no provision of successful fish paths in the existing hydro electric projects which has drastically reduced the populations of migrants in upper reaches
% of aquatic biota that are exotic	4	4	4	At least 3 exotic species are present in the reach. One of 100 worst invasive species of world 'brown trout' occurs in the river
Aquatic species' relative richness	5	5	5	Compared to availability of area, richness of fish species is very high
Human population density as % of that in the main floodplains	2	2	2	Compared to other parts, this stretch has moderate population
Overall water quality	5	5	5	Based on the data of CPCB, the quality of water is still very high
Sum of indicator scores	47	47	47	
Maximum possible sum of scores	65	65	65	
% of Maximum possible sum of scores	72	72	72	
Probable environment management class	C	C	C	The habitats and dynamics of the biota of these rivers have been disturbed, but basic ecosystem functions are still intact. Some sensitive species are reduced in extent (as per Smakhtin et al. 2007)

data at selected sites on Satluj River (i.e. Rampur, Suni, and Kasol) was used to estimate the environmental flows. There are six EMCs (A, B, C, D, E, F) are used in GEFC, fluctuating from "Unmodified" to "Critically modified". Each EMC is expressed by its exclusive FDC. In the present study, daily discharge data for Satluj River at different location as Rampur, Kasol and Suni site were available and these data are converted into average monthly data series. As estimated in Section (i), all the three sites in the study area belong to EMC 'C', hence, the final outcomes on environmental flows of Satluj river at Rampur, Suni and Kasol are pertaining to the results obtained by GEFC software for EMC 'C'. The environmental flows of Satluj river at Rampur, Suni and Kasol have been presented in Table 13.2 and Fig. 13.2. It is clear from the table and figure that the environmental flows of Rampur, Suni and Kasol sites vary from 63.06 m^3/s (February) to 397.92 m^3/s (July), 71.32 m^3/s (January) to 431.62 m^3/s (July) and 77.78 m^3/s (January) to 455.45 m^3/s (July) respectively.

Table 13.2 Environmental flows of Satluj river for EMC 'C'

Month	Ref flow	Environmental flow (m^3/s)		
		Rampur	Suni	Kasol
June	746.21	258.75	285.01	285.67
July	978.98	397.92	431.62	455.45
August	858.64	324.08	389.29	437.94
September	432.91	137.77	148.39	163.26
October	190.70	104.01	111.74	118.74
November	132.70	88.38	95.00	99.48
December	106.93	73.10	79.47	84.07
January	93.68	64.49	71.32	77.78
February	90.92	63.06	71.92	79.09
March	104.26	71.58	82.43	88.05
April	169.27	99.40	107.16	113.21
May	370.01	128.55	133.85	144.35
MAR (MCM)	11299.16	4784.01	5302.49	5673.07
% of MAR	100.00	42.34	41.50	41.49

13.5 Conclusions

For the present study, Satluj River has been selected for the assessment of its ecological status and environmental flows at its three sites (Rampur, Suni, and Kasol). The ecological status has been assessed in terms of Environmental Management Class described by the International Water Management Institute (Smakhtin et al. 2007). The assessment is based on 13 parameters related to the ecosystem of that particular area. Each parameter is given a score of 1–5 depending up on its status. As per the assessment, the ecological status of Satluj River is representative of Environmental Management Class 'C'. The Global Environmental Flow Calculator (GEFC) developed by International Water Management Institute (IWMI) has been used in the present study. The environmental flows of Rampur, Suni and Kasol sites are found to vary from 63.06 m^3/s (February) to 397.92 m^3/s (July), 71.32 m^3/s (January) to 431.62 m^3/s (July) and 77.78 m^3/s (January) to 455.45 m^3/s (July) respectively is shown in Table 13.1 and Fig. 13.2.

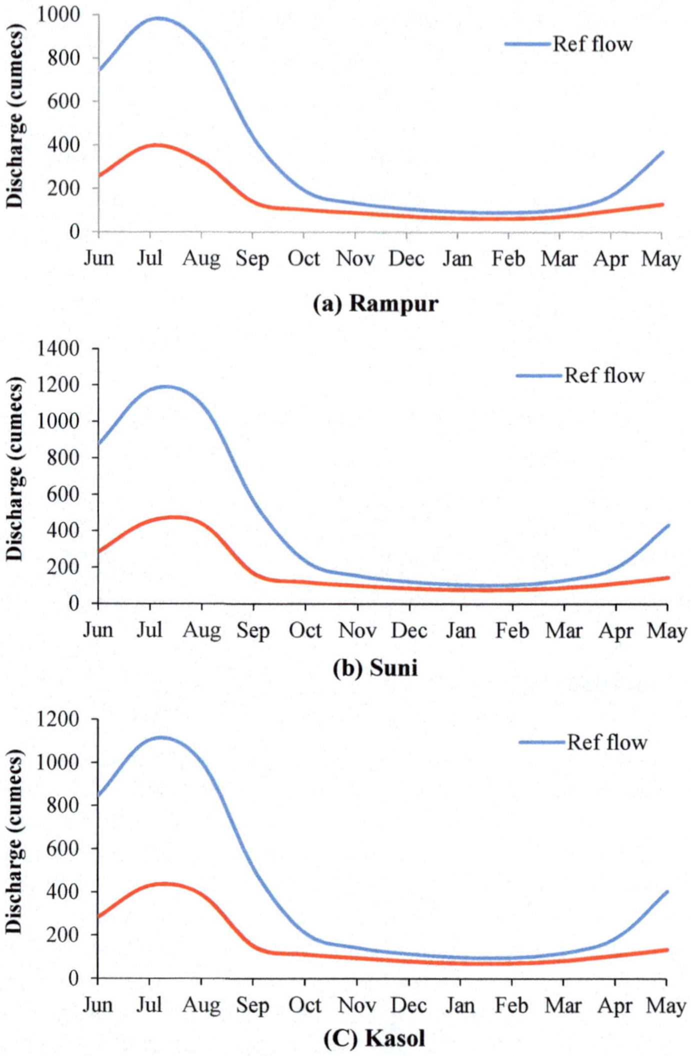

Fig. 13.2 Average monthly environmental flows of Satluj river

References

Beecher HA (1990) Standards for instream flows. Rivers 1(2):97–109

DWAF (1997) White paper on a national water policy for South Africa. Department of Water Affairs and Forestry, Pretoria, South Africa

Puckridge JT, Sheldone F, Wlaker KF, Boulton AJ (1998) Flow variability and ecology of large rivers. Mar Freshw Res 49:55–72

Smakhtin VU, Anputhas M (2006) An assessment of environmental flow requirements of Indian river basins. IWMI research report 107, International water management institute, Colombo, Sri Lanka, p. 42

Smakhtin V, Arunachalam M, Behera S, Chatterjee A, Das S, Gautam P, Joshi GD, Sivaramakrishnan KG, Unni KS (2007) Developing procedures for assessment of ecological status of Indian river basins in the context of environmental water requirements. IWMI research report 114, International water management institute, Colombo, Sri Lanka, p. 40

Smakhtin VU, Erivagama N (2008) Developing a software package for global desktop assessment of environmental flows Environmental Modelling & Software. Elsevier Science Publishers B. V. Amsterdam, the Netherlands 23(12):1396–1406

Chapter 14
Impact of Environmental Flow on Hydro Power Projects—A Case Study

Naresh Dongre and Vivek Gupta

Abstract Large number of river valley Hydro Power Projects (HPP) in Himalayan region are in different stages of implementation. Many of these Hydro Power Projects include diversion structures with long Head Race Tunnel. This results in reduction of river flow in downstream reaches. To keep the river live and to ensure perennial environmental flow in the river, a provision for release of minimum flow called "Environmental Flow" as per the Ministry of Environment & Forest (MOEF) norms are being considered while estimating the design energy of Hydro Power Projects. In India, many Hydro Power projects in the Himalayan region got MOEF clearance with 10% of minimum flow for 90% dependable year as Environmental Flow. As per the latest environmental clearances for Hydro Power Projects from MOEF, the release of Environmental Flows recommended are 30, 25 and 20% of 90% dependable year for Monsoon season, Non-Monsoon-Non-lean season & Lean season respectively. In this paper, a power potential studies have been carried out for Kolodyne HPP-Stage-II (460 MW), Rupsiabagar Khasiyabara HPP (261 MW), Loharinag Pala HPP (600 MW) and Lata Tapovan HPP (171 MW) for earlier and revised Environmental Flow guidelines. With the revised releases of Environmental Flows, the Design Energy for Hydro Power Projects reduces drastically and makes projects unviable in some cases.

14.1 Introduction

Environmental Flows are defined as the spatial as well as temporal variations of flow not only in quantity, but also in quality of water required to ensure downstream environmental, social and economic benefits, including the river ecology. The Environmental Flows should carry bed load as well as suspended load in nearly the same proportions as present in virgin Flow. Minimum flow is a general term used to describe a flow required to maintain same features of a river eco-system. In rivers without any man made controls, the minimum flow will be a function of the prevailing

N. Dongre (✉) · V. Gupta
NTPC Ltd, Noida, Delhi, India
e-mail: nareshdongre@ntpc.co.in

A. Pandey et al. (eds.), *Climate Impacts on Water Resources in India*, Water Science and Technology Library 95, https://doi.org/10.1007/978-3-030-51427-3_14

meteorological and hydrological conditions. Where changes to the natural flow in a river are brought about at some location, it is necessary to provide for a certain flow to sustain the downstream requirements. Mostly, the minimum flow guidelines emphasis on quantity of flow but quality factors should also be not ignored. While there is no disagreement on the need to maintain a certain minimum flow in the river as environmental flow for ecological objectives, there is no agreement on how to quantify this minimum flow.

14.2 Available Guidelines on Environmental Flows

14.2.1 HPSEP & PCB Guideline

As per the Environment Protection and Pollution Control Board of Himachal Pradesh State (HPSEP & PCB) minimum flow of water as 15% threshold value of the minimum flow observed in lean season.

14.2.2 Report of Working Group Constituted to Advise Water Quality Assessment Authority (WQAA) Constituted by Government of India (July 2007)

As per the guidelines outlined by the working group in respect of minimum flows are as under:

(1) Methodology similar to Tenant's methodology may be adopted for estimation of minimum flows.
(2) For maintaining the in-stream environment, the minimum flows may be considered as minimum flows with 99% exceedance.
During flood period, a range of minimum flows may be recommended as flushing flows.
(3) For summer season recommendation may be different for the rivers originating in the Himalayas, as the Himalayan rivers are carrying large snow melt component.

(a) **Himalayan Rivers**
The Minimum flow (m^3/s) is not less than 2.5 of 75% dependable annual flow. For monsoon season, minimum flushing peak (m^3/s) not less than 250 of 75% dependable annual flow.

(b) **Other Rivers**
Observed ten daily flow with 99% exceedence may be considered as Minimum flow. In case of non-availability of ten daily flow data, this may be taken as 0.5% of 75% dependable annual flow.

14.2.3 *The Expert Appraisal Committee (EAC) for River Valley and Hydroelectric Projects Constituted Under the Provisions of EIA Notification 2006*

As per the Minutes of the 61st Meeting, the committee suggested the environmental flows (Table 14.1) for Lara Sumta HPP (104 MW) Project in Lahaul and Spiti District in Himachal Pradesh.

(a) **Environmental flow as per 65st EAC Meeting**

As per the 65th Meeting of meeting, the committee has recommended the environmental flows for Hutong, Kalai & Demwe HPP's (Table 14.2).

(b) **Environmental flow as per 91st EAC Meeting**

The environmental flow for the Satluj River Basin Study in Himachal Pradesh mentioned in Minutes of the 91st Meeting of the EAC for River Valley and Hydroelectric Projects held on 8–9th February, 2016 (Table 14.3).

Table14.1 Environmental flow as per 61st EAC Meeting

S. No	Month/Season	Environmental flows for 90% dependable year (%)
1	Monsoon Season	30
2	Non-monsoon non-lean season	20–30
3	Lean season	20

Table 14.2 Environmental flow as per 65st EAC Meeting

S. No	Month	Environmental flows for 90% dependable year (%)
1	May–September	30
2	October and April	25
3	November–March	20

Table 14.3 The environmental flows as per cumulative impact assessment and carrying capacity study (CIA and CCS) of Satluj River Basin study in Himachal Pradesh

HPPs in the non fish occurring zone		
Months	EFR requirements	Remarks
December–February	For lean season, 20% of the mean flow	
March–May, October and November	20% of the flow	
June–September	30% of the total flow	
HPPs in the fish occurring zone		
December–March	20% of the mean lean season flow	
April, May, October and November	20% of the inflow or 25 cm of water depth whichever is more	These are the months in which the fishes migrate from lower reaches to higher reaches or from higher reaches to lower reaches
June–September	30% of the cumulative flow	

14.2.4 Environmental Flow as Cumulative Impacts of Hydroelectric Projects on Aquatic and Terrestrial Biodiversity

Based on assessment of Cumulative Impacts of Hydroelectric Projects on Aquatic and Terrestrial Biodiversity in Alaknanda and Bhagirathi Basins, Uttarakhand by Wildlife Institute of India, there are mainly four seasons (Table 14.4).

Based on this classification of seasons, Mean Seasonal Flow (MSF) was suggested as 21.8 and 14.5% of MSF for a stretch that falls in the Mahseer and Trout zones and 'No Fish Zone' respectively.

Table 14.4 Type seasons as per cumulative impacts of hydroelectric projects on aquatic and terrestrial biodiversity

S. No	Season type	Season	Period
1	Season I	Monsoon Season	May–September
2	Season II	Average flow period	October
3	Season III	Low or lean or dry flow season	November–March
4	Season IV	Average flow period	April

Table 14.5 The environmental flows as per the Gazete notifications 10th October 2018 Government of India

S. No	Season type	Season	Period
1	High flow Season	June to September	30[a]
2	Lean season	October, April and May	25
3	Dry Season	November to March	20

[a]30% of monthly flow of High flow season

14.2.5 The Gazete Notification, 10th October 2018 by Ministry of Water Resource, River Development and Ganga Rejuvenaton (National Mission for Clean Ganga), Government of India

The environmental flows (Table 14.5), as per the Gazete notifications Government of India, for Upper Ganga river basin stretch.

14.3 The Projects Undertaken for Studies

14.3.1 Kolodyne HPP (460 MW)

Kolodyne project is a storage scheme, located on the river Kolodyne in Lawngtali and Saiha district of Mizoram (India). The installed capacity of the project is 460 MW (4 × 115 MW) having Francis wheel turbines with overall efficiency of 0.92. The Net gross and design discharge of the project is 131.7 m and 490 m^3/s. It will feature a 129 m high Dam consisting of 9 gate of size 8 m × 10 m having total discharging capacity of 19,426 m^3/s. The Full Reservoir Level (FRL) and Minimum Draw Down Level (MDDL) of the project are at El. 230 m and at EL 207 m having Gross and live storage of 1350 MCM and 694.05 MCM respectively. The River will be diverted into a 3.5 km long HRT of 10.5 m. There are 2 nos. of intake having discharge capacity of 490 m^3/s.

14.3.2 Rupsiyabagar Khasiabara HPP (271 MW)

Rupsiabagar Khasiabara project is planned as a run-of-river project with peaking operation, located on the river Goriganga in Pithoragarh district of Uttarakhand (India). The installed capacity of the project is 261 MW (3 × 87 MW) having Pelton wheel turbines with overall efficiency of 0.90. The Gross head and design discharge of the project is 449.8 m and 69.12 m^3/s respectively. It will feature a 62 m high Dam consisting of 3 nos. of sluice spillway of gate size 9 m × 14 m and 1 no of Log chute having Gate size 3 × 3 m. The FRL and MDDL of the project are at EL. 1720 m and at EL 1700 m. The River will be diverted into a 7.47 km long HRT. There are 2 nos. of intake, followed by desilting basin. The HRT discharge is 69.13 m^3/s and flushing discharge is around 6.9 m^3/s.

14.3.3 Loharinag Pala HPP (600 MW)

Loharinag Pala HPP is a run of river scheme, located on the river Bhagirathi in Uttarkashi district of Uttarakhand (India). Installed capacity of the project is 600 MW (4 × 150 MW) having Pelton wheel turbines with overall efficiency of 0.90. It will feature a barrage consisting of 4 bays of 15 m each. The barrage consists of 4 nos. of radial gate of size 15 m × 8 m having Maximum and Minimum Pond level at EL 2147.50 m and at EL 2145.00 m. The River will be diverted into a 13.85 km long HRT. There are 3 nos. of intake followed by desilting basin. The HRT discharge is 152.6 m^3/s and Gross head of 475.67.

14.3.4 Lata Tapovan HPP

Lata Tapovan HPP is a run of river scheme, located on the river Dhauliganga in Chamoli district of Uttarakhand (India). Installed capacity of the project is 171 MW (3 × 57 MW) having Francis wheel turbines with overall efficiency of 0.93. The gross head and design discharge of the project is 298 m and 67.15 m^3/s. It will feature a 16 m high barrage consisting of 12 bays of 11 m each, having MDDL at EL 2099.5 m and FRL at 2103.00 m. The Gross and live storage of the project is 1.4 MCM and 0.52 MCM respectively.

The River will be diverted into a 7.51 km long HRT. There are 2 nos. of intake having design discharge of 79 m^3/s, followed by desilting basin. The HRT discharge is 67.15 m^3/s.

14.4 Power Potential Studies

14.4.1 Kolodyne Stage-II HPP (460 MW)

Initially, the power potential studies for Kolodyne Stage-II HPP has been envisaged based on carry over of live storage to the next water year. Further, it is observed that releases throughout the year are required for movement of ships in downstream reaches and hence the carry over is not possible. In view of the above, the reservoir operation based on annual storage has been simulated and Power Potential studies

Table 14.6 Environmental flows for 90% dependable year

S. No	Month	Environmental flows (%)	Environmental flows (m^3/s)
1	May–September	30	191.8
2	October and April	25	30.2
3	November–March	20	13.8

have been carried out. A minimum discharge of 80 m^3/s is required to be released from the reservoir, so as to ensure the necessary water depth of flow for movement of ships in downstream reaches. This aspect has been considered for both, planning as well as reservoir operation of Kolodyne Stage-II HPP.

An inflow series of 19 years i.e. from year 1986–1987 to year 2004–2005 is has been used for the present study having monthly flow data for 90% dependable year.

The 90% dependable year flow series for the purpose of Design Energy computations has been worked out. Aquatic release of 2.647 m^3/s (10% of minimum flow of 90% dependable year) throughout the year directly from the dam has been considered in the power computations. The Annual Energy generation and Design Energy generation for Kolodyne Stage-II HPP are 1842 GWh (annual Load Factor is 45.7%) and 1808 GWh respectively.

Environmental Flows recommended for Hutong, Kalai and Demwe HPP's have been adopted for the present study as the Kolodyne Stage-II HPP is located in North East region. Accordingly, the season wise Environmental flows works out (Table 14.6).

By considering above revised Environmental flows and DPR flow series, the revised power potential studies for Kolodyne Stage-II HPP has been carried out. The revised Annual Energy generation and Design Energy generation for Kolodyne Stage-II HPP are 1455 GWh (annual Load Factor is 36.1%) and 1420 GWh respectively.

14.4.2 Rupsiyabagar Khsiabara HPP (271 MW)

Rupsiabagar-Khasiyabara Hydro Power Project was planned to harness the hydro power potential of river Goriganga, by utilizing a drop of about 456.5 m in the river stretch upstream of its confluence with Sarda River. The project involves construction of a 62 m high concrete dam to divert the water into HRT after desilting arrangements. The gorge upstream the dam is narrow and steep, resulting in creation of small storage only to meet the diurnal peaking requirements. The project is therefore planned as a run-of-river project with peaking operation.

An inflow series of 27 years i.e. from year 1977–1978 to year 2003–2004 is has been used for the present study and having monthly flow data for 90% dependable year (Table 14.7).

Table 14.7 Monthly flow data for 90% dependable year for Rupsiyabagar Khasiabara HPP

Month	June	July	August	September	October	November	December	January	February	March	April	May
Flow (m^3/s)	65	83	92	68	29	18	15	12	13	20	37	63

Table 14.8 Environmental flows as per Minutes of 61st EAC

S. No	Month	Environmental flows (%)	Environmental flows (m^3/s)
1	June–September	30	23.1
2	October and May	25	11.4
3	November–April	20	3.9

Table 14.9 Summary of Revised power potential studies for Rupsiyabagar Khasiyabara HPP

Particulars	90% dependable year
Revised installed capacity (MW)	234
Annual energy generation (GWh)	1029
Design energy (GWh)	1013
Average annual load factor	50.2

The power potential studies have been carried out for Rupsiabagar-Khasiyabara Hydro Power Project, situated on river Goriganga in Uttarakhand State. The stretch of the river proposed to be exploited for development of hydro power by way of this project is immediately downstream of the proposed Sirkari Bhyol-Rupsiabagar Project. The Rupsiabagar-Khasiyabara Project would thus be a part of the cascade development of river Goriganga and form an integral constituent of the overall plan for the development of Sarda Basin.

The project has been planned to provide diurnal pondage for peaking of about 2 h twice a day. Diurnal pondage, if provided in the upstream schemes would facilitate additional peaking at Rupsiabagar-Khasiyabara Hydro Power Project also, in tendom with upstream schemes. However, minimum 1.09 m^3/s of discharge has been considered during non-monsoon period (October–May) as the aquatic release for downstream river reach. For silt flushing, 20% of design discharge has been considered during the active monsoon period (June–September). The Annual Energy generation and Design Energy generation for Rupsiyabagar Khasiyabara HPP are works out to be 1283 GWh (annual Load Factor is 56.13%) and 1257 GWh respectively.

As per "Minutes of the 61st Meeting of the Expert Appraisal Committee for River Valley and Hydroelectric Projects", environmental flows (Table 14.8) for monsoon season, Non-monsoon Non Lean season and Lean season as 30, 25 and 20%.

Considering DPR flow series and above environmental flow released, the revised power potential studies (Table 14.9) for Rupsyabagar Khasiabara HPP has been carried.

14.4.3 Loharinag Pala HPP (600 MW)

The stretch of the Bhagirathi river is lies between the tail waters of proposed Harsil dam on the upstream side and head waters of Pala Maneri Project on the downstream side. The Loharinag Pala Project would thus form integral part of the overall plan visualized for the Development of Bhagirathi Basin. This is a run-of-river type development without any storage and would utilize a net head of water of 438.32 m. The plant would thus operate as base load station.

An inflow series of 14 years i.e. from year 1989–1990 to 2003–2004 has been used for the present study and having monthly flow data for 90% dependable year (Table 14.10).

MoEF has accorded clearance to Loharinag Pala HPP with the minimum flow to be released as 0.86 m^3/s. Accordingly, the power potential studies has been carried out with environment flow of 1 m^3/s. The Annual Energy generation and Design Energy generation for Loharinag Pala HPP are works out to be 2412 GWh (annual Load Factor is 45.9%) and 2353 GWh respectively.

As per "Minutes of the 61st Meeting of the Expert Appraisal Committee for River Valley and Hydroelectric Projects" the environmental flows for monsoon season, Non-monsoon Non Lean season and Lean season as 30, 25 and 20% and worked out (Table 14.11).

With DPR flow series and above environmental flow released, the revised power potential studies for Loharinag Pala HPP has been carried. The revised Annual Energy generation and Design Energy generation for Loharinag Pala HPP are works out to be 1966 GWh (annual Load Factor is 37.40%) and 1925 GWh respectively.

14.4.4 *Lata Tapovan HPP (171 MW)*

The location of the Lata Tapovan HPP site was decided in conjunction with the upstream and downstream projects envisaged to be developed and future peaking operations, which will utilize the effective storage held against the diversion barrage. MDDL at the barrage is fixed from the minimum submergence depth criteria for preventing air entry into the headrace tunnel. Based on this the MDDL has been fixed at EL 2099.50 m The FRL of the barrage has been fixed at EL: 2103 m mainly from consideration of free flow requirement of tailrace discharge at the proposed upstream development in future. During normal operation, the water levels will vary between FRL and MDDL. The storage provided between MDDL an FRL is 0.52 MCM, which will allow 2 h each diurnal peaking operation. Design energy is calculated based on the energy projected in a 90% dependable year with 95% machine availability of the installed and firm energy is calculated based on average inflows during lean period in the 90% dependable year, in line with the definitions adopted by the Central Electricity Regulatory Commission (CERC). Flushing discharge for the desilting chamber is assumed as 15% of the power draft for all monsoon months and zero for non-monsoon months. However, environmental/aquatic release of minimum 1 m^3/s has been considered for throughout the year.

This flow series is available for a period of 31 years from the year 1971–1972 to 2001–2002 has been utilized for estimation of power potential having monthly flow data for 90% dependable year (Table 14.12).

Table 14.10 Monthly flow data for 90% dependable year for Loharinag Pala HPP

Month	June	July	August	September	October	November	December	January	February	March	April	May
Flow (m^3/s)	152	277	282	139	51	36	25	20	19	19	26	72

Table 14.11 Environmental flows as per minutes of 61st EAC for Loharinag Pala HPP

S. No	Month	Environmental flows (%)	Environmental flows (m^3/s)
1	June–September	30	63.8
2	October, November, April and May	25	11.5
3	December–March	20	4.2

Accordingly, the power potential studies has been carried out for with environment flow of 1 m^3/s for non-monsoon season (Oct–May) & silt flushing discharge of 15% of inflow for monsoon season (June–sept). The Annual Energy generation and Design Energy generation for Lata Tapovan HPP are works out to be 890 GWh (annual Load Factor is 58%) and 869 GWh respectively.

As per "Minutes of the 61st Meeting of the Expert Appraisal Committee for River Valley and Hydroelectric Projects" the environmental flows for monsoon season, Non-monsoon Non Lean season and Lean season as 30, 25 and 20% has been worked out (Table 14.13).

With DPR flow series and above environmental flow released, the revised power potential studies for Lata Tapovan HPP has been carried out. The revised Annual Energy generation and Design Energy generation for Lata Tapovan HPP are works out to be 769 GWh (annual Load Factor is 50.2%) and 752 GWh respectively.

14.5 Conclusions

- A simple figure for environmental flow could be misleading. Therefore, season wise Environmental Flows has been adopted for power potential study of Kolodyne Stage-II HPP, Rupsiyabagar Khasiabara HPP, Loharinag Pala HPP and Lata Tapovan HPP. The comparative statement of power potential studies for these HPP's is summarised (Table 14.14).
- If the Environmental Flows approved by MOEF for some HPP's in North Eastern region has been used for Kolodyne Stage-II HPP, there will be 21.4% loss of Design Energy and project may be unviable.
- If the Environmental Flows approved by MOEF for some HPP's in Northern region has been used for Rupsiyabagar Khasiabara HPP, Loharinag Pala HPP & Lata Tapovan HPP, there will be 19.4, 18.2 and 13.5% loss of Design Energy. This loss of Design Energy due Environmental Flow will results to increases the Levellised Tariff of HPP's.
- This loss of Design Energy due Environmental Flow will results to increases the Levellised Tariff of HPP's. To compensate in some extent loss of energy due to revised Environmental flows, there should be different tariff for peaking energy.

Table 14.12 Monthly flow data for 90% dependable year for Lata Tapovan HPP

Month	June	July	August	September	October	November	December	January	February	March	April	May
Flow (m^3/s)	354	313	335	208	122	84	62	60	54	55	74	141

Table 14.13 Environmental flows as per Minutes of 61st EAC for Lata Tapovan HPP

S. No	Month	Environmental flows (%)	Environmental flows (m^3/s)
1	June–September	30	30.2
2	October, November, April and May	25	8.8
3	December–March	20	3.8

- The Loss of Design Energy can be compensated to some extent by utilising Environmental Flow for power generation without any adverse impact on the environmental flow (by providing small capacity Power House at d/s of diversion structure).

14.6 Recommendations

- The term 'Environmental Flow' or 'Minimum Flow' needs to be re-defined. It should be "Flow Management" considering additional discharge available from tributaries in the downstream of diversion structure. Also, we need to prescribe other hydraulic variables like velocity of flow, flow variability, depth of flow, submerged width of river bed, etc. These are important factors for aquatic habitat, silt flushing, water quality, aesthetic river condition, etc.

Table 14.14 The comparative statement of power potential studies

Particulars	Kolodyne Stage-II HPP		Rupsiyabagar Khasiabara HPP		Loharinag Pala HPP		Lata Tapovan HPP	
	With approved E-flow	With adoptedE-flow	With approved E-flow	With adopted E-flow	With approved E-flow	With adopted E-flow	With approved E-flow	With adopted E-flow
IC (MW)	460	460	261	234	600	600	171	171
Annual energy (GWh)	1842	1454	1283	1029	2412	1966	890	769
Design energy (GWh)	1808	1420	1257	1013	2353	1925	869	752
Annual load factor (%)	45.7	36.10	56.13	50.2	45.90	37.40	58.01	50.2
Loss of design energy (%)	21.4		19.4		18.2		13.5	

Bibliography

Assessment of Cumulative Impacts of Hydroelectric Projects on Aquatic and Terrestrial Biodiversity in Alaknanda and Bhagirathi Basins, Uttarakhand by wildlife institute of India

DPR's of Kolodyne Stage-II HPP, Rupsiyabagar Khasiabara HPP, Loharinag Pala HPP & Lata Tapovan HPP

Kumar P, Chaube UC, Mishra SK (2007) Environmental flows for hydropower projects—a case study international conference on small hydropower-hydro Sri Lanka, 22–24 Oct 2007

Minutes of the 61st Meeting of the Expert Appraisal Committee for River Valley and Hydroelectric Projects constituted under the provisions of EIA Notification 2006, held on 12–13th October, 2012 at SCOPE Complex, New Delhi

Minutes of the 65th Meeting of the Expert Appraisal Committee for River Valley and Hydroelectric Projects constituted under the provisions of EIA Notification 2006, held on 22nd–23rd March, 2013 at SCOPE Complex, New Delhi

Minutes of the 91st Meeting of the Expert Appraisal Committee for River Valley and Hydroelectric Projects held on 8–9 th February, 2016 at Teesta Meeting Hall, 1st Floor, Vayu Wing, Indira Paryavaran Bhawan, Jor Bagh Road, New Delhi, 110003

Report of working group constituted to advise Water Quality Assessment Authority (WQAA) constituted by Government of India (July 2007)

Report of High Level Expert Group on Examination of Various Technical Issues Involved in Ensuring the Required Flow in River Bhagirathi to Keep the River Alive and to Ensure Perennial Environmental Flow in the River, Feb 2009

Chapter 15
Improved Wastewater Treatment by Using Integrated Microbial Fuel Cell-Membrane Bioreactor System Along with Ruthenium/activated Carbon Cathode Catalyst to Enhance Bio-energy Recovery

G. D. Bhowmick, M. M. Ghangrekar, and R. Banerjee

Abstract A two-stage integrated microbial fuel cell (MFC)-aerobic membrane bioreactor (MBR) based wastewater treatment technology with ruthenium/activated carbon (Ru/AC) as cathode catalyst in MFC was investigated. The results showcased the maximum volumetric power density and coulombic efficiency were nearly 1.4 and 1.5 times higher in case of MFC-MBR system with Ru/AC as cathode catalyst (2.7 W.m^{-3} and 12.8 ± 1.2%) than the control MFC-MBR system without Ru/AC (2.0 W.m^{-3} and 8.2 ± 0.6%), respectively. Moreover, these integrated MFC-MBR systems offered chemical oxygen demand (COD) removal efficiency of more than 96% from the synthetic wastewater having initial COD of around 1 g.L^{-1}. This integrated MFC-MBR technology offers an immense potential to be developed as full-scale application to offer reliable wastewater treatment along with bio-energy recovery.

15.1 Introduction

Energy intensive conventional aerobic treatment processes involve huge financial involvement, additional operational cost and also cost for handling and disposal of excess sludge. Conversely, these aerobic treatment processes are helpful in reducing organic matter and nutrient concentration, thus demonstrating higher treatment efficacy with better effluent quality. On the other hand, rather than consuming higher amount of energy, anaerobic treatment processes can recover energy in a more useful

G. D. Bhowmick · R. Banerjee
Department of Agricultural and Food Engineering, Indian Institute of Technology Kharagpur, Kharagpur 721302, India

M. M. Ghangrekar (✉)
Department of Civil Engineering, Indian Institute of Technology Kharagpur, Kharagpur 721302, India
e-mail: ghangrekar@civil.iitkgp.ac.in

A. Pandey et al. (eds.), *Climate Impacts on Water Resources in India*, Water Science and Technology Library 95, https://doi.org/10.1007/978-3-030-51427-3_15

way and can provide a greener solution, except the fact that it requires post treatment for meeting the disposal standards for the effluent (Grady et al. 1999; Oh and Logan 2005). Hence, to attain higher treatment efficacy, appropriate sequence or combination of advance wastewater treatment systems is essential to provide reliable treatment performance and produce effluent suitable for reuse (Oh et al. 2010).

The bio-electrochemical systems, like microbial fuel cell (MFC) has the potential to harvest the bioenergy from the wastewater while treating it. Research on MFC has extended a lot in the last few decades by adopting novel reactor configurations, developing suitable low-cost catalyst materials for electrodes, operational regime modification, etc. (Martinez-Huitle and Ferro 2006; Zhou et al. 2013; Bhowmick et al. 2018). However, sluggish nature of the oxygen reduction reaction (ORR) and higher cathode overpotential losses result in serious voltage depletion in MFC (Bhowmick et al. 2019c; Türk et al. 2018) and cause a major hindrance for its practical applications. Another approach to overcome existing barriers to optimize the process performance of MFC and to reduce the footprint of the wastewater treatment system was done by combining MFC with membrane filtration units elsewhere (Wang et al. 2012; Ma et al. 2015; Bhowmick et al. 2019a).

The aerobic membrane filtration process, like membrane bioreactor (MBR), which is a suspended-biomass growth process; where organic matter from the wastewater gets biodegraded in combination with a filtration unit for higher treatment efficiency (Van Dijk and Roncken 1997; Côté et al. 1997; Rosenberger et al. 2002). Advantages of the MBR over the conventional activated sludge process is like the small footprint with effective treatment efficacy and susceptible to higher mixed liquor suspended solid (MLSS) concentrations, therefore less reactor volume is required to achieve same or even higher treatment performance reliably (Yoon et al. 2004; Hasan et al. 2012). Thus, combination of MFC with the MBR can be used to further accomplish almost complete eradication of biological organic matter from the wastewater along with an advantage of bioenergy recovery (Cha et al. 2010; Min and Angelidaki 2008; Ray et al. 2016; Yuan and He 2015; Bhowmick et al. 2019b).

In this investigation, a two-stage MFC-MBR integrated system was investigated to achieve higher efficiency in terms of wastewater treatment and energy recovery by implementing ruthenium/activated carbon (Ru/AC) composite as novel cathode catalyst. Synthetic wastewater was introduced to the integrated MFC-MBR system consisting of Ru/AC as cathode catalyst and its performance was compared with another MFC-MBR system without any cathode catalyst. This integrated MFC-MBR system was targeted to yield higher power recovery along with significantly reduced organic matter concentration of the final effluent.

15.2 Materials and Methods

15.2.1 Electrodes Fabrication

The electrodes were made with carbon cloth having a projected surface area of 25 cm^2. Ruthenium (Ru) (on 10% activated carbon powder, Alfa-Aesar, USA) powder with a loading of 0.5 $mg.cm^{-2}$ of cathode surface area was coated along with four coatings of diffusion layers in the MFC-MBR combined system, named as M-1 here onwards. The purpose of using diffusion layer on the cathode of MFC was to improve the cathode kinetics significantly by increasing the active oxygen site for ORR, while simultaneously reducing the loss of water through the ceramic separator used as proton exchange membrane (PEM) made of red soil with 20% montmorillonite. Whereas, another MFC-MBR system without any cathode catalyst was fabricated in order to compare the performance and named as M-2 here onwards. The procedure of Ru/Activated carbon catalyst layer coating is stated in details in subsequent sections.

15.2.1.1 Application of the Carbon Base Layer

The AC, with loading of 1.56 mg cm^{-2} of the cathodic surface area, was used along with 20 μl of 30% polytetrafluoroethylene (PTFE) solution for each mg of AC and then the whole solution was put into the plastic sample vial with 8–10 glass beads to vortex it for about 20 s. The suspension was coated over the carbon cloth using small paintbrush and air dried for at least 2 h followed by heating at 370°C for about 20 min.

15.2.1.2 Application of the Diffusion Layers

The 60% PTFE solution was used to coat the ink uniformly over the already coated carbon base layer and heated for 15 min at 370°C in a muffle furnace. After cooling the pieces of carbon cloth, this process was repeated for another three times to provide the optimum diffusion layer thickness over the cathode.

15.2.1.3 Application of the Catalyst Layer

The procured catalyst material (Ru, 10% on activated carbon powder) was applied corresponding to the loading of 0.5 $mg.cm^{-2}$ of cathode surface area and added with 0.83 μl of distilled water for every 1 mg of catalyst powder in a drop-wise fashion into the vial along with 6–8 glass beads and then vortex for couple of seconds. A 6.7 μl of Nafion for every mg of catalyst powder was also added and it was vortex for another 20 s. Thus the mixture paste formed was then coated over the carbon cloth on the side opposite to the diffusion layers maintaining proper consistency using

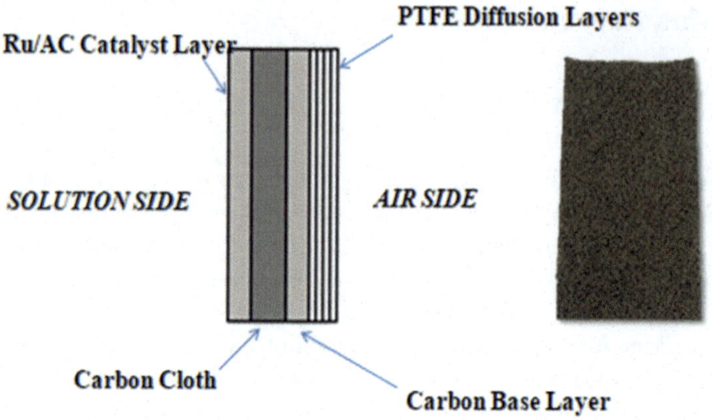

Fig. 15.1 Side-view of Ru/AC cathode composite with a four-coatings of PTFE diffusion layer

paintbrush and it was allowed to air-dry for at least 24 h. The catalyst coated carbon cloth was thus made ready for further application (Fig. 15.1).

15.2.2 Fabrication and Operation of MFC-MBR Systems

Two identical MFC-MBR combined systems were developed to investigate the efficacy of Ru-AC as cathode catalyst. Glass bottles with 4 cm of internal diameter and working volume of 140 mL were used to fabricate the aqueous-cathode MFCs. The wall of the glass bottles was cut in a rectangular pattern (5 cm × 5 cm) using the glass cutter to make space for the membrane-electrode assembly. The PEM, made up of red soil material with 20% of montmorillonite, and carbon cloth as electrodes were used in this membrane-electrode assembly. The operating parameters like operating voltage (OV), current response was measured over an external resistance of 100 Ω. All the MFCs were inoculated with 20 mL of anaerobic mixed-sludge collected from the septic tank (volatile suspended solids, VSS of 19.8 $g.L^{-1}$) after heat pretreatment was done at 100 °C for 15 min to suppress the methanogenesis. Sucrose based synthetic wastewater was prepared (Jadhav and Ghangrekar 2009) with organic matter concentration of around 1000 mg of chemical oxygen demand $(COD).L^{-1}$ after adjusting the pH to 7.4. The MFCs were operated in one day fed-batch mode.

The effluent from the MFCs was fed continuously to the aerobic MBR (70 mL), which was run with a liquid retention time of 12 h. Hollow-fibre polyethylsulfone (PES) based UF membrane with an average pore size of 80 nm, external diameter of 1 mm, and 0.8 mm of inner diameter with an exposed surface area of 100 cm^2 was procured for this investigation. The permeate was collected by applying vacuum pressure of 0.1 bar periodically (twice a day) (Fig. 15.2). The aeration was done by air-flow pump (SOBO air pump, China) in the MBR, using ceramic diffusers by placing it at the bottom of it for adequate aeration. The membrane assembly was

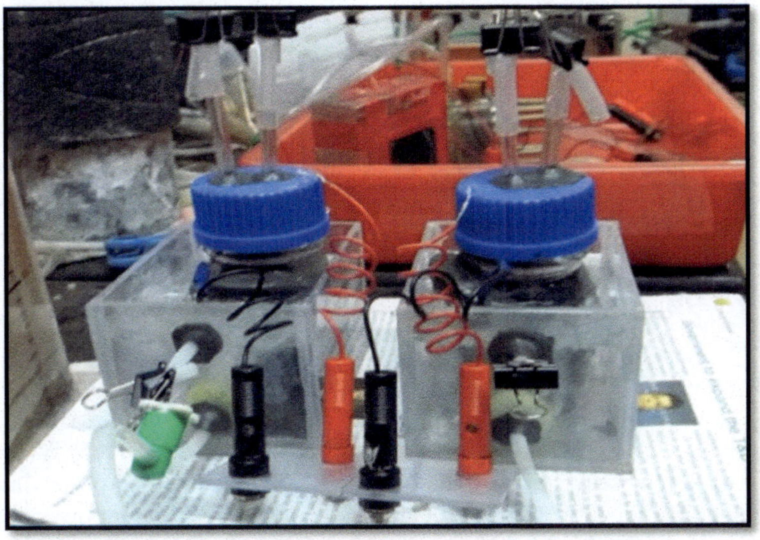

Fig.15.2 Integrated MFC-MBR set-up used in this experiment

cleaned after every two weeks by back-flushing in 0.5% NaOCl solution for 1 h in a different container to mitigate the membrane biofouling.

15.2.3 Electrochemical and Data Acquisition

The voltage and current produced by the MFCs were measured by digital multi-meter (Agilent Technologies, Malaysia) with a data acquisition unit to finally measure the generated power (Eq. 15.1).

$$P = I * V \tag{15.1}$$

where, P is the power in W; I is current in A; and V is voltage in V. The maximum volumetric power density normalized to the anodic chamber volume of MFC was determined from the polarization curve, ranging the applied resistances value from 40,000 Ω to 10 Ω. The COD and solids concentration of the samples were analysed according to the Standard Methods (APHA 1998). The coulombic efficiency (CE) was estimated by calculating the amount of coulombs actually recovered to the amount of coulombs theoretically existing in the wastewater being treated (Oh and Logan 2005) as shown in Eq. 15.2:

$$\mathrm{CE} = \frac{M_s \int_0^t I.dt}{b.\mathrm{F}.V_{an}.\Delta\mathrm{COD}} \tag{15.2}$$

where, M_s is the molecular weight of the oxygen in g mol^{-1}, F is Faraday's constant, i.e. 96,485 C mol^{-1}, b is number of electrons exchanged with per mole of oxygen, V_{an} is the anodic chamber volume in m^3 and ΔCOD is the change in the value of COD over time t.

15.3 Results and Discussion

15.3.1 Wastewater Treatment and Power Generation

The electrical output for both the systems was measured after reaching the steady-state condition with proper microbial attachment over the anodes. The maximum OV for M-1, with Ru/AC as cathode catalyst was found to be 169 mV. Whereas, the reactor M-2, in absence of any cathode catalyst exhibited 93 mV of maximum OV during the operational period. Through polarization, a maximum power density of 2.7 $W.m^{-3}$ was obtained for M-1, which was 1.4 times higher than the maximum power density of 2.0 $W.m^{-3}$ obtained for M-2 (Fig. 15.3). These results indicated that M-1 produced higher power output than M-2 because of the presence of Ru/AC as cathode catalyst and PTFE diffusion layer in the composite cathode, which might help in accelerating the ORR to enhance the power production. Also from the Fig. 15.3, polarization curves revealed the internal resistance, which was found to be slightly less in case of M-1 (128 Ω) than the M-2 (145 Ω). The maximum open circuit voltage (OCV) during the stable phase of operation was 609 mV in case of M-1, which was also higher than the M-2 (511 mV) following the same trend of improved performance of M-1 due to the application of Ru/AC as cathode catalyst.

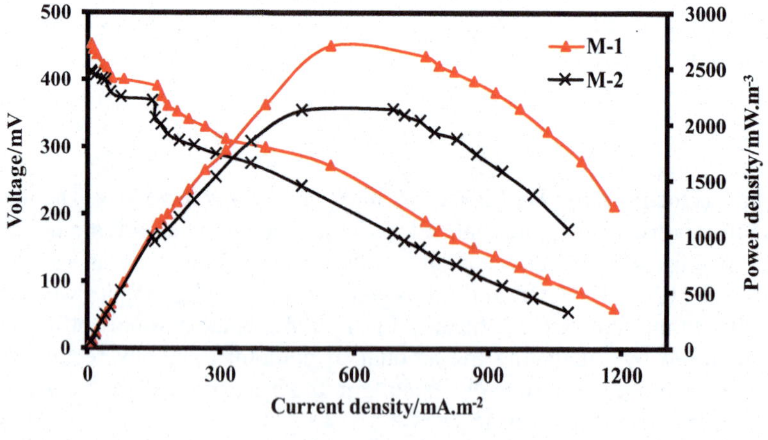

Fig. 15.3 Power density versus current density and polarization plots for the MFCs of M-1 (with Ru/AC) and M-2 (without Ru/AC)

The COD removal efficiency for the MFC part of the M-1 and M-2 was found to be 87.5 ± 2.4% and 75.2 ± 2.5%, respectively (Fig. 15.4a). The respective average CE of M-1 was 12.8 ± 1.2%, which was 1.5 times that of M-2 (8.2 ± 0.6%) (Fig. 15.4b). Therefore, from the overall discussion it can be concluded that M-1 was capable to perform better ORR than control MFC due to the presence of Ru/AC as catalyst and PTFE diffusion layer in the composite cathode.

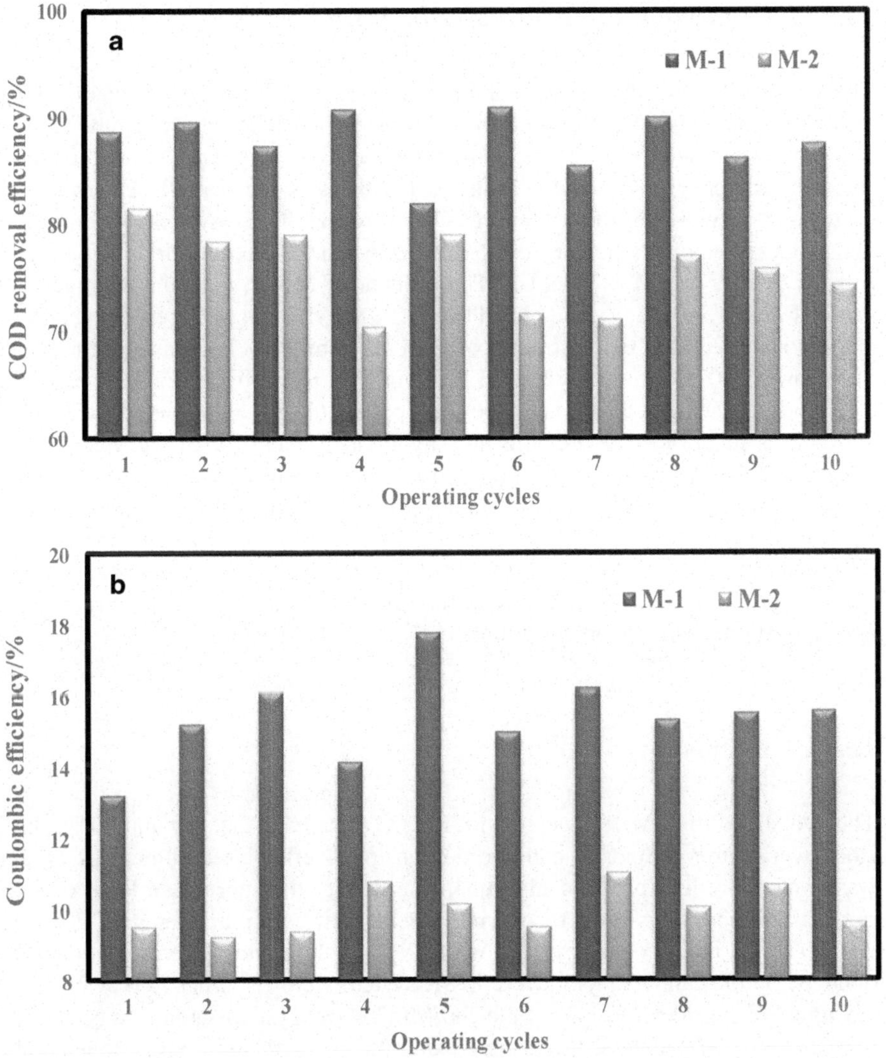

Fig. 15.4 **a** COD removal efficiency and **b** Coulombic efficiency at each batch cycle for M-1 (with Ru/AC) and M-2 (without Ru/AC)

Table 15.1 Performance evaluation in terms of COD ($mg.L^{-1}$) for M-1 (with Ru/AC) and M-2 (without Ru/AC) at each stages of operation

Treatment stage	M-1	M-2
MFC influent	999.6 ± 23.7	999.6 ± 23.7
MFC effluent	125 ± 5	248 ± 11
Permeate (Final effluent)	40 ± 4	76 ± 7
Overall removal efficiency (%)	**96.0 ± 0.3**	**92.4 ± 0.2**

15.3.2 Treatment of Wastewater in MBR

The performance of MFC part of the combined systems has already been discussed; however, as there was no difference in between the MBR structural regime for both of the systems (M-1 and M-2), the results showed slight difference in organic matter removal efficiency on a whole. Table 15.1 elaborates the overall performance at each stage of treatment of M-1 and M-2 on a whole. These systems evidenced an overall COD removal efficiency of 96.0 ± 0.3% and 92.4 ± 0.2% in final permeate, respectively for the M-1 and M-2. The effluent of MFCs, with the average COD value of 125 ± 5 $mg.L^{-1}$ and 248 ± 11 $mg.L^{-1}$ for M-1 and M-2, respectively, were continuously fed to MBR. Permeate of both the composite systems had final COD of as low as 40 ± 4 $mg.L^{-1}$ and 76 ± 7 $mg.L^{-1}$, respectively for M-1 and M-2. The physico-chemical features of the final permeate of the combined MFC-MBR system was superior to the previous investigations (Su et al. 2013; Wang et al. 2012; Li et al. 2014). The reason of this might be the pre-treatment done in MFC and thus it helped to increase in the overall organic matter removal from the system similar to a previous investigation by Tian et al. (2015). Therefore, this research work exposed a MFC–MBR combined technology for effective organic matter removal from the wastewater and produced effluent can therefore be used for various purposes like agro-irrigation, aesthetic enhancement, etc.

15.4 Conclusion

The two-stage MFC-MBR system with Ru/AC as cathode catalyst and PTFE diffusion layer in the composite cathode demonstrated effective treatment of organic wastewater. It could produce high-quality effluent with more than 96% of COD removal efficiency vis-à-vis recovering bioelectricity with an average CE of 12.8 ± 1.2% and a power density of 2.7 $W.m^{-3}$. The volumetric power density and CE obtained by the catalyst laden MFC-MBR system were 1.4 times (2.0 $W.m^{-3}$) and 1.5 times (8.2 ± 0.6%) higher than the MFC-MBR system without any catalyst, respectively. Nevertheless, the energy requirement for this system is considerably higher than the energy recuperated by the MFC; hence efforts are required to be given to increase the energy recovery from the MFC, to convert this overall system energy neutral. After such sustainability investigations, the MFC-MBR combined system can offer enormous possibility of pilot-scale application of the same.

Acknowledgement Department of Biotechnology, Government of India (BT/IN/INNO-INDIGO/28/MMG/2015–16) has supported this research project by providing the financial assistance. The help extended by Ms. Sreemoyee Ghosh Ray during experimentation planning and execution level is duly acknowledged.

References

American Public Health Association (APHA) (1998) Standard methods for the examination of water and wastewater. Am Public Heal Assoc Water Work Assoc Environ Fed 552

Bhowmick GD, Noori MdT et al (2018) Bismuth doped TiO_2 as an excellent photocathode catalyst to enhance the performance of microbial fuel cell. Int J Hydrogen Energy 43:7501–7510

Bhowmick GD, Chakraborty I et al (2019a) TiO_2/Activated carbon photo cathode catalyst exposed to ultraviolet radiation to enhance the efficacy of integrated microbial fuel cell-membrane bioreactor. Biores Technol Rep 100303

Bhowmick GD, Das S et al (2019b) Improved wastewater treatment by combined system of microbial fuel cell with activated carbon/TiO_2 cathode catalyst and membrane bioreactor. J Instit Eng (India) Series A 100:675–682

Bhowmick GD, Kibena-Põldsepp E, Matisen L, Merisalu M, Kook M, Käärik M, Leis J, Sammelselg V, Ghangrekar MM, Tammeveski K (2019c) Multi-walled carbon nanotube and carbide-derived carbon supported metal phthalocyanines as cathode catalysts for microbial fuel cell applications. Sustain Energy & Fuels 3(12):3525–3537

Côté P, Buisson H et al (1997) Immersed membrane activated sludge for the reuse of municipal wastewater. Desalination 113(2):189–196

Cha J, Choi S et al (2010) Directly applicable microbial fuel cells in aeration tank for wastewater treatment. Bio-electrochemistry 78:72–79

Grady CPL, Daigger GT et al (1999) Biological Waste Water Treatment. Marcel Dekker, New York

Hasan SW, Elektorowicz M et al (2012) Correlations between trans-membrane pressure (TMP) and sludge properties in submerged membrane electro-bioreactor (SMEBR) and conventional membrane bioreactor (MBR). Biores Technol 120:199–205

Jadhav GS, Ghangrekar MM (2009) Performance of microbial fuel cell subjected to variation in pH, temperature, external load and substrate concentration. Biores Technol 100(2):717–723

Li J, Ge Z et al (2014) Fluidized bed membrane Bio-electrochemical reactor for energy-efficient wastewater treatment. Biores Technol 167:310–315

Ma J, Wang Z et al (2015) Long-term investigation of a novel electrochemical membrane bioreactor for low-strength municipal wastewater treatment. Water Res 78:98–110

Martinez-Huitle CA, Ferro S (2006) Electrochemical oxidation of organic pollutants for the wastewater treatment: direct and indirect processes. Chem Soc Rev 35(12):1324–1340

Min B, Angelidaki I (2008) Innovative microbial fuel cell for electricity production from anaerobic reactors. J Power Sources 180:641–647

Oh S, Logan BE (2005) Hydrogen and electricity production from a food processing wastewater using fermentation and microbial fuel cell technologies. Water Res 39:4673–4682

Oh ST, Kim JR et al (2010) Sustainable wastewater treatment: how might microbial fuel cells contribute. Biotechnol Adv 28(6):871–881

Ray SG, Bhowmick GD et al (2016) Advances in wastewater treatment by combined microbial fuel cell-membrane bioreactor. In: 13th IWA specialized conference on small water and wastewater systems, pp 1–3
Rosenberger S, Kruger U et al (2002) Performance of a bioreactor with submerged membranes for aerobic treatment of municipal waste water. Water Res 36:413–420
Su X, Tian Y et al (2013) Performance of a combined system of microbial fuel cell and membrane bioreactor: wastewater treatment, sludge reduction, energy recovery and membrane fouling. Biosens Bioelectron 49:92–98
Tian Y, Li H et al (2015) In-situ integration of microbial fuel cell with hollow-fiber membrane bioreactor for wastewater treatment and membrane fouling mitigation. Biosens Bioelectron 64:189–195
Türk KK, Kruusenberg I, Kibena-Põldsepp E, Bhowmick GD, Kook M, Tammeveski K, Matisen L, Merisalu M, Sammelselg V, Ghangrekar MM, Mitra A, Banerjee R (2018) Novel multi walled carbon nanotube based nitrogen impregnated Co and Fe cathode catalysts for improved microbial fuel cell performance. Int J Hydrogen Energy 43(51):23027–23035
Van Dijk L, Roncken GCG (1997) Membrane bioreactors for wastewater treatment: the state of the art and new developments. Water Sci Technol 35(10):35-41
Wang YP, Liu XW et al (2012) A microbial fuel cell-membrane bioreactor integrated system for cost-effective wastewater treatment. Appl Energy 98:230–235
Yoon SH, Kim HS et al (2004) Incorporation of ultrasonic cell disintegration into a membrane bioreactor for zero sludge production. Process Biochem 39:1923–1929
Yuan H, He Z (2015) Integrating membrane filtration into bio-electrochemical systems as next generation energy-efficient wastewater treatment technologies for water reclamation: A review. Biores Technol 195:202–209
Zhou M, Wang H et al (2013) Recent advances in microbial fuel cells (MFCs) and microbial electrolysis cells (MECs) for wastewater treatment, bioenergy and bio-products. J Chem Technol Biotechnol 88(4):508–518

Chapter 16
Acclimation and Treatability Studies on Slaugter House Wastewater by Hybrid UASB Reactor

R. Loganath and Debabrata Mazumder

Abstract The slaughterhouse wastewater (SWW) is highly influenced by carbonaceous and nitrogenous organic matter aside from significant quantity of dissolved inorganics. There are different streams of SWW, composite of which is exceptionally perfect for secondary biological treatment. The amount of biodegradable organic substances present in the slaughterhouse stream recommends that the anaerobic treatment process could be actualized as the best decision. In this study, the biological treatment of SWW was observed using the Hybrid Upflow Anaerobic Sludge Blanket Reactor (HUASBR) at a laboratory scale setup under continuous mode operation. In this study the reactor was subjected to various Chemical Oxygen Demand (COD) loading rates in the range of (2.95–37.98) Kg COD/m^3/d under the Hydraulic Retention Times (HRT) between 24 and 6 h. The outcome of the HUASB Reactor shows that maximum COD removal efficiency of 94% can be achieved at the OLR of 18.29 kg COD/m^3/d under 10 h HRT.

16.1 Introduction

In most populated countries like India there are various kinds of industries to fulfil their everyday purposes and therefore the industrial development is also increasing every ever. As a cost of non-renewable fuels most of the large-scale commercialised industries are moving towards the anaerobic digestion as an alternative of the industrial wastewater treatment as well as for generating good renewable sources from the organic matter present in the wastewater (Eder et al. 2018; Pedroza et al. 2017; Schroyen et al. 2017; Tsapekos et al. 2017; Yasar et al. 2017). The high pollutant concentrations in the wastewater is a threat to the environmental impacts on the water bodies and it's difficult to treat in the municipal sewers. Stringent discharge standard and public awareness regarding environmental impacts on water bodies, now-a-days compel industries to treat their effluent onsite for reuse or treatment

R. Loganath (✉) · D. Mazumder
Department of Civil Engineering, Indian Institute of Engineering Science and Technology, Shibpur, Howrah, India
e-mail: loganath.rk@gmail.com

A. Pandey et al. (eds.), *Climate Impacts on Water Resources in India*, Water Science and Technology Library 95, https://doi.org/10.1007/978-3-030-51427-3_16

before its discharge to the sewers (Loganath and Mazumder 2020a, b; Loganath and Mazumder 2018a, b, c; Pradhan et al. 2017; Schaider et al. 2017; Waclawek et al. 2017). Out of available technologies of wastewater treatment, anaerobic digestion is observed to be viable for reducing the high organic loading on the receiving water bodies and sewers (Un et al. 2009; Loganath and Mazumder 2020a, b; Rajeshwari et al. 2000; Ravinadranath et al. 2010; van Lier et al. 2008; Zhang et al. 2014).

The economic impact of anaerobic wastewater processing results from the production of biogas, which contains about 50–75% methane. Capture of this gas would provide an additional source of energy in the day-to-day life, where the gap between supply and demand is wide (Antwi et al. 2017; Bazrafshan et al. 2012; Bustillo-Lecompte and Mehevar 2017; Flugaur 2003, Loganath and Mazumder 2018b, c, a). The usage of methane in vehicles and public transport systems could be encouraged due to the economy of lubricating oils, longer engine life, considerable reduction in emission of pollutants. Production and application of renewable energy, reduction of wastes, and utilization of by-products as fertilizers from anaerobic digestion has increased the attractive application of anaerobic digestion (Pedroza et al. 2017; Loganath and Mazumder 2018a, b, c; Lenz et al. 2009; El-Mashad and Zhang 2010; Verstraete 1983).

The two important concerns for SWW treatment are removal of organic matter and suspended solids. Application of high-rate anaerobic digestion like Upflow Anaerobic Sludge Blanket (UASB) has become popular for treatment of SWW over last few decades (US-EPA 2002, 2004; Johns 1995). Further modification of UASB in form of hybrid UASB is also under operation worldwide for high strength wastewater.

In last 20 years, significant research investigation has been made on HUASB reactor dealing with diverse kinds of domestic and Industrial wastewater towards removal of BOD/COD and nutrients. Since later 1990s HUASB reactor has been extensively used for up-gradation of the existing treatment plants to improve their efficiency in terms of BOD/COD and nutrients from domestic and industrial wastewater (Torkian et al. 2003; Speece 1996; Rajakumar and Meenambal 2008). In recent times HUASB reactor has also been extensively used for treatment of variety of industrial wastewater including food processing, pulp and paper, tannery, pharmaceutical, oily and petrochemical, slaughterhouse, distillery wastewaters etc. (Rajakumar et al. 2011; Rajakumar et al. 2012; Loganath and Mazumcer 2018). Although a lot of information regarding laboratory and pilot scale HUASB reactor operation is available, there is a very few literatures on HUASB reactor treating Industrial wastewater especially SWW without any special attention to explore the effects of various operating parameters on the same. Along these lines, aim of this present study is set to evaluate the performance efficiency of the HUASB Reactor with respect to the COD removal efficiency and suspended solids from SWW.

16.2 Materials and Methods

16.2.1 Fabrication of Reactor Setup

"A lab scale Hybrid Up-flow Anaerobic Sludge Blanket (HUASB) reactor" of 13.5 L liquid volume was fabricated for performance evaluation. It was fabricated utilizing acrylic fiber tube with diameter of 15 cm, height of 60 cm and thickness of 5 mm. An outlet was given at the highest point of the bio-film media, which was associated with the effluent tank. On the highest point of the reactor a three-phase separator was provided to isolate gas and solid raised because of the upward movement of the feed. The gas outlet was associated through elastic tubing to the liquid displacement system to measure the gas generation. The measure of gas produced is equivalent to the measure of liquid displaced and consequently gas produced was estimated at certain interval of time. A bio-film media of height 15 cm made of Polypropylene reticulated rings was given. Around 50 biofilms were filled inside the reactor in staggered way to facilitate microbial development and gas–liquid-solid separation. The individual surface area of each ring was 7.1 cm^{-1} and the aggregate surface region possessed by the packing was 6700 m^2/m^3. The reactors were worked at the temperature of 37 ± 2 °C. A schematic representation of HUASB reactor is appeared in Fig. 16.1.

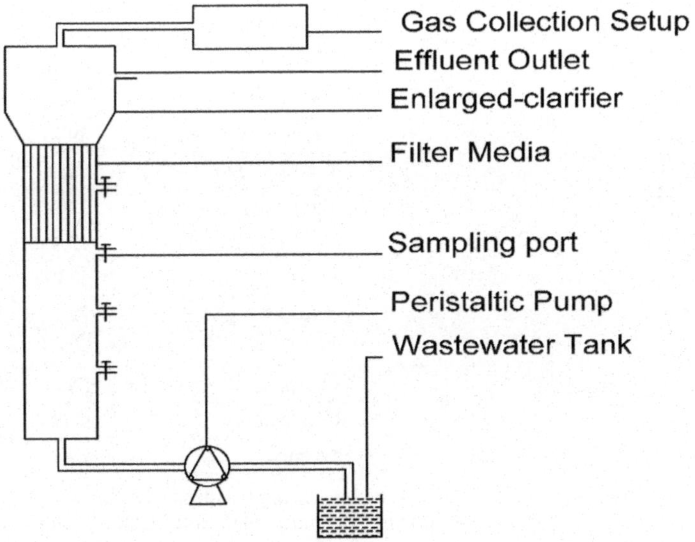

Fig. 16.1 HUASB reactor used in this study

16.2.2 Start-Up of HUASB Reactor

The HUASBR was initially filled with the anaerobic biomass from a conventional biogas plant during start-up. This AD sludge was obtained from the biogas plant located at Narandrapur, Kolkata. This plant was a traditional single stage AD plant, where generates methane from dairy animal manure slurry. AD biomass was withdrawn from the base of the sludge outlet tank. The collected sludge characteristics showed pH 6.9, Total COD 2200 mg/L & TSS 33,087 mg/L.

The reactor was fed with slowly increasing amount of SWW during acclimation phase. The SWW was collected monthly twice in the amount of 20 L/time and stored in 4 °C. Table 16.1 representing characteristics of the SWW used in this study.

The HUASB reactor was firstly fed under batch mode with a uniform retention period of 3 days over 240 days for the sake of acclimation so that it could withstand with maximum COD load of the real raw SWW. During this stage a constant temperature of (37 $\pm$ 2) °C was sustained by means of temperature controller. The first phase of HUASB reactor operation SWW was loaded with COD of 1730 mg/L and final COD concentration was increased up to 5900 mg/L with the stepwise increment.

Table 16.1 Characteristics of SWW used in this study

S. No	Measured parameters	Observed value**
1	pH	7.39
2	Alkalinity	239
3	Total COD	8426
4	Soluble COD	6294
5	Total solids	16,262
6	Total dissolved solids	8980
7	Total suspended solids	7281
8	Total fixed solids	2767
9	Total volatile solids	13,495
10	BOD	5730
11	Nitrogen	72
12	TKN	426
13	Phosphate	25
14	Conductivity	9.47
15	Salinity	2.84

Note All the values in mg/L except pH, salinity (psu), conductivity (ms/cm) and Alkalinity (mg/L as $CaCO_3$)
**Average of Six Samples

16.2.3 Analytical Procedures

All the mentioned parameters used in Table 16.1 were measured according to the Standard Methods (APHA 2005).

16.2.4 Experimental Procedure

After the batch study was over, HUASB Reactor was operated under continuous mode after the biomass was acclimated. It was run using raw SWW with variable COD concentration as well as hydraulic retention time (HRT). Thus, the organic loading rate (OLR) was increased gradually by decreasing HRT to evaluate the COD removal efficiency of the reactor. Hence, the reactor was loaded with OLR in the range of (3.13–38.11) kg COD/m^3/d on account of variation in HRT from 24 to 6 h during the continuous study.

16.3 Results and Discussion

16.3.1 Start-Up of HUASB Reactor

During start-up cum acclimation, the performance of the HUASB reactor was monitored by measuring pH, Alkalinity, initial and final COD, MLSS growth and VFA formation. Both the carbonaceous synthetic wastewater and the raw SWW were used for this purpose. During this period, a mixed solution was fed to the reactor, where the SWW was varied in the ratio of 1/20, i.e. the feed consisted of 50 ml of the SWW and 950 ml of the synthetic wastewater to make a total volume of 1000 ml. Accordingly, the amount of SWW was increased @ 50 ml in total feed volume of 1000 ml for each batch acclimation run. Under batch mode the reactor was operated over 8 months with 3 days batch retention time. The feed ratio to the reactor under batch mode operation is shown in Fig. 16.2.

The inflow SWW pH was adjusted between 6.9 and 7.2 throughout the acclimation study. Hence, the COD concentration was varied between (1730–6390) mg/L. While acclimation period after the treatment COD was observed between 950 to 720 mg/L with respect to this the COD removal efficiency was 45–88%. These observations show that the seed sludge acclimation was accomplished with the presence of raw SWW. The detailed results are shown in Fig. 16.3 and 16.4.

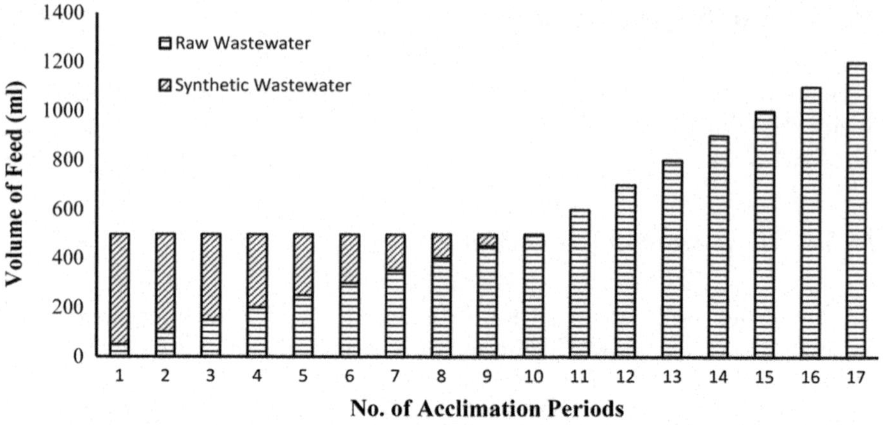

Fig. 16.2 Proposition of raw SWW and synthetic wastewater during acclimation process

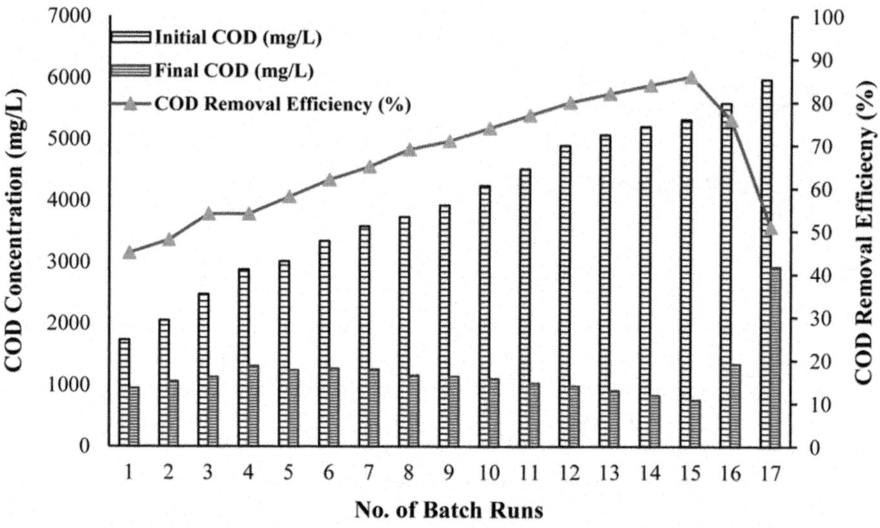

Fig. 16.3 Profile of COD removal in HUASB Reactor during acclimation

16.3.2 COD Removal Efficiency Under Continuous Mode Operation

The continuous mode of operation with the HUASB Reactor was performed with the varying OLR with respect to the HRT. The raw SWW was diluted and fed to the reactor to overcome the shock loading to the HUASB Reactor. The steady state condition of each OLR and HRT was confirmed with the three consecutive repetitive results. After every successful steady state condition, OLR was further risen with

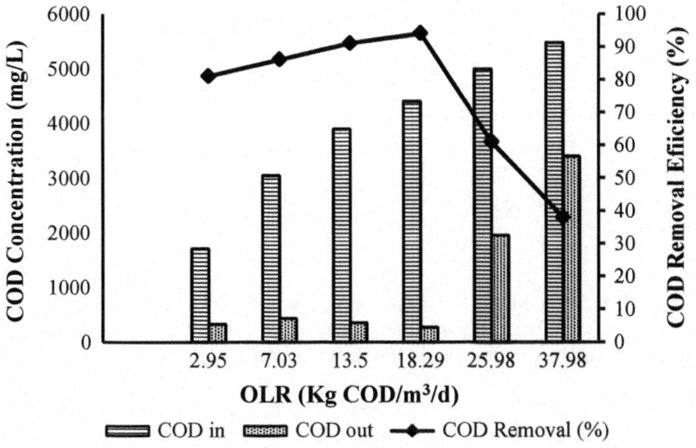

Fig. 16.4 COD removal Performance of HUASB Reactor treating Raw SWW

decreasing HRT and the reactor was monitored. In the first steady state condition OLR to the reactor was 2.95 kg COD/m^3/d with respect to the 24 h HRT. During final stage of reactor operation, the OLR to the reactor was 37.98 kg COD/m^3/d under 6 h HRT.

The maximum COD removal efficiency was observed to be about 94.8%, for the influent COD concentration of about 4390 mg/L at 10 h HRT. The optimum OLR was noted as 18.29 kg COD/m^3/d in the present HUASB Reactor under continuous mode of operation, which exhibited more than 94% COD removal efficiency. From the earlier studies, it is already proved that HUASB Reactor can handle the maximum OLR of 9.27–13.27 kg COD/m^3/d at 10 h HRT (Rajakumar et al. 2011). In this present study due to the proper acclimation of sludge with the raw SWW the maximum COD efficacy of 94% was achieved at the OLR of 18.29 kg COD/m^3/d. Therefore, compared to other studies the OLR could be increased to around 30% due to the well acclimated sludge and the addition of more surface area of attached growth media in the present treatment process.

16.4 Conclusions

The SWW contains a high amount of organic matter, representing a good source of organic carbon. HUASB reactor used in the study was satisfactorily acclimated under SWW using the anaerobic digested sludge. During acclimation phase, COD removal efficiency reached to as high as 88% even in presence of raw SWW entirely. It also performed very well towards removal of COD from SWW under continuous mode of operation. The maximum COD removal efficiency was attained up to 94%, under the influent COD concentration of about 4390 mg/L and HRT of 10 h. The optimum OLR of 18.29 kg COD/m^3/d was provided to the HUASB reactor, which showed more than 94% COD removal efficiency. Hence, the present HUASB reactor can be recommended for COD removal from high-strength wastewater generated from slaughterhouse operation.

References

Antwi P, Li J, Boadi PO, Meng J, Quashie FK, Wang X, Ren N, Buelna G (2017) Efficiency of an upflow anaerobic sludge blanket reactor treating potato starch processing wastewater and related process kinetics, functional microbial community and sludge morphology. Biores Technol 239:105–116

APHA (2005) Standard methods for the examination of water and wastewater, American Public Health Association (APHA): Washington, DC, USA

Bazrafshan E, Mostafapour FK, Farzadkia M, Ownagh KA, Mahvi AH (2012) Slaughterhouse wastewater treatment by combined chemical coagulation and electrocoagulation process. PLoS ONE 7:e40108

Bustillo-Lecompte C, Mehrvar M (2017) Slaughterhouse wastewater: treatment, management and resource recovery. In: Physico-chemical wastewater treatment and resource recovery, InTech

Eder A, Mahlberg B (2018) Size, subsidies and technical efficiency in renewable energy production: the case of Austrian biogas plants. Energy J 39

El-Mashad HM, Zhang R (2010) Biogas production from co-digestion of dairy manure and food waste. Biores Technol 101:4021–4028

Flugaur NJ (2003) Wastewater effluent treatments and control technologies in the beef processing industry, University of Wisconsin–Stout

Johns M (1995) Developments in wastewater treatment in the meat processing industry: a review. Biores Technol 54:203–216

Lenz M, Enright AM, O'Flaherty V, van Aelst AC, Lens PN (2009) Bioaugmentation of UASB reactors with immobilized Sulfurospirillum barnesii for simultaneous selenate and nitrate removal. Appl Microbiol Biotechnol 83:377–388

Loganath R, Mazumder D (2018a) Determination of kinetic co-efficients of anaerobic treatment of slaughterhouse wastewater. J Indian Chem Soc 95:467–473

Loganath R, Mazumder D (2018b) Performance study of hybrid upflow anaerobic sludge blanket reactor treating slaughterhouse wastewater. J Indian Chem Soc 95:365–370

Loganath R, Mazumder D (2018c) Performance study on organic carbon, total nitrogen, suspended solids removal and biogas production in hybrid UASB reactor treating real slaughterhouse wastewater. J Environ Chem Eng 6:3474–3484

Loganath R, Mazumder D (2020a) Development of a simplified mathematical model for anaerobic digestion. In: Sustainable waste management: policies and case studies 2020, Springer, Singapore, pp 571–578

Loganath R, Mazumder D (2020b) Performance study on enlarged-clarifier hybrid upflow anaerobic sludge blanket reactor for treating the slaughterhouse wastewater. Water Environ J

Pedroza M, Souza J, Vieira G, Bezerra M (2017) Bio-oil and biogas from the pyrolysis of sewage sludge, and non-isothermal degradation on Usy Zeolite. Brazilian J Petrol Gas 11

Pradhan D, Sukla LB, Sawyer M, Rahman PK (2017) Recent bioreduction of hexavalent chromium in wastewater treatment: a review. J Ind Eng Chem 55:1–20

Rajakumar R, Meenambal T (2008) Comparative study on start-up performance of HUASB and AF reactors treating poultry slaughterhouse wastewater. Int J Environ Res 2

Rajakumar R, Meenambal T, Banu JR, Yeom I (2011) Treatment of poultry slaughterhouse wastewater in upflow anaerobic filter under low upflow velocity. Int J Environ Sci Technol 8:149–158

Rajakumar R, Meenambal T, Saravanan P, Ananthanarayanan P (2012) Treatment of poultry slaughterhouse wastewater in hybrid upflow anaerobic sludge blanket reactor packed with pleated poly vinyl chloride rings. Biores Technol 103:116–122

Rajeshwari K, Balakrishnan M, Kansal A, Lata K, Kishore V (2000) State-of-the-art of anaerobic digestion technology for industrial wastewater treatment. Renew Sustain Energy Rev 4:135–156

Ravindranath E, Kalyanaraman C, Begum SS, Gopalakrishnan AN (2010) Effect of recirculation rate on anaerobic treatment of fleshing using UASB reactor with recovery of energy

Schaider LA, Rodgers KM, Rudel RA (2017) Review of organic wastewater compound concentrations and removal in onsite wastewater treatment systems. Environ Sci Technol 51:7304–7317

Schroyen M, Van Hulle SW, Holemans S, Vervaeren H, Raes K (2017) Laccase enzyme detoxifies hydrolysates and improves biogas production from hemp straw and miscanthus. Biores Technol 244:597–604

Speece RE (1996) Anaerobic biotechnology for industrial wastewaters. In: Anaerobic biotechnology for industrial wastewaters

Torkian A, Eqbali A, Hashemian S (2003) The effect of organic loading rate on the performance of UASB reactor treating slaughterhouse effluent. Resour Conserv Recycl 40:1–11

Tsapekos P, Kougias P, Treu L, Campanaro S, Angelidaki I (2017) Process performance and comparative metagenomic analysis during co-digestion of manure and lignocellulosic biomass for biogas production. Appl Energy 185:126–135

Ün ÜT, Koparal AS, Öğütveren ÜB (2009) Hybrid processes for the treatment of cattle-slaughterhouse wastewater using aluminum and iron electrodes. J Hazard Mater 164:580–586

US-EPA (2002) Environmental assessment of proposed effluent limitations guidelines and standards for the meat and poultry products industry point source. In: Office of water, US Environmental Protection Agency, Washington, DC, EPA-821-B-01008

US-EPA (2004) Effluent limitations guidelines and new source performance standards for the meat and poultry products point source category. In: U. S. Environmental Protection Agency (US EPA) Federal Register 69 (173)

van Lier JB, Mahmoud N, Zeeman G (2008) Anaerobic wastewater treatment, biological wastewater treatment, principles, modelling and design, pp 415–456

Verstraete I (1983) Biomethanation of wastes: perspectives and potentials. In: International conference on the commercial applications and implications of biotechnology. vol 1, pp 725–742

Wacławek S, Lutze HV, Grübel K, Padil VV, Černík M, Dionysiou DD (2017) Chemistry of persulfates in water and wastewater treatment: a review. Chem Eng J 330:44–62

Yasar A, Rasheed R, Tabinda AB, Tahir A, Sarwar F (2017) Life cycle assessment of a medium commercial scale biogas plant and nutritional assessment of effluent slurry. Renew Sustain Energy Rev 67:364–371

Zhang B, Liu Y, Tian C, Wang Z, Cheng M, Chen N, Feng C (2014) A bibliometric analysis of research on upflow anaerobic sludge blanket (UASB) from 1983 to 2012. Scientometrics 100:189–202

Chapter 17
Development of Biofilters for the Treatment of Greywater

Sujata S. Kulkarni, Basavaraj S. Hungund, Raghavendraprasad Suryavanshi, Geeta C. Bellad, and M. R. Patil

Abstract The depletion of water resources is increasing due to the excessive usage of surfactants and polyphosphates. These surfactants and polyphosphates promote the growth of nuisance causing microorganisms by providing a source of phosphate. Greywater contributes approximately 70–75% of the total domestic wastewater. Therefore, it is necessary to treat greywater and reuse it for gardening, irrigation, etc. The treated water can also be released into natural water bodies. Two biofilters were developed using activated carbon and wood chips in which coconut coir acted as support material. The average grain size of activated carbon and wood chips was found to be 1.0 mm–2.0 mm and 1.0 cm–2.0 cm respectively. The thin sheets of coconut fibers were collected from the local market whose density and permeability were measured. The activated carbon and wood chips of thickness 5.0 cm and 10.0 cm were sandwiched between the coconut fibers of 3.0 cm thick in two separate filter setups. A study was conducted for phosphate degradation in greywater using both the biofilters in the batch scale. Synthetic greywater was prepared by adding a known amount of detergent and liquid soap to potable water. Phosphate content in the greywater was analyzed using UV–Vis Spectrophotometer and chemical oxygen demand (COD) was determined by the closed reflux method. A batch of 5.0 L of water is retained in each of the filters with the retention time of eight days and the filtrate was analyzed for phosphate concentration and COD after every 24 h. Activated carbon removed 80% of phosphate and 72% of COD whereas woodchips could remove 45% of phosphates and 42% of COD when retained for 24 h. The removal capacity of

S. S. Kulkarni
Department of Civil Engineering, B.V.B. College of Engineering and Technology, Hubballi 580031, India

B. S. Hungund (✉)
Department of Biotechnology, KLE Technological University, Hubballi 580031, India
e-mail: hungundb@gmail.com

R. Suryavanshi
Sustainable Biosolutions, Hubballi 580031, India

G. C. Bellad · M. R. Patil
School of Civil & Environmental Engineering, KLE Technological University, Hubballi 580031, India

A. Pandey et al. (eds.), *Climate Impacts on Water Resources in India*, Water Science and Technology Library 95, https://doi.org/10.1007/978-3-030-51427-3_17

biofilters was compared with the removal capacity of conventional sand filter, which reduced the phosphates and COD by 38% and 45% respectively. Studies were also conducted to assess the effect of hydraulic retention time on the removal of phosphate and COD concentration.

17.1 Introduction

Water is found almost everywhere on earth but very less water is fit for consumption. Increased population and industrialization have resulted in the contamination of water bodies posing severe environmental problems (Proshad et al. 2018). The World Health Organization says, "a large proportion of the overall disease burden, 3.3% of global deaths and 4.6% of global disability-adjusted life years (DALYs), was attributed to quantifiable effects of inadequate water, sanitation and hygiene (WASH)" (WHO 2019). This makes it the leading cause of deaths worldwide making it very necessary to implement effective methods of water purification. Most of the regions of the world suffer from water scarcity, therefore nowadays the development of the techniques which can make the wastewater reusable is of much importance. The reuse of wastewater after the treatment may be advantageous as it reduces the water scarcity in that region and also reduces the burden on wastewater treatment plants. Greywater is the wastewater generated from sinks, bathrooms, and dishwashers excluding sewage. It constitutes a major part of domestic waste. It contains a negligible amount of fecal matter and hence can be easily treated and reused for irrigation, gardening, and other non-potable purposes.

Greywater is typically characterized by very high concentrations of xenobiotic compounds and other residues from soaps and detergents (Edwin et al. 2014). It is reported that the biological treatment of the greywater needs a risk free reuse of greywater for various purposes other than potable water (Nolde 1999). Soaps and detergents can be collectively called as surfactants which act to clean the surface of the material. The major constituent of detergent which is responsible for the efficient performance is polyphosphate. Phosphates chelate calcium and magnesium ions and increase the efficiency of surfactant in hard water. Excess use of phosphates may be disadvantageous as they remain in the wastewater and find their way into natural bodies causing an elevation in the nutrient content of water bodies. This leads to eutrophication and algal blooms. Due to the serious problems posed by the polyphosphates on the environment, most of the countries including the United States and European Union have restricted the use of phosphates. But in India, there is no restriction on the use of phosphates in detergents. This makes the treatment of greywater inevitable before disposing it into the environment. Conventionally, sand filters are used for the treatment of wastewater on a larger scale which has certain disadvantages like lesser efficiency and non-portability.

Therefore, in the present work, a comparative study is carried out with an attempt to develop a suitable biofilter more efficient than a sand filter to reduce phosphate and chemical oxygen content (COD) from greywater. Dalahmeh and team (2012) studied the efficiency of bark, activated charcoal, foam and sand filters in reducing pollutants from greywater. The results were encouraging with activated charcoal being the most suitable for improving greywater quality. Erikson and co-workers (2002) identified Zn, Pb, etc. as heavy metals most likely to be present in greywater with their maximum values being 1.6 mg/L, 0.15 mg/L respectively. This is assumed due to the corrosiveness of local water supplies and the composition of household infrastructure. Some personal care products like whitening toothpaste, cosmetics, etc. also have heavy metals in them that find their way into wastewater. If present in greywater, the heavy metals cause long term effects on crops and soil as determined by Rattan and co-workers (2005). Therefore, the heavy metals must be removed from greywater to make it safe for its reuse. In the present work, the relative reduction capacity of a sand filter is evaluated with respect to activated carbon filter for the removal of lead and nickel.

17.2 Materials and Methods

17.2.1 Characterization of Materials

For the construction of biofilters coconut coir, sand, gravel, and activated carbon have been used. The dry coconut coir was procured from the local market with a density of 0.10 g/cc and its permeability being 0.011 cm/sec. The average grain size of activated carbon and wood chips was found to be 1.0 mm–2.0 mm and 1.0 cm–2.0 cm respectively. Iodine number of activated carbon was found to be 9.0. The top layer of the sand filter had the average grain size ranging from 0.3–0.6 mm and uniformity coefficient and porosity being 3% and 30% respectively. The gravel of size 20 and 6 mm was used.

17.2.2 Development of Biofilters

Two of the biofilters containing activated carbon and wood chips separately along with coconut coir fibers were developed to study the phosphate degradation in wastewater in batch scale. The thin sheets of coconut fibers were used as supporting material. The activated carbon and wood chips each of thickness 10 cm were sandwiched between the coconut fibers of 3.0 cm thick in two separate filter setups. To carry out the comparative study, a sand filter was also developed in our laboratory considering the design specification of previous work (Verma et al. 2017). Sand and gravel were washed twice using tap water followed by air drying. After that, they were soaked in the solution of 5% HCl and 5% NaOH for one hour followed by oven drying. The cross-section of the filters is given in Fig. 17.1.

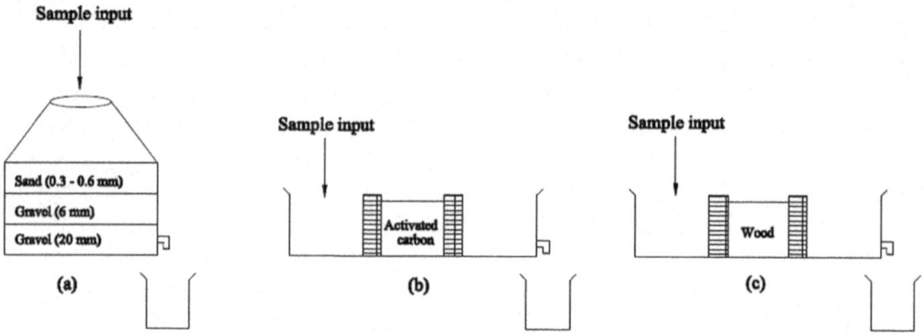

Fig. 17.1 Figure depicting cross-sections of filters. (**a**) Sand filter. (**b**) Activated carbon filter. (**c**) Wood filter

17.2.3 Batch Studies for Treatment Process

Batch studies were conducted to remove phosphate ions from greywater samples, prepared in the laboratory. Greywater samples containing 1% detergent was fed into the filters and analyzed for the reduction in phosphates and also chemical oxygen demand (COD). The analysis for phosphate ionic content was done using a spectrophotometer at 450 nm as defined by the American Public Health Association (APHA, Ed: Eaton 2005). The COD was also estimated by the closed reflux method as given by APHA. During the process, 5.0 L of greywater sample was fed initially and allowed to retain for eight days at ambient.

Temperature. Samples were collected and analyzed after every 24 h. Similarly, synthetic water samples containing lead and nickel at a concentration of 10 ppm each were fed into the filter bed and allowed to retain for five days. Further samples were collected after every 24 h and analyzed for nickel and lead content using atomic absorption spectroscopy (AAS) referring to APHA guidelines.

17.3 Results and Discussion

The synthetic greywater samples containing 1% detergent prepared in the laboratory were fed into filters and analyzed for the reduction in phosphates and chemical oxygen demand (COD).

Figure 17.2 depicts the details of phosphate content with respect to time. Initial phosphate content in the greywater was 40 mg/L at 1% detergent. After 24 h of retention, the phosphate content in the sand filter was found to be 34 mg/L indicating around 15% removal. After eight days of retention, the phosphate content was found to be 11.88 mg/L, indicating 70% removal of the phosphate. The study performed by Erickson and co-workers (2012) on capturing phosphates with iron enhanced sand filtration reported 88% removal of phosphates. Figure 17.3 depicts the details of

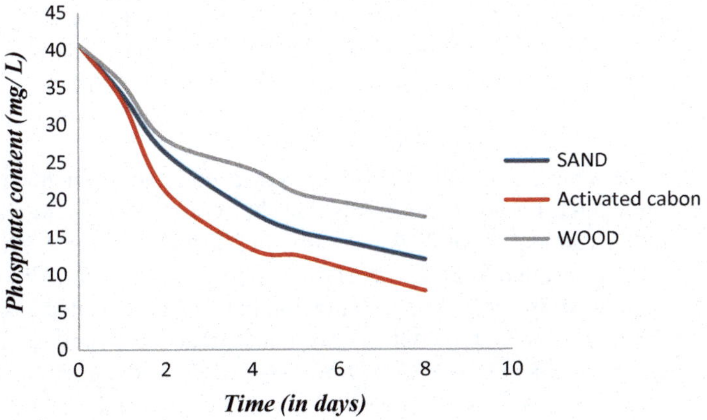

Fig. 17.2 Profile of phosphate content in greywater during treatment process

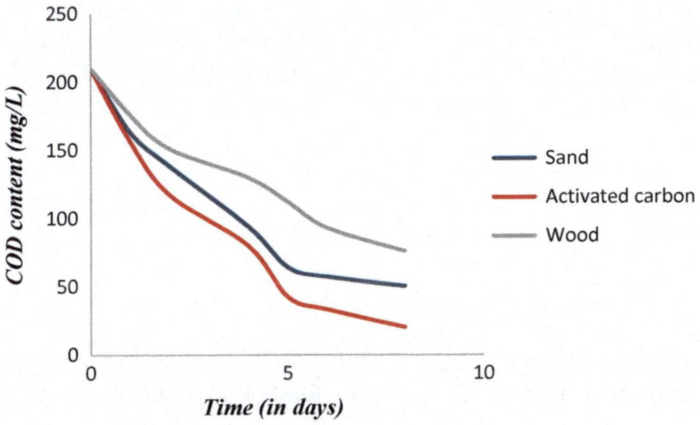

Fig. 17.3 Profile of COD level in greywater during treatment process

COD level in greywater during treatment process where the initial COD content was noted as 210 mg/L. After 24 h, COD was reduced to 163 mg/L. After eight days of retention time, the sand filter was able to remove 74.76% of COD by reducing it to 53 mg/L. The biofilm developed in sand filter due to the presence of native microorganisms might have helped in the degradation of the greywater sample.

The development of biofilm was slow in the initial days which has resulted in the slower degradation process. After two days, the increase in removal efficiency was noticed probably due to the sufficient growth of phosphate reducing microorganisms in the sand filter. The initial concentration of heavy metals in the synthetic sample was maintained at 10 ppm. Sand filter removed 60% of lead and 35% of nickel after the retention time of five days. Sand is very effective in removing lead as compared to nickel at lower concentrations. These results are in agreement with the work evidenced by Pappalardo and the team (2010). Anderson and co-workers (1985) summarized that some degree of physical, chemical, and biological treatment processes are responsible for the treatment in sand filters. Adsorption of metal ions on sand can be explained based on interactions between the surface functional groups and the metal ions (Awan et al. 2003).

The filter containing activated carbon proved to be most efficient in removing phosphates and COD. Figure 17.2 depicts that the phosphate content was reduced to 7.73 mg/L. This shows about 80% reduction in phosphate content. The previous studies have shown around 9–87% reduction in phosphate content (Erickson et al. 2017). The present study shows 90.4% reduction in COD in activated carbon filters. Hami and associates (2007) studied the performance of activated carbon in the water filters and evidenced a decrease in COD from 72–92.5% for the inlet values of 110–200 mg/L. Adsorption on activated carbon takes place through three basic steps: substances adsorb to the exterior of the carbon granules; substances move into the carbon pores; substances absorb to the carbon. Adsorption mainly depends on the properties of adsorbents like surface area, pore size distribution, surface functional groups, etc. as well as the experimental condition (Dong et al. 2016). Activated carbon filter removed 73% of lead and 67% of nickel after the retention time of five days with the initial concentration being maintained at 10 mg/L. These results are in agreement with the work carried out by Pappalardo and the team (Pappalardo et al. 2010).

As indicated in Figs. 17.2 and 17.3, the wood chips showed the least reduction in phosphates and COD being 53.1% and 63.8% respectively. The literature shows that the wood chips have certain active sites for the adsorption of phosphates and the wood chips treated with iron salts have enhanced removal capacity due to an increased number of active sites (Eberhardt et al. 2006). But, the wood chips used in the present study are the waste produced by the furniture industries. It was used without pretreatment which is expected to possess lesser active sites and lesser surface area for adsorption. This can be the reason for their reduced performance.

17.4 Conclusion

The activated carbon utilized adsorption mechanism and was found to be efficient in removing phosphate content from greywater. The ionic interactions between the active sites of adsorbent and chemical pollutants may be responsible for the treatment of greywater. The present study also showed that both adsorption and biofiltration techniques are useful in the treatment of greywater. Both activated carbon and sand filter are efficient in removing the traces of heavy metals from greywater. The present study has a wide scope for future work. The effect of temperature and pH on the removal efficiency can be studied as both the parameters significantly control the rate of adsorption and also affect the rate of bioremediation. Future work may include scale-up studies to assess exhaust time, breakthrough time of adsorbents, column experiments, etc.

Acknowledgement The authors acknowledge Dr. Ashok Shettar, Honorable Vice-Chancellor, KLE Technological University, Hubballi for his constant support and encouragement. The authors would like to acknowledge the Technical Education Quality Improvement Program (TEQIP) for financial support.

References

Anderson DL, Siegrist RL (1985) Technology assessment of intermittent sand filters. Municipal Environmental Research Laboratory, EPA

Awan MA, Qazi IA, Khalid I (2003) Removal of heavy metals through adsorption using sand. J Environ Sci 15(3):413–416

Dalahmeh SS, Pell M, Vinnerås B, Hylander LD, Öborn I, Jönsson H (2012) Efficiency of bark, activated charcoal, foam and sand filters in reducing pollutants from greywater. Water Air Soil Pollut 223(7):3657–3671

Dong L, Liu W, Jiang R, Wang Z (2016) Study on the adsorption mechanism of activated carbon removing low concentrations of heavy metals. Desalin Water Treat 57(17):7812–7822

Eaton AD (ed) (2005) Standard methods for the examination of water and waste water 21. American Public Health Association, Washington, D.C

Eberhardt TL, Min SH, Han JS (2006) Phosphate removal by refined aspen wood fiber treated with carboxymethyl cellulose and ferrous chloride. Biores Technol 97(18):2371–2376

Edwin GA, Gopalsamy P, Muthu N (2014) Characterization of domestic gray water from point source to determine the potential for urban residential reuse: a short review. Applied Water Science 4(1):39–49

Eriksson E, Auffarth K, Henze M, Ledin A (2002) Characteristics of grey waste water. Urban water 4(1):85–104

Erickson AJ, Gulliver JS, Weiss PT (2012) Capturing phosphates with iron enhanced sand filtration. Water Res 46(9):3032–3042

Erickson AJ, Gulliver JS, Weiss PT (2017) Phosphate removal from agricultural tile drainage with iron enhanced sand. Water 9(9):672

Hami ML, Al-Hashimi MA, Al-Doori MM (2007) Effect of activated carbon on BOD and COD removal in a dissolved air flotation unit treating refinery wastewater. Desalination 216(1–3):116–122

Nolde E (1999) Greywater reuse systems for toilet flushing in multi-storey buildings–over ten years' experience in Berlin. Urban water 1(4):275–284

Pappalardo L, Jumean F, Abdo N (2010) Removal of cadmium, copper, lead and nickel from aqueous solution by white, yellow and red United Arab Emirates sand. Am J Environ Sci 6(1):41

Proshad R, Kormoker T, Mursheed N, Islam MM, Bhuyan MI, Islam MS, Mithu TN (2018) Heavy metal toxicity in agricultural soil due to rapid industrialization in Bangladesh: a review. Int J Adv Geosci 6(1):83–88

Rattan RK, Datta SP, Chhonkar PK, Suribabu K, Singh AK (2005) Long-term impact of irrigation with sewage effluents on heavy metal content in soils, crops and groundwater—a case study. Agr Ecosyst Environ 109(3–4):310–322

Verma S, Daverey A, Sharma A (2017) Slow sand filtration for water and wastewater treatment–a review. Environ Technol Rev 6(1):47–58

World Health Organization (2019) Safer water, better health

Chapter 18
Greywater Treatment by Two-Stage Bioreactor

P. Naresh Kumar and Arun Kumar Thalla

Abstract The present study was carried out with an objective to treat Greywater by a two-stage biological process i.e. anaerobic followed by aerobic each with a working volume of 3 L. It emphasises the results obtained during its operation for the last three months. Reactor was started with an HRT of 10 h and gradually reduced to 6 h in stages. The overall average efficiency by the system in terms of COD reduction, nutrients i.e. total nitrogen and phosphorous was around 83, 70 and 71% respectively. With the trend being followed by the reactors in the last three months, it is noticed that the system efficiency is gradually increasing and so we are evident that the system is gaining stability and thus may lead to better quality effluent with time. Investigations are being done with Photocatalysis as post treatment to further enhance the quality of water.

18.1 Introduction

Domestic wastewater is majorly divided into two categories based on their origin. One is black water which consists of wastewater from toilets and urinals and the second one is greywater which is composed of the wastewater generated by handwash basin, shower, kitchen, washing machine. Greywater is normally diluted and only contains low level of contaminating pathogens, nitrogen and other contaminants. Hence greywater is more suitable for recycle and reuse (Li et al. 2009) for toilet flushing, irrigation or other recreational purposes which has no human contact, by this about 30–40% reduction in the demand of portable water can be achieved (Jefferson et al. 2000).

Different approaches have been applied for greywater treatment including physical, chemical and biological operational methods. Even though physical and chemical processes can effectively remove suspended solids, organic materials and surfactants, they are not cost-effective for removing the full array of dissolved components in greywater (Bani-Melhem and Smith. 2012). The study conducted by Elmitwalli

P. N. Kumar · A. K. Thalla (✉)
Department of Civil Engineering, National Institute of Technology, Mangalore, Karnataka, India
e-mail: thallpce@gmail.com

A. Pandey et al. (eds.), *Climate Impacts on Water Resources in India*, Water Science and Technology Library 95, https://doi.org/10.1007/978-3-030-51427-3_18

and Otterpohl (2007), Hernández Leal et al. (2007) and other researchers used anaerobic treatment for the greywater and achieved COD by 70–85%, phosphorus by 22–33% and of 15–22% nitrogen removal was achieved, hence suggesting that anaerobic treatment can significantly reduce the organic load from the greywater. Similarly, the study conducted by Ghunmi et al. (2010) showed that aerobic bioreactor can reduce about 86% of COD from the greywater.

In this study we are combining both anaerobic and aerobic (moving bed bioreactor) process to increase the efficiency of the biological treatment. However, post treatment process such as photocatalysis can be used so as to degrade the remaining organic matter and also disinfect the water to meet recyclable water requirement standards.

18.2 Material and Methods

18.2.1 System Configuration

Two cylindrical containers of 4 L capacity, having working volume of 3 L each, was used as reactors for both anaerobic and aerobic processes followed by a settling tank of capacity 3 L. The reactor were fabricated from an acrylic glass cylinder. Two reactors are then operated in sequential continuous reactor mode. The reactors are equipped with two peristaltic pumps one for feeding the anaerobic reactor and the other for activated sludge recycling in aerobic reactor from settling tank. Anaerobic reactor was capped at the top to prevent atmospheric air from entering in. A provision at top is provided for gases to escape out produced during anaerobic digestion. Nitrogen gas is passed periodically into the reactor to maintain anaerobic condition in the reactor. Aerobic reactor is provided with diffusers for continuous oxygen supply as an external source. The reactor was operated at hydraulic retention time (HRT) of 10 h at the start-up of the experiment and gradually brought down to 6 h (Fig. 18.1).

18.2.2 Formulation of Synthetic Greywater

Synthetic Greywater was prepared by dissolving arrowroot powder, urea, potassium dihydrogen phosphate, detergent and hand wash in water to best imitate real greywater.

Fig. 18.1 Laboratory scale setup of two-stage bioreactor

18.2.3 Analytical Procedures

Chemical oxygen demand (COD), biochemical oxygen demand (BOD), nutrient such as ammonia nitrogen and total phosphorus and mixed liquor suspended solids (MLSS) were analyzed according to standard methods. Phosphorus as phosphates was analyzed with UV spectrometer at 690 nm according to moly do-phosphoric method. Nitrogen as Ammonia–Nitrogen was analyzed with UV spectrometer at 420 nm according to direct nesslerization method. Samples for analysis were filtered through 0.45 μm to remove suspended solids, if present, prior to being fed to UV spectrometric analyses.

18.2.4 Operation of Two-Stage Bioreactor

The anaerobic reactor was seeded using active anaerobic sludge collected from Kavoor treatment plant and aerobic sludge was obtained from NITK treatment plant for start-up. Before the start of experiments, the reactor had been operated using $K_2Cr_2O_7$ for tracer study analysis. Tracer study response curve showed that our laboratory scale reactor setup was close to ideal CSTR and the operation of reactor was started at room temperature without control.

The laboratory scale reactor setup was then operated in a continuous mode for studying its performance in removing carbon and nutrients from greywater. The mixed liquor suspended solids (MLSS) concentration were maintained between 1500 and 2500 mg/l. The sludge recycling is maintained periodically to maintain the mixed liquor suspended solids in the aerobic reactor for best removal of organic loading and nutrients.

The settling properties were also evaluated using the sludge volume index (SVI). SVI is a good indicator of the system performance. Low SVI values i.e., value less than 100–150 mL/g indicate good sedimentation characteristics of the sludge yielding high biomass concentration in the aerobic reactor, whereas high SVI values greater than 150 mL/g reflect bulking sludge and low biomass concentration in the aerobic reactor. During experiments, SVI value of sludge were good and did not exceed 150 mL/g.

18.3 Results and Discussion

18.3.1 Characteristics Study of Synthetic Greywater

To determine the feasibility of the synthetic greywater produced, their characteristics were examined in laboratory and compared with previous literature of real greywater and is presented in Table 18.1.

Table 18.1 Composition of synthetic greywater

Characteristics	Synthetic greywater			Greywater, literature		
				Jefferson et al. (2001)	Eriksson et al. (2002)	Frielder (2004)
	Max	Min	Mean	Mean	Mean	Mean
pH	7.6	6.7	7.2	7.41	7.18	7.47
Turbidity (NTU)	48	36	42	93	–	101
COD (mg/L)	294.4	224	256	339	364	451
BOD (mg/L)	170	110	140	142	240	146
NH_4–N (mg/L)	23.16	20.87	21.866	–	–	–
PO_4–P (mg/L)	6.597	6.192	6.463	–	–	–

18.3.2 Performance of Two-Stage Bioreactor for Organic Matter (Cod) and Nutrient Removal

18.3.2.1 Organic Matter Removal

Figure 18.2 shows the variation in influent and effluent COD concentration and their removal efficiency when the system was operated at 6 h HRT.

The experimental results shows that removal efficiency gradually increased from 60 to 83% over two months of reactor operation at HRT of 6 h. The COD removal efficiency in individual anaerobic & aerobic reactor were 51% and 26% respectively. Thus, representing the increased removal efficiency of the system by the addition of the aerobic reactor after anaerobic reactor. Similar results were found in the study of Hernández Leal et al. (2010) where anaerobic reactor was capable of reducing only 25% of COD during first 15 days and after 180 days about 80% COD removal efficiency was achieved. Similar results were observed from the study of Hernández Leal et al. (2007) on greywater reuses for irrigation using anaerobic–aerobic reactor and found that this combination was cable of reducing about 85% of organic load, but it was unable to efficiently reduce the e-coli from the influent.

Huge amount of COD is consumed by denitrifiers and polyphosphate accumulating organisms, with the simultaneous removal of nitrogen and phosphorus.

Similarly, the BOD of the effluent from the combined reactor varied between 26 and 40 mg/l. A maximum removal of 77% BOD loading is observed from the effluent after two months of operation time. Efficiency of organic reduction of the combined reactor is expected to further improve with the operation time.

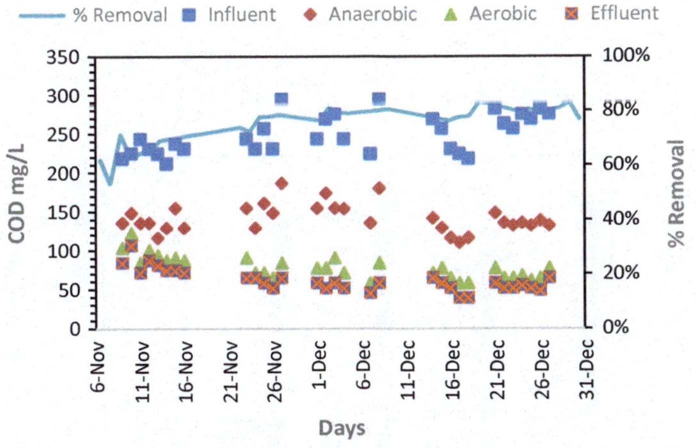

Fig. 18.2 COD removal profile of the exp. Setup

18.3.2.2 Ammonia Nitrogen Removal

The variations of ammonia nitrogen (NH_3–N) concentration of influent, effluent & total removal efficiency during the complete operation period after the system reached steady state condition were measured and illustrated in Fig. 18.3. The average influent ammonia nitrogen concentration was about 21.18 mg/L. From the figure, it can be observed that the anaerobic–aerobic system along with the setting tank was capable of reducing about 70% of the total NH_3–N load, having the effluent concentration as 6.325 mg/L. Major reductions of NH_3–N i.e. about 40% was achieved only by the anaerobic reactor with its effluent concentration being 12.704 mg/L. Some researchers such as Li et al. (2009), Hernández Leal et al. (2010) reported that the greywater has lower concentrations of ammoniaa nitrogen as they did not consider the presence of urine in the greywater. This was different from the present study as we consider the presence of urine in the greywater.

Similarly, by the study of Couto et al. (2014) on anaerobic degradation of greywater over a period of 6 months showed that the anaerobic reactor was capable of reducing about 60% of the ammonia nitrogen, this reduction in nitrate was explained as the consequence of the denitrification process, which takes place predominantly in anoxic environments.

During the biological nitrification process, the aerobic autotrophic bacteria present in the reactor oxidises ammonia (NH_4–N) into nitrite (NO_2^-) and then into nitrate. In settling tank under anoxic condition, denitrification process by heterotrophic bacteria takes place where the final output of nitrification process nitrate could be reduced to nitrogen (N_2) gas. Generally 2.86 g of BOD is sufficient for the reduction of 1 g of nitrate.

The minimum sludge age required to acheive complete nitrification in the reactor is said to be 16 days (Balakrishnan and Eckenfelder 1969) and is maintained as such

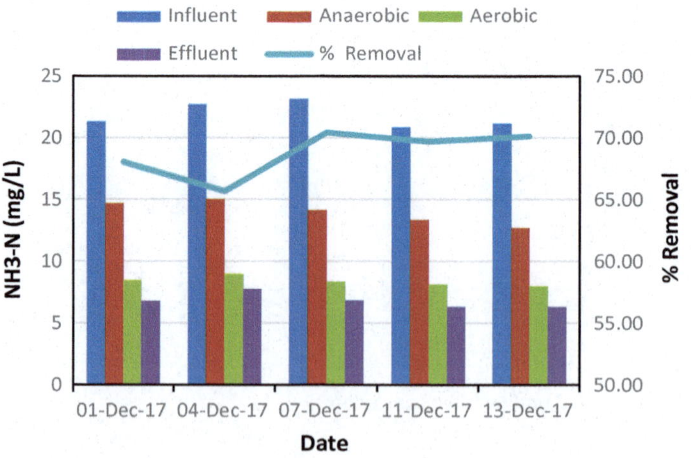

Fig. 18.3 Ammonia–Nitrogen removal profile

in the aerobic reactor unit. Nitrification is a pH sensitive process. Its rate goes down as the pH value drops to below 6.8. Hence it is important to maintain pH in the range between 6.8 and 7.3.

18.3.2.3 Total Phosphorus Removal

The variations of total phosphorus (TP) of influent, effluent and total removal efficiency during complete cycle of two-stage bioreactor system after the system reached steady state condition is illustrated in Fig. 18.4. The influent concentration of total phosphorus that is fed into the reactor was about 6.338 mg/L. It can be observed that only a minor reduction in the total phosphorus was observed by the anaerobic reactor where the concentration reduced from 6.338 to 5.015 mg/L. In other words, there were no noticeable removal of total phosphorus concentration in the effluent from anaerobic process. Major reduction was observed from the aerobic stage of the setup, where the concentration of total phosphorus further reduced to 1.851 mg/L achieving about 71% reduction in the concentration of total phosphorus (TP).

The phosphorus in the integrated system is removed by two different population of microorganism, one is Phosphate accumulating organisms (PAOs), that is capable of accumulating polyphosphates under aerobic condition (type (A) PAOs) and other is PAOs which are capable of accumulating polyphosphates under both anoxic and aerobic condition (DNPAOs). In aerobic condition, the activity of phosphate uptake depends on both the activity of type (A) PAOs and that of DNPAOs (Tsuneda et al. 2006). In order to stimulate the growth of phophate accumulating organisms (PAOs) in an activated sludge system, an anaerobic–aerobic sequence and presence of short chain fatty acids in the anaerobic reactor system are required (Wentzel and Ekama 1997). PAOs use the energy that are released from the hydrolysis of inter cellular polyphosphate under anaerobic conditions, to transport volatile fatty acids (especially

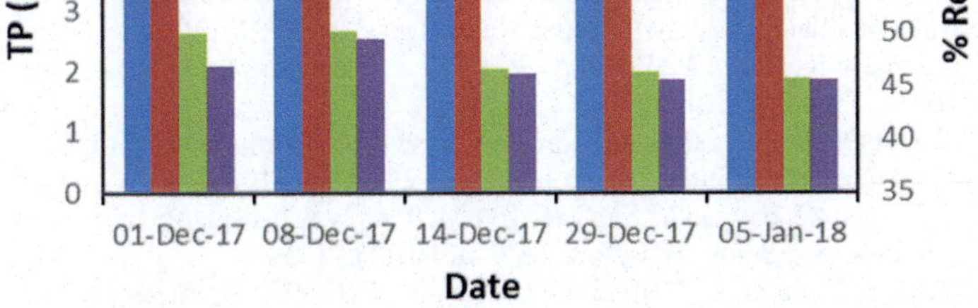

Fig.18.4 Phosphorus removal profile

acetic acid) across membranes of the microbial cell producing polyhydroxybutyrate (PHB). The phosphate are released in anaerobic unit in connection with storage of organic matter. PHB serves as an energy source for cell growth in the aerobic unit as a storage of polyphosphate (Kuba et al. 1994).

In anaerobic system, mostly nitrification and denitification takes place. Hence the denitrification bacteria consumes a portion of substrate before which can be used by phosphorus removing bacteria in the system. Then in the aerobic reactor, the PHB stored disintegrates and degraded for glycogen restoration, thus phosphorus removal takes place. Hence the performance to reduce phosphorus is limited in anaerobic phase (Kuba et al. 1994). So, the removal of phosphorus was therefore limited to aerobic stage.

18.4 Conclusion

Two-stage lab scale anaerobic–aerobic hybrid reactor was able to achieve 83% COD, 70% ammonia nitrogen and 71% phosphorus from greywater when operated at 6 h of HRT. As the removal efficiency gradually increased over time, it is expected to increase further in future days.

Acknowledgement Greywater treatment by two-stage bioreactor experiments was conducted in the present study is a preliminary work done as a part of research project titled "Small scale and sustainable household greywater recycling (S3HWR)". Grant no. IMPRINT 5670 funded by MHRD and MOUD Government of India.

References

Balakrishnan S, Eckenfelder WW (1969) Nitrogen relationships in biological treatment processes—I. Nitrification in the activated sludge process. Water Res 3(1):73–81

Bani-Melhem K, Smith E (2012) Greywater treatment by a continuous process of an electrocoagulation unit and a submerged membrane bioreactor system. Chem Eng J 198:201–210

Do Couto D, Popescu SM, Suc JP, Melinte-Dobrinescu MC, Barhoun N, Gorini C, Jolivet L, Poort J, Jouannic G, AuxietreJL (2014) Lago Mare and the Messinian salinity crisis: evidence from the Alboran Sea (S. Spain). Mar Pet Geol 52:57–76

Elmitwalli TA, Otterpohl R (2007) Anaerobic biodegradability and treatment of grey water in upflow anaerobic sludge blanket (UASB) reactor. Water Res 41(6):1379–1387

Eriksson E, Auffarth K, Henze M, Ledin A (2002) Characteristics of grey wastewater. Urban water 4(1):85–104

Friedler E (2004) Quality of individual domestic greywater streams and its implication for on-site treatment and reuse possibilities. Environ Technol 25(9):997–1008

Ghunmi LA, Zeeman G, Fayyad M, van Lier JB (2010) Grey water treatment in a series anaerobic–aerobic system for irrigation. Bioresour Technol 101(1):41–50

Hernández Leal L, Zeeman G, Temmink H, Buisman C (2007) Characterisation and biological treatment of greywater. Water Sci Technol 56(5):193–200

Hernández Leal L, Temmink H, Zeeman G, Buisman CS (2010) Comparison of three systems for biological greywater treatment. Water 2(2):155–169
Jefferson B, Laine A, Parsons S, Stephenson T, Judd S (2000) Technologies for domestic wastewater recycling. Urban Water 1:285–292. https://doi.org/10.1016/S1462-0758(00)00030-3
Jefferson B, Burgess JE, Pichon A, Harkness J, Judd SJ (2001) Nutrient addition to enhance biological treatment of greywater. Water Res 35(11):2702–2710
Kuba T, Wachtmeister A, Van Loosdrecht MCM, Heijnen JJ (1994) Effect of nitrate on phosphorus release in biological phosphorus removal systems. Water Sci Technol 30(6):263–269
Li F, Wichmann K, Otterpohl R (2009) Review of the technological approaches for grey water treatment and reuses. Sci Total Environ 407(11):3439–3449
Luostarinen S, Sanders W, Kujawa-Roeleveld K, Zeeman G (2007) Effect of temperature on anaerobic treatment of black water in UASB-septic tank systems. Biores Technol 98(5):980–986
Tsuneda S, Ohno T, Soejima K, Hirata A (2006) Simultaneous nitrogen and phosphorus removal using denitrifying phosphate-accumulating organisms in a sequencing batch reactor. Biochem Eng J 27(3):191–196
Wentzel MC, Ekama GA (1997) Principles in the design of single-sludge activated-sludge systems for biological removal of carbon, nitrogen, and phosphorus. Water Environ Res 69(7):1222–1231

Chapter 19
Waste Water Management in Super Thermal Power Stations of NTPC

Sudarsan Chakrabarti, S. Padmapriya, and Anirudh Sood

Abstract Water is the priceless endowment of nature which sustain life for billions of species on mother earth. We must make all efforts to preserve the precious gift of earth for sustainable existence in future. Main sources of natural fresh water supply are rivers, lakes, springs, ground water, rain water wells which are depleting rapidly with the increased pace of industrialization and due to the increasing pressure of burgeoning population on earth. While various water conservation Technologies in industries and particularly in Power Plants have been or are being employed to strive towards attainment of least amount of consumption of fresh water from earth surface, these measures alone are not enough, leaving the scope for something radical and novel practice or business model in entire water management chain for humans and industries including Power industries. Treated and recycled sewage water employing appropriate advanced processes for treatment of various types of waste water e.g. domestic sewage as well as industrial effluents, are also being growingly used now-a-days for various domestic & industrial applications, thereby reducing the pressure on fresh water. The present paper is intended for reading and practicing by modern power plant engineers towards attaining the goal of utmost water economy in Power Plants by adopting the sustainable practice of using recycled and treated municipal/urban sewage water which is otherwise disposed-off as waste water (treated or untreated) eventually finding its way to the river and natural water bodies on earth surface—thus contaminating continuously the pristine sources of water on one hand and indirectly helping the unabated escape of precious water to atmosphere through Cooling Tower and adsorption in solid ash after burning of coal in furnace. The water vapours will probably in course of time escape from earth surface rendering this wonderful compound a rare and scarce thing on earth.

S. Chakrabarti (✉) · S. Padmapriya · A. Sood
Patratu Vidyut Utpadan Nigam Limited (A Subsidiary of NTPC Ltd, Patratu, Dist. Ramgarh, Jharkhand, 829119)/NTPC Ltd, Scope Complex 110003, New Delhi, India
e-mail: sudarsanchakraborty@ntpc.co.in

A. Pandey et al. (eds.), *Climate Impacts on Water Resources in India*, Water Science and Technology Library 95, https://doi.org/10.1007/978-3-030-51427-3_19

19.1 Introduction

Waste water is generated in power plants mainly by way of generation of high TDS blowdown from closed cycle operation of condenser cooling water circuit with cooling tower, wet ash handling, coal water slurry formed as a result of water sprinkling for coal dust suppression in coal handling plant transfer points, crusher house, coal stack-yard, neutralised waste water generation from Demineralisation plant which is producing power cycle make-up water, formation of clarifier sludge from water pre-treatment (PT) plant, PT plant Filter back wash water, UF (ultra-filtration)/RO (reverse osmosis plant) reject water, oil–water mixing in Fuel oil handling area, etc. Ash water overflow from ash dyke, usually at a good distance from power plant and its toe-drain also may find its way to adjoining land if not recirculated and recycled back to power plant for handling of ash generated from coal burning. The typical water accounting/balance diagrams have been studied for two mega power projects—one mega Power Project (having 2 nos. 660 MW units) with water cooled condenser at Tanda-II, India and other mega Power Project Phase-I (having 3 nos. 800 MW units) with air cooled condenser at Patratu, India (Water Balance Diagrams for Tanda-II STPP, Pre-Feasibility Report for Patratu Phase-I STPP). These power projects are having latest state-of-the-art units with super-critical parameters for highest-in-class energy efficiency. While in Tanda-II two number units have common condenser cooling system circulating approximately 136,400 m^3/h at COC (cycle of concentration) of 5.5 in closed cycle with evaporation quantity of 2390 m^3/h and blowdown quantity of 450 m^3/h from its cooling towers, overall plant make-up water quantity from nearby Saryu canal, a natural water body, is 3300 m^3/h. The high TDS blow-down is completely used for various consumptive graded water use in different processes of power plant,most significant of which is ash handling water make-up. It may be seen evaporation in cooling tower is approximately 72.4% whereas blowdown from cooling tower is approximately 13.6% of total plant water make-up. For Patratu phase-I, 3 nos. supercritical units of 800 MW each, in contrast has the total plant water make-up of only 2000 m^3/h at most and most optimum consumption will be of order of 1800 m^3/h, as these units do not have closed cycle cooling water system for condenser cooling on account of their using Air cooled condensers and Dry ash handling system provision. The facts are conspicuous that cooling tower losses are the prime losses of surface water in water cooled power plant.

Keeping the waste water generation sources in context to a large size mega thermal power plant in mind, the major step which has been taken in India in recent times, is implementing the concept of Zero Liquid Discharge (ZLD) in power station which will not only arrest such waste water effluent discharge from power plants but also will completely treat waste water, recycle and reuse it for various process applications based on the requrements of various quality grades of water. NTPC Ltd. India as a responsible corporate citizen has proactively committed itself to the conservation and rejuvenation of water and has taken a number of steps towards waste water management to reduce the water consumption by using 3Rs (Reduce, Recycle and

Reuse) as guiding principle (NTPC Water Policy 2017). A brief gloss-over in what follows will help throw some deeper insights.

NTPC stations have mostly wet Ash handling systems and have installed Ash Water Recirculation System (AWRS) at ash dykes/ponds for recycling the decanted water for ash slurry system make up that will substantially reduce the intake of fresh water make-up and zero spill-over to adjoining natural field. Similarly Toe-Drain Recirculation (TDR) system in ash dyke/pond has been adopted for ensuring zero discharge and zero seepage to adjoining field. HDPE lining are used in Ash-Dyke/Pond, Raw Water Reservoir to arrest leaching of ash water to earth ground water reserve, thereby the loss of water. The journey from lean ash slurry system (containing 25–30% ash by weight) to High concentration ash slurry disposal system (HCSD system containing more than 60% ash by weight) has also been very successful in almost all recent plants since 2008, which drastically cut fresh water consumption in handling both ESP (electrostatic precipitator) fly ash and furnace bottom ash since HCSD maintain a very meagre water to ash ratio by weight compared to that in conventional lean slurry system, an old practice. In complete Dry furnace bottom ash system and Dry fly ash (ESP hopper fine ash) transportation system there is no usage of water at all as dry fly ash are fluidized thru compressed air and conveyed thru pipelines and crushed furnace-bottom ash are conveyed thru metallic/vulcanised rubber conveyor respectively to near-by ash silos at plant outskirt from where these ash are taken away by various industries like cement, ash-aggregate, road/highway making, brick manufacturing, other construction industries etc. depending on fineness requirements, thru bulkers/closed trucks, which is a win–win situation for both power generators and ash consumers. However, some amount of dust suppression water (consumptive) would be needed which will be typically of the order of 150–250 m^3/h for a typical 1500 MW power station. NTPC has also adopted Dry bottom ash handling first time in its Patratu mega power project mentioned above which is an example of complete dry ash handling system in true sense.

High Cycles of Concentration (COC) operation in Circulating Water (condenser cooling) System is another endeavour by NTPC to reduce the water consumption and thereby generation of waste water of CT blow-down quality by targeting its coal based stations with COC beyond 5.5 while for its Gas stations, high COC operation even upto 10 is being attempted. This calls for rigorous chemical treatment regime for condenser cooling water closed cycle involving condensers, pumps and cooling towers. Recent advances in non-chemical techniques like galvanic dispersion of precipitating dissolved solids at condenser end, electrode insertions employing electro-chemical methods at Cooling tower basin end have also been reported from some power stations with high COC.

NTPC has also been a pioneer in India in adopting Air Cooled Condenser (ACC) based mega super-critical coal based units in North Karanpura (3 × 660 MW) and Patratu (3 × 800 MW). Both the projects are under construction and North Karanpura is in advanced stage. Air Cooled Condenser will not require any water for cooling the power cycle steam for power generation in condenser. It will need only atmospheric air to cool the steam in condenser by sacrificing cycle efficiency marginally, however with Turbine inlet temperature (TIT) at or above 600C with improved metallurgy of

Boiler Tubes, Main Steam lines, reheat lines and Turbine high pressure modules, such marginal lowering of efficiency can be well compensated with the gain in cycle efficiency due to higher TIT.

High efficiency Drift-Eliminator in cooling towers of 0.001% (max. drift loss) in place of 0.05% drift loss are being targeted recent projects of NTPC. NTPC has also recently embarked upon the practice of using hot side blow down from condenser cooling water closed circuit instead of cold side, which will reduce the evaporation loss from cooling tower thereby conserving the fresh water. This is easily established from the famous relationship between evaporative loss, blow-down loss, drift loss and cycle of concentration in close cooling circuit employing cooling towers.

Recycle and Reuse of plant STP (sewage treatment plant) water is implemented in horticulture, gardening and upkeep of various facilities at stations. Usually in NTPC scenario Moving bed bio-reactors (MBBR) having high COD and BOD removal capability are used for township STPs.

Adoption of efficient effluent management scheme, consisting of collection, treatment through Effluent treatment plant (ETP), recycling/reuse of treated effluents in NTPC stations are targeted to achieve Zero Liquid Discharge (ZLD). Water meters for consumptive measurement and monitoring at strategic points in water distribution network in NTPC plants are being installed for taking pre-emptive action for efficient use of water. Usually, plant level effluent treatment plants of capacity approximately 150–350 m^3/h with or without RO plant depending on ionic loading are employed in NTPC to treat the collected effluents and recycle back to various process use of varied water quality grade.

Second major step recently taken by Government Of India through its gazette notification on the revised (Electricity) Tariff policy dated 28th January, 2016 has made it mandatory for all power plants to use treated sewage water from Municipal Sewage Treatment Plants if they exist within 50 km radius of the power plants as a prime source of supply of make-up water using existing fresh raw water from natural water bodies only as a operational back-up source for uninterrupted generation of power-indeed a noble concept (Conference of Power 2017). The above two measures, viz. zero liqid discharge and Treated sewage water use in power plant together will help cut down use of fresh pristine raw water from natural sources and thereby free up natural water for drinking water and irrigation purpose. NTPC has taken a lead for making use of treated sewage water from Municipality wherever available near vicinity of power plants resulting in saving of fresh water and currently making attempts for its plants like Dadri (2 nos. units each of 490 MW), Patratu (3 nos. units each of 800 MW), Solapur (2 nos. each of 660 MW), Mouda (2 nos. units each of 660 MW and 2 nos. units each of 500 MW), Meja (2 nos. each of 660 MW), etc.

The above two approaches are the classic examples of waste water management in the context of power plant water use in India in recent days.

19.2 Zero Liquid Discharge in Power Plants-NTPC Experience

To achieve Zero Liquid Discharge (ZLD), separation of storm water drain from plant drain and treating the plant drain for recycling/reuse in the first phase, is under implementation in various NTPC projects. For implementing the second phase, namely treatment/clarification process and complete recycle/reuse strategy for coal based plant & gas based plant shall be different.

In coal based plants, after segregating plant effluent water from storm water, the effluents collected in common basin can be treated and reused to a large extent in Ash handling plant (AHP), Coal handling plant (CHP), FGD (flue gas de-sulphurisation plant make-up water), Service water etc. This will be achieved by a combination of measures like increased COC (Cycle of Concentration) of CW system, Optimized ash water ratio, AWRS (Ash Water Recirculation System-70% ash waters recycled back from dykes (CEA report 2012), ash dyke toe drain recycle and plant township STP (sewage treatment plant) water recycle for horticulture etc. (Effluent collection and disposal scheme Darlipalli STPP (2 × 800 MW).

More is the COC, less is the blow down from condenser closed cooling circuit (CEA report 2012) hence less burden on Zero liquid discharge target. It may be noted that blow down water from CW system is used for various service applications like FGD make up, area wash, ash handling supplementary water, Fly ash conditioning in dry ash disposal silos. Less ash water ratio similarly need less handling water. Present day NTPC is handling ash in most of its stations in High Concentration slurry transportation mode (HCSD)-a recent technology requiring very less handling water compared to conventional lean slurry mode as explained earlier. Clarifier sludge, DM plant neutralisation waste water, CPU waste water, Boiler and ESP area wash water laden with ash is pumped back to Ash water sump or effluent treatment plant sump (as applicable). Oil is skimmed off from oil water mix in drain pit and recycled back to drain oil tank of FO (fuel oil) plant. For collection of run-off water from coal stack yard after dust suppression, there are coal slurry settling ponds at the end of stack yard which constantly store and help in recycling black water back for dust suppression purpose thru a set of dedicated pumps on a stand-alone basis.

For gas based plants, ZLD will be achieved through measures like higher COC than that for coal based plants (at times with molybdate salt addition apart from usual corrosion inhibitors, scale inhibitors, ant-scale dispersants bio fouling inhibitors, bio-dispersants and aqueous sulphuric acid solution in CW circuit to prevent precipitation of dissolved solids as well as prevent chloride attack on stainless steel in condenser tubes which cannot tolerate more than 300 ppm chloride concentration happening at very high COC), very low blow-down from CW system, use of effluent in housekeeping, horticulture in plant and township, leaving balance water for natural evaporation in an already existing reservoir or a new one created in available areas. Evaporators, concentrators and crystallizers also are employed in certain exceptional cases where there is a severe space constraint particularly in a compact gas based thermal power plant layout with advanced class Gas turbine machines and that too after weighing the option whether high efficiency Reverse Osmosis (HERO) plant could be employed upstream of Evaporator/concentrator/crystalliser train to reduce ionic load in waste water and thereby economise on both capital cost of evaporators (single effect or multi-effect), concentrators and crystallizers and their concomitant power consumptions (Chakrabarti 2016). Gas based plants in this respect is in disadvantage because there is no any available sink like ash handling plant of coal based units, to require treated waste water for its handling and also it does not have a robust FGD plant to require treated waste water make up.

19.3 Use of Treated Sewage Water in Power Plants and the Genesis of the Concept

19.3.1 Need for Treatment of Sewage Water and Converting It to Industry Usable Quality

Treatment of sewage is absolutly necessary for making our rivers & water boidies clean. Partially or untreated sewage is the single major source for deterioation of surface or ground water quality. It contributes 70% of the pollution load to streams or water bodies of India. The untreated wastewater causes severe soil and water pollution by coming in contact with ground, ground water, rivers and natural drainage system causing pollution in downstream catchment area (Conference of Power 2017). The rivers, lakes, streams, canals, etc. are used by human being for fulfillment of their daily needs. The consumption of polltuted water has adverse impact on human health as well as on aquatic life. While the total estimated sewage generation in the country is 62,000 MLD (approximately), only 23,277 MLD of treatment capacity have been developed as per Central Pollution Control Board, Govt.of India estimate in year 2015 (Conference of Power 2017). Together, India's largest cities generate more than 38,254 MLD of sewage as per an estimate. Of this, less than 30% collected sewage undergo treatment. Furthur, it is noteworthy that 80% of household water is released as wastewater. With the increasing population and growth of per capita water consumption over the years,sewage treatment has already assumed the proportions of full-fledged industry in its own right and has thrown open a big challange to the water authorities/municipal authorities/urban local bodies both in terms of adequate treatment capacity creation and sustainable business model.

19.3.2 Need of Recycle and Reuse of Treated Sewage Water

The large volume of sewage offers tremendous potential for municipal bodies/Urban Local Bodies to recycle water for industrial and/or residential use and reduce their reliance on bulk fresh water sources. On the other hand, it is seen that with the increasing population and rapid urbanisation, there is more and more pressure on existing fresh surface water sources like rivers/lakes/reservoirs/canals etc. and even ground water acquifers for meeting the needs of drinking as well as irrigation water (Conference of Power 2017). Thus, it has become imperative to conserve the fresh waters sources to the utmost extent for priority use of drinking and irrigation water as a principle, particularly in water-stressed and water-stretched areas of India (Conference of Power 2017). Keeping this in mind, the water policy of nation and state has set the priority first for potable use followed by agricultural use and then industrial use as far as usage of fresh water is concerned.

19.3.3 Use of Treated Sewage Water in Power Plants

In view of given scenario as experienced above, the idea of industrial use of treated sewage water particularly by power plants has come into being, which will serve the noble cause of freeing up equivalent quantity of fresh surface water for drinking and agricultural purpose. In order to conserve fresh water and meet the requirement of households and agriculture, Government of India has included a very forward looking provision in the revised Tariff Policy dated 28th January, 2016. It has been made mandatory to use treated sewage water by power stations wherever power plants are at a distance up to 50 km from the STPs.

19.3.4 Mapping of Functional STPs with the Power Plants in the Country (Including NTPC Power Plants)

Since the notification of the revised tariff policy, the activity of mapping for power station within the vicinity of 50 km from STPs across the country has started in all seriousness and the central/state power utilities are already engaged in the process of active negotiation with the municipal/urban local bodies for supply of treated sewage water for their power stations. The central utility NTPC has already taken a lead role in this regard and has identified five (5) coal based power stations namely Solapur, Mouda, Meja, Dadri and up-coming Patratu power station for use of treated sewage water to cater to the makeup water requirement. Similarly, for other NTPC stations, active negotiations with respective municipality and parallel process of finalisation of detail scheme and DPR are going on. As regards state power utilities, Mahagenco has already tied up with Nagpur Municipality Corporation through a long term agreement for supply of treated sewage water to their Koradi Power station in Maharashtra (Conference of Power 2017). However, looking at the present status of utilisation of treated sewage water in the country, combined effort from both central as well as state power utilities including private sector utilities together with the all-out efforts from municipal/urban local bodies towards creation of new

STPs and up-gradation/modernization of existing STPs are the need of the hour. For giving adequate impetus to present policy of treated sewage water utilisation by industry including power plants, Government of India has brought out various funding/financing schemes like "Amrut Yojna" to assist the state governments in setting up new STP facilities which will invigorate the financially constrained municipal/urban bodies and formed National Mission for Clean Ganga under Namami Ganga Program with a mandate to set up STPs in various towns situated in the Ganga River basin. Mapping of STPs and power plants along the course of Ganga river has already been set into motion (Conference of Power 2017).

19.3.5 Quality and Quantity of Treated Sewage Water

As per the CPCB report 2015, out of the 816 installed STPs inventory, only 522 STPs were operational as on March 2015. The capacity utilisation was 18,883 MLD out of 23,277 MLD. It means that even the existing STPs are either defunct or non-performing. Some of the STPs are either having poor/dilapidated collection network or have inadequate O&M system. This trend needs to be arrested if treated sewage water is to be used by down end industries as per their requirements. Of the functional STPs, sizable number do not maintain the standard for effluent discharge (STP outlet quality 2017) for sewage treatment plants and consequently poor quality effluents from STPs join the river stream or fresh water bodies around thus defeating the basic objective of installing a STP.

It is therefore imperative that the existing as well as proposed upcoming STPs must maintain the environment discharge quality standards/norms which will also facilitate ascertaining the quantum of further treatment to be done to improve the quality of treated water for use of the same in the receiving industry. Also since quantity of sewage generation varies with the time of the day, the treated sewage water quantity also will vary over the day and hence sufficient storage facility needs to be created at STP end to cater to such fluctuation and maintain a uniform supply round the day on 24 × 7 basis (Conference of Power 2017).

Based on the data available, it is necessary to provide tertiary treatment plant in addition to secondary treatment being done at STPs, for achieving required power plant water quality. As such, additional investment needs to be made for the tertiary treatment plant and treated water conveying pipe line from STPs boundary up-to the door step of power plants (Conference of Power 2017). Typical quality of make-up water in power plant after tertiary treatment is akin to that of clarified water with MPN of Fecal coliform as nil and total nitrogen (as N) less than 10 ppm, total sulphur (as S) less than 0.1 ppm, BOD less than 5 ppm, COD less than 10 ppm, Ammonia as NH3 less than 2 ppm, Phosphate as PO4 less than 1 ppm, TDS less than 400 ppm (NTPC MoU for Dadri Plant 2016).

19.3.6 Business Model for CAPEX and OPEX Towards Usage of Treated Sewage Water by Power Plants

Business model to be adopted by municipalities and power utilities for usages of treated sewage water can be based on either PPP (private public partnership) model with full or partial investment from PPP partner of municipalities/utilities which are public bodies, OR, on EPC (engineering, procurement and construction) model with full investment made by utility alone or municipality alone or both (Conference of Power 2017). However, issues regarding CERC (central electricity regulatory commission) and SERC (state electricity regulatory commission) approval for such EPC investment under project cost of utility needs to be addressed, particularly for existing or old power stations where project cost has already been capitalized. Federal/central ministry of Power, Govt. of India on 4th March 2020 has published typical MOU guidelines between Municipal body/urban local body and Power utility for reuse of sewage water discharged by Municipalities or ULBs, which envisages entire capital cost of cross country pipeline and pumping system at Municipality STP end as well as tertiary treatment plant at power plant end to be borne by Power utility and same will be capitalized as project cost of power station to be charged as fixed cost portion of electricity tariff. Recurring cost for pipeline and pumping O&M is also to be borne by Power utility. Municipality /ULBs will bear both capital cost of sewage collection network, storage and STPs upto secondary treatment and also their O&M cost (MOP 2016). MahaGenco's Koradi thermal power station, Maharashtra, India use treated sewage water from Nagpur municipal STP which is about 40 km away from power station in its condenser cooling system basing their long-term business model on PPP concept and same is running successfully without any problem. Process of selection of private partner by public utility should however be transparent based on bidders' rich experience and preferably on discovered price basis with tariff quoted per cubic metre.

19.3.7 NTPC Endeavour for Use of Treated Sewage Water in Its Plants

Treated sewage water will be used for CW system makeup in place of conventional clarified water makeup. Outlet quality of treated water from STPS need to be further refined through some filtration process viz. media filters or UF systems before discharging into CW channel for closed cycle condenser make up usually with COC 5 or more.

NTPC has identified five projects viz. Dadri, Patratu, Solapur, Meja and Mouda where STPs exist/are being constructed within 50 km radius of the power plants. For Dadri power station, NTPC has already signed in-principle MOU with NOIDA authority (Municipal/civic body) for utilization of 80 MLD treated sewage water from NOIDA STPs. Further legal agreement and start of survey work of pipe corridor and preparation of DPR (detailed project report) by NOIDA/UP Jal Nigam (NOIDA's execution arm) is under way. For Patratu power project, NTPC through its JV Company PVUNL has approached Ranchi MC (municipal commissioner) and secretary water resource deptt. in advance and consent was obtained for supply of 37 MLD treated sewage water from Ranchi STP for use in FGD make up, Dry Ash conditioning, Auxiliary equipment cooling water makeup, CHP (coal handling plant) dust suppression and various service applications. Very soon legally binding agreement is going to be signed between power utility Patratu Vidyut Utpadan Nigam Limited (PVUNL) and Ranchi Municipality for supply of this treated sewage water for Patratu phase-1 super thermal power project comprising three numbers of 800 MW coal based units mentioned earlier (Mandatory use of treated sewage water by the Thermal Power Plants 2020). For Solapur, Mouda & Meja projects, discussions are on with respective MCs for supply of treated sewage water on the same lines. For Solapur 52 MLD treated sewage water is being tied up whereas for Mouda and Meja agreed MOU envisage 150 MLD and 75 MLD of treated sewage water supply respectively from Solapur Municipal Corp, Nagpur Municipal Corp and Allahabad Municipal Corp. (Naini). Typical flow diagram and block boundary depicting the scheme is given in Fig. 19.1.

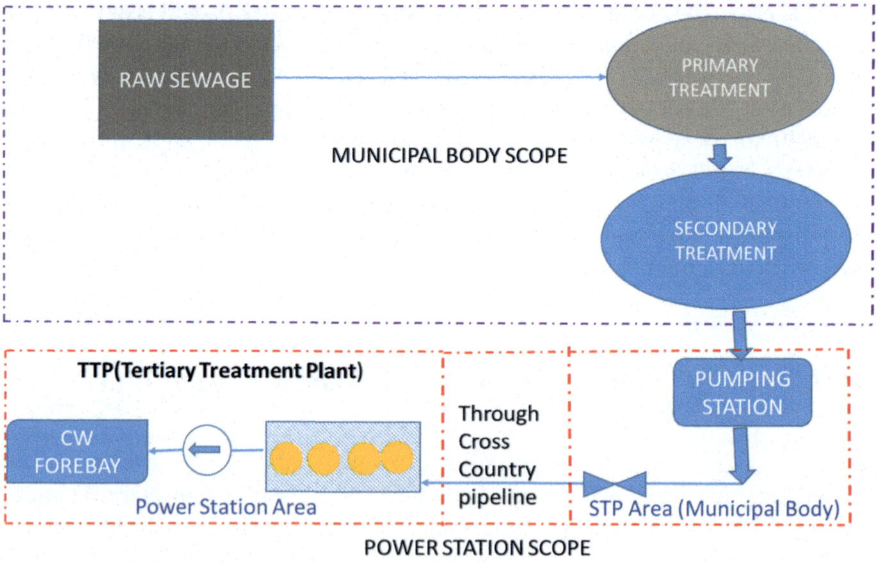

Fig. 19.1 Process flow diagram

19.4 Conclusion

Fossil fuel based Power generation should aim for latest plant fresh water consumption norm of 2.5 m^3/h/MW (as per gazette notification dated 07.12.2015 of Govt. Of India) though it is a difficult target to achieve for water cooled condenser based power generating units employing closed cycle cooling with Cooling towers further aggravated by new additional requirement of Make-Up water for FGD and DENOX systems to be introduced for stricter ambient quality for controlling SOX, NOX and Hg. Coal firing units using Direct Air Cooled Condenser cooling method and Dry bottom ash and Dry fly ash collection and transportation system have the potential to cut down fresh plant make up water consumption to only one third of that for closed cycle water cooled condenser based units that run on wet ash handling system. This fact exhorts technologists and engineers to go for technology shift in favour of air cooled condensers and dry bottom and dry fly ash system, if less waste water generation is the objective. But air cooled condenser (ACC) overuse will lead to average ambient temperature tending to go high as exhaust air temperature leaving ACC will attain temperature not less than 70 °C in peak summer in tropical country like India. Use of advanced technological practices like Adiabatic cooling and Dephlegmeter tubes in ACC can lower outlet air temperature of ACC in summer albeit with marginal increase in usage of make-up water.

The twin concepts of zero liquid discharge and treated sewage water use as make up to various closed cycles applications can be adopted by thermal power stations in India wherever possible to conserve surface water and act as a model for other industries to follow the suit.

With the present cost (as estimated for financial year April, 2012–March, 2013 in filing tariff petition of NTPC to Central Electricity Regulatory Commission) of fresh water varying (depending on location and state of Union of India) from approximately Rs. 0.8/m^3 to approximately Rs. 14.84/m^3 (NTPC Water Charges 2015) which is likely to go up further in coming days to not less than approximately Rs. 40/m^3 as estimated by planners, use of treated sewage water will definitely weigh very favorably for use in Power Plant.

References

Chakrabarti S (2016) Water and energy security: an approach towards sustainable economic development, presented in a National Seminarat TERI, India (May 2016)

Conference of Power, renewable energy and mines minister of states and UTs. (2017). (https://powermin.gov.in/sites/default/files/webform/notices/Agenda_Booklet_for_PMC_on_3_to_4th_May_0.pdf)

Effluent collection and disposal scheme Darlipalli STPP (2 × 800 MW) and P&ID diagram for effluent collection and disposal—SG & BOP area Darlipalli super thermal power Plant (2 × 800 MW). https://www.ntpc.co.in/download/darlipali-stpp-stage-i-2x800-mw

https://energy.economictimes.indiatimes.com/news/power/ntpc-inks-pact-with-noida-authority-to-use-treated-sewage-water-at-dadri-plant/64603885

Mandatory use of treated sewage water by the thermal power plants (2020) GOI Gazette notification published on dated 4 Mar 2020 https://www.cea.nic.in/reports/others/thermal/tppd/treatment%20sewage%20water.pdf

Minimisation of water requirement in coal based thermal power stations. CEA report published in 12 Jan. http://www.indiaenvironmentportal.org.in/content/355428/report-

Ministry of Power (MOP) (2016) Mandatory use of treated sewage water by the thermal power plants as per the provisions of the tariff policy 2016. https://www.cea.nic.in/reports/others/thermal/tppd/treatment%20sewage%20water.pdf

NTPC Water charges (2015) NTPC affidavit dated 30 Apr 2014 to CERC. https://www.cercind.gov.in/2015/orders/SO121.pdf

NTPC Water Policy (2017) https://www.ntpc.co.in/sites/default/files/downloads/WaterPolicy2017.pdf

Pre-Feasibility report for Patratu phase-I STPP (3 × 800 MW), Jharkhand (India) with air cooled condenser. https://environmentclearance.nic.in/writereaddata/Online/TOR/0_0_23_Oct_2015_1017175831FinalPreFRPatratuSTPPPh1.pdf

Sewerage Treatment Plant (STP) outlet quality (2017) GOI Gazette notification published on dated 13 Oct 2017. https://www.indiaenvironmentportal.org.in/files/file/Sewage%20Treatment%20Plants_2.pdf

Water balance diagrams for Tanda-IISTPP (2 × 660 MW), Uttar Pradesh (India) with water cooled condenser. https://www.cea.nic.in/reports/others/thermal/tetd/min_of%20water_coal_power.pdf

Chapter 20
Removal of Dyes and Iron Using Eco-Friendly Adsorbents

S. L. Devika, P. Nimitha, Venkatesh Muganur, and S. Shrihari

Abstract This paper examines the efficiency of lemon peel adsorbents for removal of dyes, and cassava peel biochar for removal of iron from synthetic wastewater under laboratory conditions. These are the eco-friendly and economical adsorbents. Experiments were conducted to find out the effects of initial concentration and pH. The biochar was prepared by pyrolysis of cassava peel at 500^0 C for 4 h. The removal of total iron was found to be dependent on adsorbate initial concentration and pH. Percentage removal increased with increase in initial concentration of adsorbate. Optimum pH for removal was found to be 5. Optimum removal efficiency of 89.84% by cassava peel biochar on iron adsorption was obtained from the batch study. The removal efficiency of dyes was found to be dependent on pH for lemon peel adsorbent. The optimum pH was obtained to be 2 and 4 for Acid Violet 17 and Remazol Brilliant Blue R respectively. The optimum initial dye concentration was 40 mg/L. The efficiency of dye removal increased with the increase in initial dye concentrations, and then decreased after the optimum value. An optimum Iron removal efficiency of 89.84% by cassava peel biochar was obtained from the batch study, but decreased in column studies (73.09%).

20.1 Introduction

The perception of public on water quality is significantly influenced by the color of water. The removal of color from wastewater is frequently more critical than the removal of the soluble colorless organic substances (Buthelezi et al. 2012). Removing dye from the textile industry wastewater is very costly (Malik et al. 2013). Dye removal from textile industry effluents is of great concern in many countries globally for both environmental and water reuse concerns (Choy et al. 1999). There are several techniques for the removal of dye which are cost effective, but due to the low biodegradability are not very useful in treating dye wastewaters. Consequently, they are normally dealt with by either physical or chemical methods (Saiful et al.

S. L. Devika · P. Nimitha · V. Muganur · S. Shrihari (✉)
N.I.T.K. Surathkal, Mangaluru 575025, India
e-mail: shrihari@nitk.edu.in

A. Pandey et al. (eds.), *Climate Impacts on Water Resources in India*, Water Science and Technology Library 95, https://doi.org/10.1007/978-3-030-51427-3_20

2005). It was reported in literature that adsorption is the best equilibrium process and effective in the removal of contaminants from the wastewater (Yagub et al. 2014). Because of its high capacity to adsorb dyes, the activated carbon is the most effective adsorbents. But due to the high cost of activated carbon other low-cost adsorbents are being tried (Bhatnagar and Minocha 2006). Many scientists have tried to use economical conventional adsorbents for the treatment of industrial effluents (Pollard et al. 1992). Adsorbents used are agricultural solid wastes such as coir pith (Namasivayam and Kavitha 2002), banana pith (Namasivayam and Kanchana 1992), coconut husk (Low and Lee 1990) and orange peel (Sivaraj et al. 2001), etc.

In this study, removal of two synthetic dyes viz. Acid Violet 17 and Remazol Brilliant Blue R, commonly used in textile coloring, has been studied. Acid violet 17 is an acidic dye (anionic dye) used commonly in textile industries, which is of violet colour. It gives bright blue purple colour in aqueous phase. The dye is very harmful and is carcinogenic and mutagenic. This dye is mainly used for the colouring of wool, and leather. Remazol Brilliant Blue R (RBBR) is an anthraquinone dye (azo dye), which is used in textile industries, which is of peacock blue colour. It gives peacock blue colour in aqueous phase. RBBR is a harmful dye and it can harm aquatic life and also vegetative life if the contaminated water is used for irrigation.

Velmurugan et al. (2011) studied the removal of dye from aqueous solution using economical adsorbents such as orange peel, neem leaves and banana peel for the removal of Methylene Blue. The isotherm evaluation showed that orange peel prepared with nominal treatment was very effective than the Neem leaves and Banana Peel. The research proved that even though activated carbon was the most effective adsorbent, other low-cost adsorbents may be used for color elimination as the final preference of the adsorbent was a matter of economics. The Acid Violet 17 removal from aqueous solutions by adsorption using activated carbon was investigated using pistachio nut shell (PNS), an agricultural waste biomass (Vijayalakshmi et al. 2010), and sunflower seed hull (Thinakaran et al. 2008). The Remzol Brilliant Blue Reactive (RBBR) dye removal from effluent was studied using watermelon rinds activated carbon adsorbent (Ahmad et al. 2014), biosorption on the macrophyte *Salvinia natans* (Pelosi et al. 2014), pirina (olive oil factory waste) pretreated with commercial activated carbon and nitric acid (Dagdelen et al. 2014), orange peel adsorbent (Namasivayam et al. 1996) with promising results. The use of lemon peels for adsorption was used for the removal of cobalt ions on raw and alkali-treated lemon peels (Singh and Shukla 2015), for zinc from waste water (Rajoriya and Kaur 2014), and for the removal of cutting oil from wastewater (Tembhurkar and Deshpande 2012). The investigation result revealed that lemon peel adsorbent could be effectively utilized as low-cost adsorbent for the removal of dye from effluents.

Hegazi (2013) studied the feasibility of agricultural and industrial wastes such as rice husk and fly ash for the removal of heavy metals viz Iron, Lead, Nickel, Cadmium and Copper. He reported that rice husk was effective in the simultaneous removal of Iron, Lead and Nickel. With the increase in the amount of absorbent concentration, removal of Iron using rice husk increased from 68.59 to 99.25%, while removal of Iron using fly Ash varied from 46.18 to 86.757%. Thermally and chemically modified banana peels were used for removing total iron by Yousaf and

Sajjad (2015). The removal of iron by adsorption using oil palm factory ash was done by Isa et al. (2004). Oil palm industry ash was capable of removing over 99% of iron from aqueous solution.

20.2 Experimental Methods

20.2.1 Adsorbent Preparation

Lemon peels were collected from the fruit stall at Surathkal, a region in Karnataka. The collected lemon peels were washed two times in tap water and two times in distilled water. Then they were cut into small pieces and oven dried at 100 °C for 24 h. Then they were powdered using a grinder. The powdered lemon peels were sieved using sieves of sizes, >0.425, 0.425–0.355, 0.355–0.212, 0.212–0.106 and <0.106 mm and were used as the adsorbent. And the lemon peel powder was stored in air tight containers for further use.

Cassava peels were collected from small scale food Industry, Kasargod, Kerala. Peels were washed several times to eliminate dust and other impurities with tap water followed by distilled water. Cleaned peels were cut in to small pieces and oven dried for 24 h at 100 0 °C to remove the moisture content. The washed and oven dried peels were crushed in to powder and filled in a silica crucible then introduced to a muffle furnace. The sample was pressed a little and closed with an aluminum foil and a lid so as to achieve maximum oxygen limited condition. The powder was pyrolyzed at a temperature of 500 °C for 4 h. This process was repeated until substantial amount of biochar was obtained. The resulting biochar was sieved to five different sizes using standard sieves.

20.2.2 Characterization of Adsorbents

The characterization of the adsorbents was done by proximate analysis. As per American Society for Testing and Materials (ASTM) D4979. The results are presented in Table 20.1.

20.2.3 Preparation and Characterization of Adsorbate

The stock solution of 1000 mg/L of the Acid Violet 17 dye was prepared by dissolving 1000 mg of acid violet dye powder in 1000 mL of distilled water. Similarly, the stock solution of 1000 mg/L of the Remazol Brilliant Blue R dye was prepared by dissolving 1000 mg of acid violet dye powder in 1000 mL of distilled water. The dyes were

Table 20.1 Characteristics of proximate analysis

Sl. No.	Parameter	Lemon peel, percentage	Biochar, percentage
1	Moisture content	7.28	1.992
2	Ash content	6.32	14.4
3	Volatile matter	34.52	8.61
4	Fixed carbon content	51.88	74.99
5	pH	6.4	10.2

scanned for its wavelength using the UV–VIS spectrophotometer. The stock solution of 100 mg/L of iron was prepared by dissolving 49.78 mg of analytical grade ferrous sulphate ($FeSO_4 \cdot 7H_2O$) in 100 ml of distilled water. The required concentration of Adsorbate solution for the study were prepared by diluting the stock solution.

20.2.4 Feasibility of Lemon Peel Powder (Adsorbent)

The adsorption of Acid Violet 17 and Remazol Brilliant Blue R on lemon peel powder was studied using the batch studies. The 50 mL of 20 ppm dye solution of both the dyes which were already prepared were taken in the BOD bottles and 600 mg of the adsorbent (lemon peel powder of size 0.355–0.212 mm) was added. The solutions of both the dyes with adsorbent were kept in a shaker at 150 rpm for 100 min. After 100 min, the bottles were taken out and the solution was centrifuged at 10,000 rpm for 20 min to separate the particles from solution. Then both dye solutions were analyzed in a UV–VIS spectrophotometer. Similarly, for Iron, the adsorption of iron on cassava peel biochar was studied using the batch studies. The efficiency of iron removal with different initial adsorbate concentration and pH were investigated. BOD bottles (300 ml capacity) containing a fixed adsorbate dosage with 50 ml of iron solution corrected to various pH were placed in a rotary shaker for a specified period of time. The bottles were then removed and the samples were filtered through Whatman filter paper NO 42. Final Iron concentration was determined using 1, 10 phenanthroline method in a UV spectrophotometer.

20.3 Results and Discussion

20.3.1 Effect of Initial Adsorbate Concentration

The initial adsorbate concentration has great effect on its removal by adsorption from the aqueous solution. The effect on the removal of Acid Violet 17 and Remazol

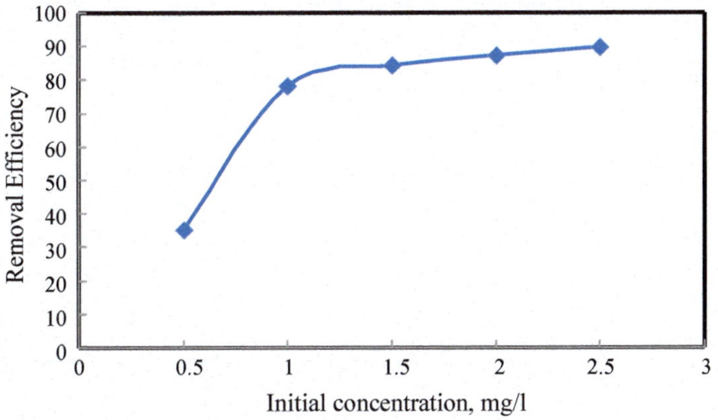

Fig. 20.1 Effect of initial iron concentration

Brilliant blue R from the aqueous solution by adsorption on lemon peel powder by varying its initial dye concentration (10, 20, 30, 40, and 50 mg/L) were studied at equilibrium conditions. The removal efficiencies of the dyes have increased as the initial dye concentration increased from 10–40 mg/L. This trend may be due to the fact that at higher concentrations of dyes, there might be high driving forces for mass transfer. Further, at higher adsorbent dosage, the active sites on the adsorbent are surrounded by a greater number of molecules which could lead to efficient adsorption. After 40 mg/L, the removal efficiency decreased considerably. It was probably due to the active sites of the adsorbent had attained saturation and there were no more active sites available for adsorption.

The Initial adsorbate concentrations was varied from 0.5 to 2.5 mg/l. Figure 20.1 shows the graph of removal efficiency with initial adsorbate concentration, and it was observed from the trend that the removal efficiency exhibited an increasing trend with increase in initial adsorbate concentration of iron. This trend could be due to the fact that at high concentrations there are higher driving forces for mass transfer. In addition to that, with the increase in metal ion concentration all the active sites in the adsorbent activate as many iron molecules were available for adsorption or might be due to the higher number of iron molecules surrounding the adsorbent which could lead to more efficient adsorption (Fig. 20.2).

20.3.2 Effect of pH

The pH of the solution plays important role in iron adsorption. The number of active sites present is governed by the pH of the solution. Figure 20.3 shows the effect of pH on the Acid Violet 17 and Remazol Brilliant Blue R dye solutions adsorption on lemon peel sample at equilibrium conditions. The pH of the solutions was adjusted by using 1N H_2SO_4 and 1N NaOH solutions. The pH was varied from 4 to 7. The

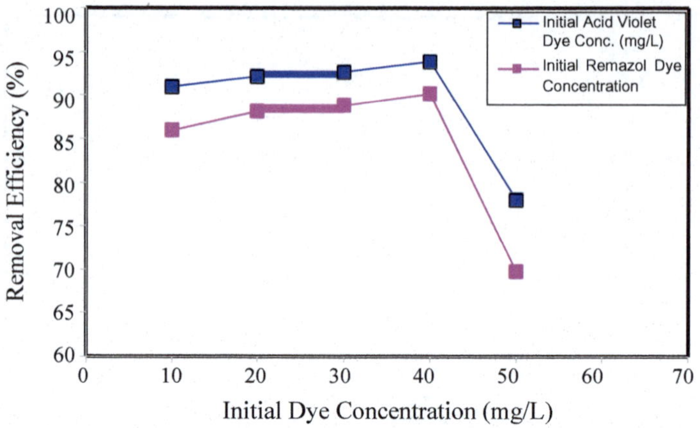

Fig. 20.2 Effect of initial dye concentration and contact times on the removal of Acid Violet 17 and Remazol Brilliant Blue R by lemon peel powder as adsorbent

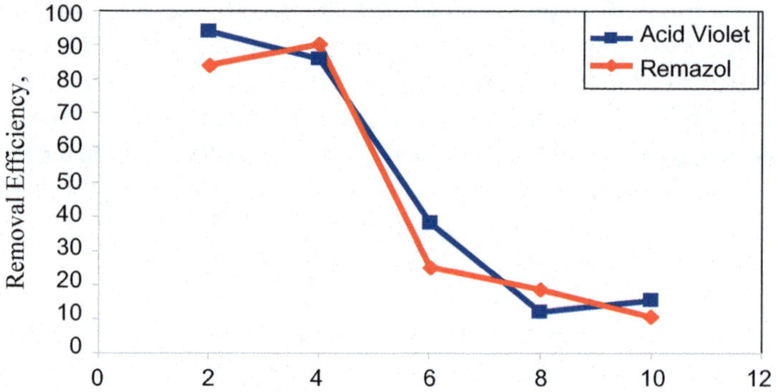

Fig. 20.3 Effect of pH on Acid Violet 17 and Remazol Brilliant Blue R adsorption on lemon peel powder as adsorbent

graph showed that at lower pH range 2–4, the removal efficiency is maximum and the removal efficiency is low in the alkaline pH range. The higher adsorption of dye at lower pH (very acidic medium) may be because of the interactions between the negatively charged dyes (anions) with surface functional groups present on lemon peels. In addition, as the pH of the solution decreases, the number of H+ ions in the system could increase, thus the number of positively charged sites could increase and the adsorption of Acid Violet 17, (anionic dye) might take place due to electrostatic attraction. For the Remazol Brilliant Blue R dye, as the pH of the solution decreases, the concentration of H+ ions in the system could increase, thus the number of positively charged sites could increase and thus negatively charged ions of the Remazol Brilliant Blue R (anthraquinone dye) could adsorb on to the positively charged sites of lemon peels due to electrostatic attraction. Lower adsorption efficiency of Acid

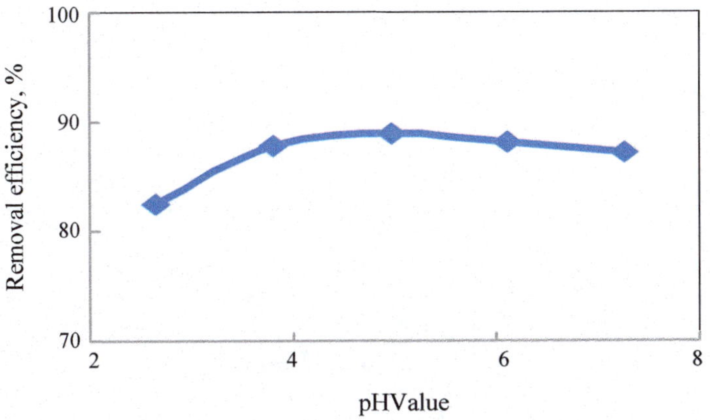

Fig. 20.4 Effect of pH on removal iron using cassava peel

violet 17 and Remazol Brilliant Blue R, at higher pH (Alkaline) is provable due to the presence of excess anions (OH−) ions interfering with the dye anions for the adsorption sites on the adsorbent, and forms of soluble hydroxyl complexes.

Figure 20.4 shows the effect of pH on iron adsorption by cassava peel biochar. The effect of pH was studied between 3 and 7. At lower pH values the removal efficiency of iron was low and with increase in pH of the solution the removal efficiency also increased. The optimal value was observed at pH 5. Beyond this optimum pH the removal efficiency decreased and this may be due to formation of soluble hydroxyl complexes.

20.3.3 Column Studies

The fixed bed column experiment was carried out to find the adsorption capacity of cassava peel biochar with continuous flow movement. The borosilicate glass tube column of 2.5 cm inner diameter and 50 cm depth was used for this purpose. Adsorption column studies were done with synthetic iron solution of 2.5 mg/l concentration at pH 5 at fixed flow rate of 1.5 ml/min. Figure 20.5 shows the Breakthrough Curve plotted between C/Co and time elapsed. With the increase in the flow time, the value of C/Co was found to increase. The increase in C/Co indicate that the concentration of iron has increased in the effluent with time. The adsorption column achieved breakpoint at 300 min. The breakthrough curve shows that there is no considerable increase in C/Co value for first 270 min which implies that there was no sudden decrease in effluent quality with passage of time.

The quantity of total metal adsorbed onto the surface of the adsorbent (q_{ad}, mg) can be deduced from the area beneath the curve between adsorbed metal concentration C_{ad} (mg/L) and time elapsed (t, minutes) and is expressed as

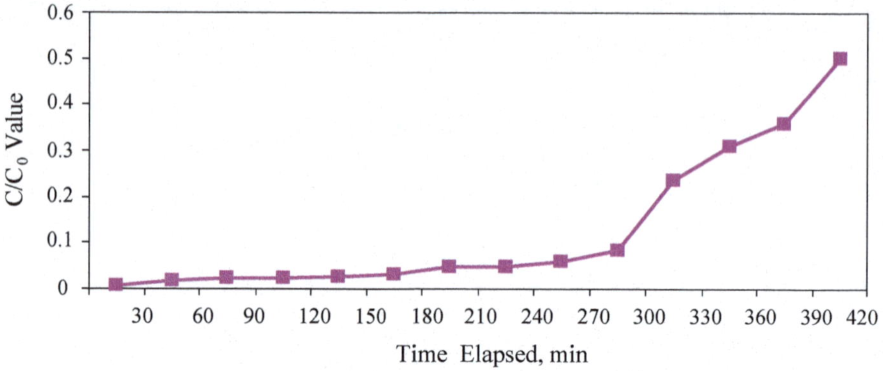

Fig. 20.5 Breakthrough curve in column studies

Table 20.2 Adsorption column performance for iron removal

Co (ppm)	q_{ad} (mg)	m_{total} (mg)	Removal (%)	Q_{eq} (mg/g)
2.5	1.398	1.9125	73.09	0.05592

$$q_{ad} = \frac{Q}{1000} \int_{t(0)}^{t(total)} C_{ad} \cdot dt$$

For a certain volumetric flow rate being maintained, the total amount of metal being sent to the column (m_{total}) is determined as:

$$m_{total} = \frac{Co \cdot Q \cdot t(total)}{1000}$$

The Total removal can then be expressed as:

$$R(\%) = \left(q_{ad} / m_{total}\right) x100$$

The total metal send to the column (m_{total}), total percentage removal from the column (%R), equilibrium metal uptake calculated is given in Table 20.2.

The total percentage removal obtained for 2.5 mg/l was 73.09% which was less than batch studies.

20.4 Conclusion

The following conclusions can be derived from the study:

- The optimum removal efficiency of Acid Violet 17 with an initial dye concentration of 40 mg/L and pH 2 was 93.59%.
- The optimum removal efficiency of Remazol Brilliant Blue R with an initial dye concentration of 40 mg/L and pH 4 was 90%.
- The efficiency of dye removal increased with the increase in initial dye concentrations, and then decreased after the optimum value. The removal efficiency of the dyes were found to be dependent on pH for lemon peel adsorbent. The optimum pH was obtained to be 2 and 4 for Acid Violet 17 and Remazol Brilliant Blue R respectively.
- From the batch study it is observed that with the increase in the adsorbent dosage, the Iron removal efficiency also increased. The maximum removal efficiency was found at a dosage of 2 g.
- Adsorption of Iron was almost pH independent. The efficiency increased with increase in pH. The maximum removal was obtained at pH 5.

References

Ahmad MA, Ahmad N, Bello OS (2014) Removal of remazol brilliant blue reactive dye from aqueous solutions using watermelon rinds as adsorbent. J Dispersion Sci Technol 36(6):845–858

ASTM D4979-19 Standard practice for physical description screening analysis in waste

Bhatnagar A, Minocha AK (2006) Conventional and non-conventional adsorbents for removal of pollutants from water—a review

Buthelezi SP, Olaniran AO, Pillay B (2012) Textile dye removal from wastewater effluents using bioflocculants produced by indigenous bacterial isolates. Molecules 17(12):14260–14274

Choy KKH, McKay G, Porter JF (1999) Sorption of acid dyes from effluents using activated carbon. Resour Conserv Recycl 27:57–71

Dagdelen S, Acemioglu B, Baran E, Kocer O (2014) Removal of remazol brilliant blue R from aqueous solution by Pirina pretreated with nitric acid and commercial activated carbon 225:1899–1908

Hegazi HA (2013) Removal of heavy metals from wastewater using agricultural and industrial wastes as adsorbents. Hous Build Nat Res Cent J 9:276–282

Isa MH, Al-Madhoun WA, Aziz HA, Asaari FAH, Sabiani NHM (2004) Fe removal by adsorption using ash from oil palm factory. AWAM, pp 1–5

Low KS, Lee CK (1990) The removal of cationic dyes using coconut husk as an adsorbent. Pertanika 13(2):221–228

Malik A, Grohmann E, Akhtar R (eds) (2013) Environmental deterioration and human health: natural and anthropogenic determinants. Springer Science & Business Media

Namasivayam C, Kanchana N (1992) Waste banana pith as adsorbent for color removal from wastewaters. Chemosphere 25(11):1691–1705

Namasivayam C, Kavitha D (2002) Removal of Congo Red from water by adsorption onto activated carbon prepared from coir pith, an agricultural solid waste. Dyes Pigm 54(1):47–58

Namasivayam C, Muniasamy N, Gayatri K, Rani M, Ranganathan K (1996) Removal of dyes from aqueous solutions by cellulosic waste orange peel. Elsevier Science Limited, vol 44, issue 2, pp 37–43

Pelosi BT, Lima LKS, Vieira MGA (2014) Removal of the synthetic dye Remazol brilliant blue R from textile industry wastewaters by biosorption on the macrophyte *Salvinia natans*. Braz J Chem Eng 31(4):256–266

Pollard SJT, Fowler GD, Sollars CJ, Perry R (1992) Low cost adsorbents for water and wastewater treatment: a review. Sci Total Environ 116:3305–3357

Rajoriya S, Kaur B (2014) Adsorptive removal of zinc from waste water by natural biosorbents. Int J Eng Sci Invention 3(6):60–80

Saiful Azhar S, Ghaniey Liew A, Suhardy D, Farizul Hafiz K, Irfan Hatim MD (2005) Dye removal from aqueous solution by using adsorption on treated sugarcane bagasse. Am J Appl Sci 2:1499–1503

Singh SA, Shukla SR (2015) Adsorptive removal of cobalt ions on raw and alkali-treated lemon peels. Int J Environ Sci Technol 13:165–178

Sivaraj R, Namasivayam C, Kadirvelu K (2001) Orange peel as an adsorbent in the removal of acid violet 17 (acid dye) from aqueous solutions. Waste Manage 21(1):105–110

Tembhurkar A, Deshpande R (2012) Powdered activated lemon peels as adsorbent for removal of cutting oil from wastewater. Am Soc Civ Eng (ASCE) 10:311–315

Thinakaran N, Baskaralingam P, Pulikesi M, Panneerselvam P, Sivanesan S (2008) Removal of acid violet 17 from aqueous solutions by adsorption onto activated carbon prepared from sunflower seed hull, vol 151, issue 2–3, pp 316–322

Velmurugan P, Rathinakumar V, Dhinakaran G (2011) Dye removal from aqueous solution using low cost adsorbent. Int J Environ Sci 1(7):1492–1503

Vijayalakshmi P, Sathya Selva Bala V, Thiruvengadaravi KV, Panneerselvam P, Palanichamy M, Sivanesan S (2010) Removal of acid violet 17 from aqueous solutions by adsorption onto activated carbon prepared from pistachio nut shell, vol 46, issue 1, pp 155–163

Yagub MT, Sen TK, Afroze S, Ang HM (2014) Dye and its removal from aqueous solution by adsorption: a review. Adv Coll Interface Sci 209:172–184

Yousaf MM, Sajjad (2015) Application of thermally and chemically modified banana peels waste as adsorbents for the removal of iron from aqueous system. Environ Anal Chem 2(3), 1–7

Chapter 21
Humic Acid Removal from Water Using Hydrophilic Polysulfone Membrane

Bharti Saini and Manish Kumar Sinha

Abstract Surrounding vegetation, animal, human and microbiological decomposition are the strong source of humic acid (HA) falling into the surface water bodies during the process of rain-generated runoff in monsoon. HA contains various functional groups such as carboxylic, phenolic, hydroxyl and quinine which are the major foulants. Contact of HA may have adversarial health issue to human being namely goiter, black foot and cancer disease. The Environmental Protection Agency (EPA) has prescribed maximum permissible limit of 2 ppm of HA content in the drinking water. Membrane technology is the prominent technology for HA removal. However, membrane fouling due to foulants is one of the major issues. In the present work hydrophilic polysulfone membrane was synthesized using water soluble polyethylene glycol monomethyl ether (PEGME) of 5000 molecular weight. Membrane was characterized for the structure, morphology, hydrophilicity, fouling and rejection percentage of HA. Blended membrane shows very high rejection percentage of HA and less fouling than plain membrane.

21.1 Introduction

Utilization of safe and clean water in daily life is still a dream in various locations across the world. One of the most common causes of polluted water is humic acid release from surrounding vegetation into the surface water. Exposure of HA results in many disease to human being namely cancer, goiter and black foot. To satisfy the increasing demand for safe water, the advancement in the area of water treatment has becomes the main obligation. Lots of efforts had been done by many researchers to enhance the quality as well as quantity of pure water. Among water treatment methods, membrane technology is one of the best promising technique around the world due to low cost, several real time applications and environmental friendly approach (Beril Melbiah et al. 2017; McCloskey et al. 2012). Based on

B. Saini (✉) · M. K. Sinha
Department of Chemical Engineering, School of Technology, Pandit Deendayal Petroleum University, Gandhinagar 382007, Gujarat, India
e-mail: bhartisaini100@gmail.com

A. Pandey et al. (eds.), *Climate Impacts on Water Resources in India*, Water Science and Technology Library 95, https://doi.org/10.1007/978-3-030-51427-3_21

pore size, membrane separation technology mainly consist four different processes such as micro-filtration (MF), ultra-filtration (UF), nanofiltration (NF) and reverse osmosis (RO). Among these processes, UF has great attention due to their effectiveness in separation of colloids, macromolecules and particles (Lee et al. 2013). For UF membranes, a broad range of polymeric materials like cellulose acetate (CA), polyvinylidene fluoride (PVDF), polysulfone (PSF), polyethersulfone (PES), polyacrylonitrile (PAN), polypropylene (PP) and polyethylene (PE) are used. Among them, PSF polymer is mostly utilized in fabrication process of UF membranes on account of good physicochemical stability, ability to resist chlorine attack and oxidation, and superior film forming capability (Sinha and Purakit 2013). However, PSF membrane easily fouled by foulants and limits the wide application in water purification system attributable to hydrophobic nature of PSF. The low fouling resistance property of the membrane results in sharp decrease in flux.

In order to overwhelm these restrictions, several hydrophilic adaptations of pure polymer membranes were employed by the researchers. Swapna et al. synthesized mixed matrix membranes by blending iron oxide (Fe_3O_4) nanoparticles (NPs) stabilized with amine and polyacrylonitrile (PAN) covered with chitosan. They found that the membranes were able to reduce the HA concentration below 2 mg/L (Panda et al. 2015). Saraswathi et al. fabricated polydopamine coated PVDF ultrafiltration membranes. They observed 89.5–95.4% solute rejection of HA from plain PVDF to modified PVDF membrane (Saraswathi et al. 2017). Hwang et al. (2013) prepared activated carbon polymer composite membranes and modified membranes achieved 80% of HA removal efficiencies. Gholamvand et al. prepared hybrid ultrafiltration PSF membranes using blends of graphine oxide-TiO_2 nanocomposite and observed that the modified membrane had significant HA separation efficiency for 10 ppm concentration (Kumar et al. 2016).

In this work, PSF membranes were successfully synthesized with polyethylene glycol monomethyl ether (PEGME) of 5000 molecular weight via immersion precipitation (phase inversion) method. The prepared PSF membranes were analyzed for functional groups, morphology, porosity and permeability. Hydrophilicity of the all prepared membranes was examined in terms of static water contact angle. In order to calculate the percentage rejection and fouling resistance ability of membrane, filtration experiments were conducted using HA solution of 100 ppm.

21.2 Experimental

21.2.1 Materials

PSF polymer, N-Methyl-2-pyrrolidone (NMP) and **polyethylene glycol monomethyl ether** of 5000 molecular weight were purchased from M/s. Sigma Aldrich, India. Humic acid (HA) was procured from Merck Limited, India. All reagents and chemicals were of analytical grade and used with no further treatment

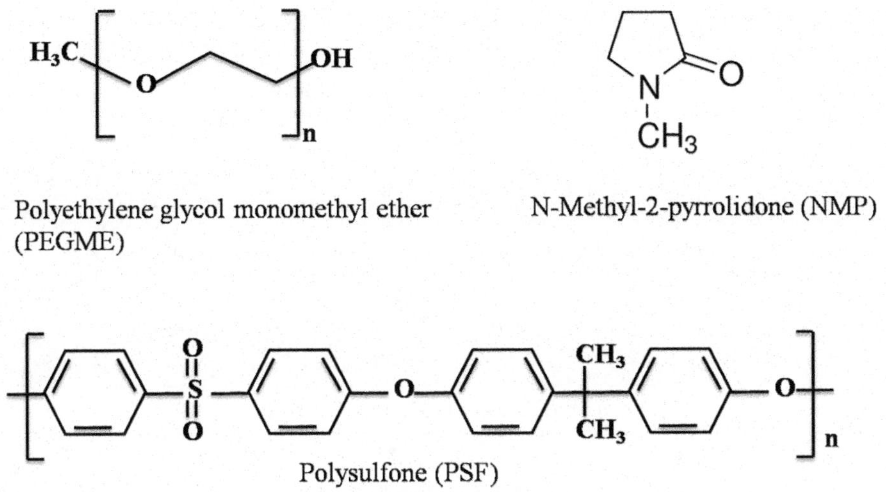

Fig. 21.1 Chemical structures of PEGME, NMP and PSF

and purification. The chemical structures of PEGME, NMP and PSF are shown in Fig. 21.1.

21.2.2 Synthesis of Membranes

Polysulfone membranes were prepared via phase inversion technique mainly focused on immersion precipitation using casting knife by maintaining 0.2 mm thickness. The membrane fabrication steps are shown in Fig. 21.2. Initially, casting solution was prepared with 15 wt% of PSF polymer and different wt% of PEGME of 5000 molecular weight (0–3 wt%) in NMP solvent by continuous stirring on magnetic stirrer under ambient conditions for 12 h. The solution was kept unstirred for 6 h for degassing then the films were casted on glass plate and immediately dip into the water bath. Precipitated films were moved into fresh water bath for 24 h for removal of traces of solvent from membrane. After that membranes were kept open for drying under ambient conditions and placed in air sealed packets.

21.2.3 Characterization of Membrane

Membranes were characterized for morphology by conducting liquid–liquid displacement porosimetry (LLDP) experiments as per procedure described in our earlier work (Saini et al. 2019). From these analysis, mean pore radius and pore size allocation were evaluated. Functional groups associated with the membranes were

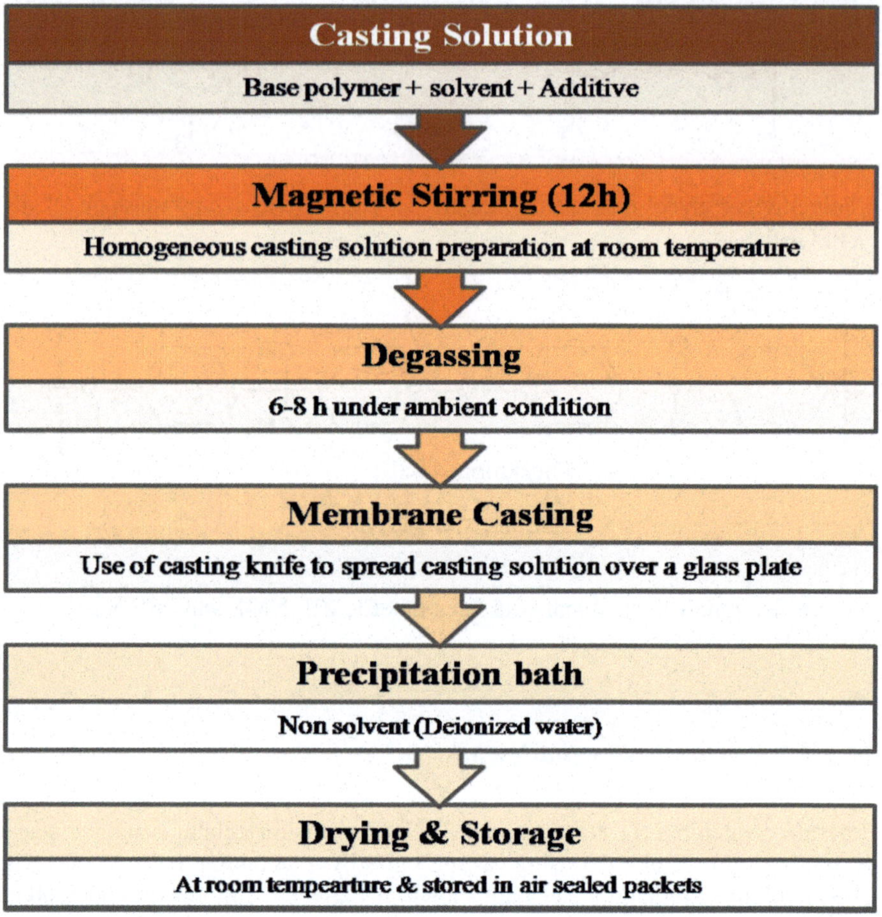

Fig. 21.2 Steps for polymeric membrane fabrication by phase inversion method

enquired by attenuated total reflectance-fourier transform infrared (ATR-FTIR) spectroscopy (Perkin-Elmer Inc., Germany). The static water contact angle was evaluated in order to determine the hydrophilicity of all fabricated membranes (Padaki et al. 2012; Saini and Sinha 2018). Porosity (Panda and De 2014) and equilibrium water content (EWC) (Qiu et al. 2009) were determined by gravimetric method using following formulas:

$$EWC(\%) = \frac{m_w - m_d}{m_w} \times 100 \tag{21.1}$$

$$\varepsilon(\%) = \frac{m_w - m_d}{\rho_w A l} \times 100 \tag{21.2}$$

where m_w and m_d are wet and dry membrane weights (g), respectively. l, A and ρ_w are the membrane thickness (cm), membrane surface area (cm^2) and water density (g/cm^3), respectively.

21.2.4 Filtration Experiments Using Pure Water and Humic Acid Solution

Filtration experiments were conducted in batch filtration setup (shown in Fig. 21.3) of 350 mL capacity (made of stainless steel) and effective filtration diameter of the membrane was 50 mm. Nitrogen gas was used to maintain the required pressure in the system. Pure water flux (PWF), HA flux and HA removal percentage (%R) was determined by following equations (Li et al. 2009; Zhao et al. 2011, 2014; Huang et al. 2015):

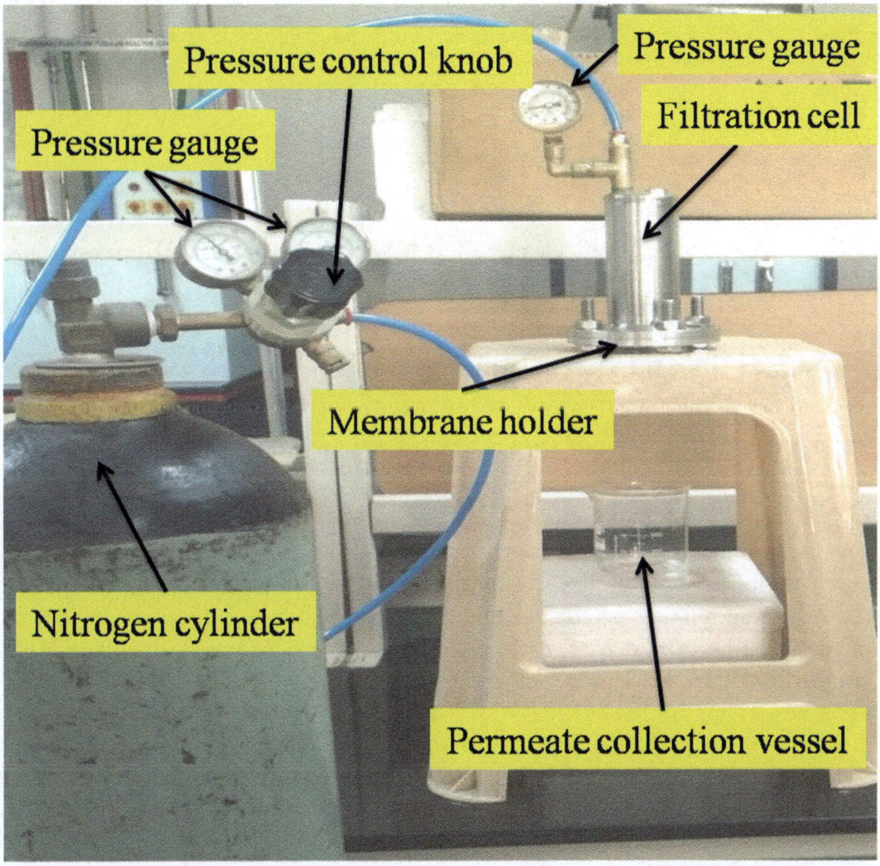

Fig. 21.3 Batch filtration experiment setup

$$J_w = \frac{V}{A\Delta T} \tag{21.3}$$

$$R(\%) = \left(1 - \frac{C_p}{C_f}\right) \times 100 \tag{21.4}$$

where J_w, V, A and ΔT indicates PWF (L/m^2 h), permeated liquid volume (L), filtration surface area of the membrane (m^2), penetration time (h), respectively. C_f and C_p shows the feed and permeate concentrations (mg/L), respectively. From the filtration experiments, some fouling parameters like flux recovery ratio ($Flux_{RR}$), irreversible fouling (F_{ir}), reversible fouling (F_r) and total fouling (F_t) were calculated by the following expressions (Shen et al. 2017):

$$Flux_{RR}(\%) = \frac{J_{pw2}}{J_{pw1}} \times 100 \tag{21.5}$$

$$F_{ir}(\%) = \left(\frac{J_{pw1} - J_{pw2}}{J_{pw1}}\right) \times 100 \tag{21.6}$$

$$F_r(\%) = \left(\frac{J_{pw2} - J_f}{J_{pw1}}\right) \times 100 \tag{21.7}$$

$$F_t(\%) = \left(\frac{J_{pw1} - J_f}{J_{pw1}}\right) \times 100 \tag{21.8}$$

where J_f, J_{pw1} and J_{pw2} indicates feed flux, PWF before and after HA ultrafiltration.

21.3 Results and Discussion

21.3.1 Morphological and Functional Group Analysis

Liquid–liquid displacement porosimetry experiments were conducted using water-isobutanol-methanol mixture (v/v) in dead end filtration setup as per the procedure mentioned in our earlier work (Saini et al. 2018) to determine the morphology (average pore number percentage and average pore size) of the plain and blended membranes. From Fig. 21.4, it can be seen that the maximum percentage of pore number was nearly 18%. For all membranes, it may also be found that most of the pores having average pore size less than 2 nm, which demonstrate that the all membranes are applicable for ultrafiltration separation process. A very few percent of pore number were observed for pore size greater than 2 nm. The trend of pore size distribution was observed to be similar in all membranes. The mean pore size is correlated with the cumulative pore numbers of the membrane. The mean pore size of the membranes was 1.33 nm for plain membrane (M0 membrane) and

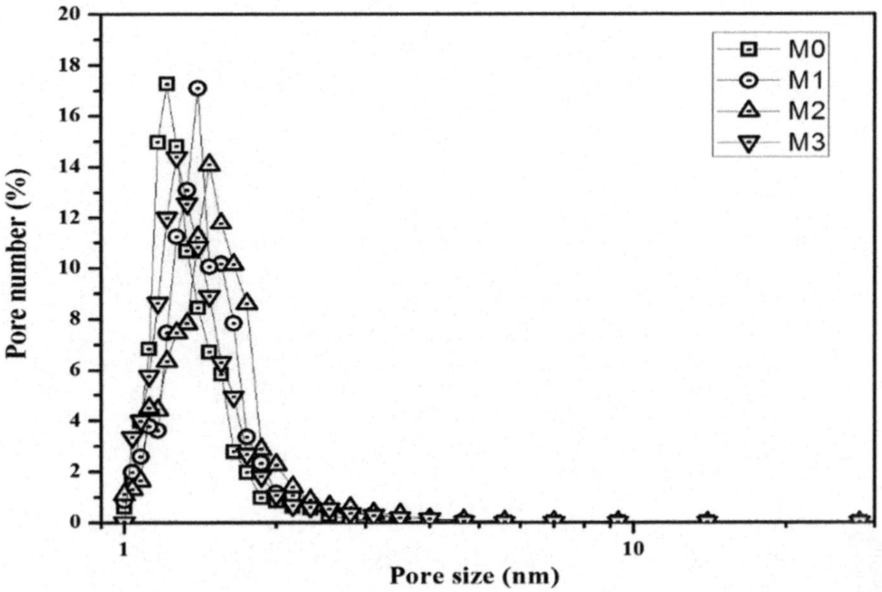

Fig. 21.4 Pore number percentage with respect to pore size for all membranes

1.49 nm for blended membrane (M3 membrane). The increment in pore size was detected as a result of hydrophilic and water-soluble nature of PEGME in the blended membrane. During preparation of membrane by immersion precipitation process, some amount of PEGME leach out into the water bath whereas some quantity remains present in the polymer film. The functional group OH present in PEGME (present in the membrane/polymer) induces the hydrophilicity in the membranes. Pore size enlargement in blended membrane is an indication of better permeability.

Figure 21.5 shows the FTIR spectra of wave number ranges from 4000 to 500 cm^{-1} for all membranes. From spectra, the standard peaks at 1153 and 1250 cm^{-1} are observed due to presence of ether (–C–O–C–) and sulfonic (–S = O) groups of PSF polymer, in that way supporting the existence of PSF polymer (Saini and Sinha 2018). In addition, the broad absorption peak at 3450 cm^{-1} is observed caused by the stretching vibration of –OH groups which confirms the presence of hydrophilic groups of PEGME additive in the membrane (Ghaemi et al. 2012). Moreover, the peak around 2974 cm^{-1} is detected due to –C–H stretching (Pakizeh et al. 2013). Apart from that some other characteristic peaks of PSF polymer can be seen in the spectra.

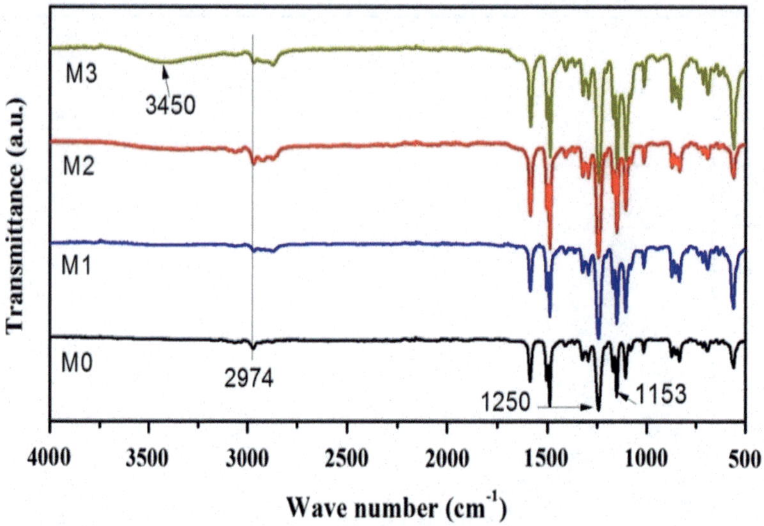

Fig. 21.5 FTIR spectra of all membranes: Transmittance (a.u.) versus wave number

21.3.2 Hydrophilicity

Hydrophilicity indicates the water attraction as well as fouling resistance property of the membrane. It was analysed by measuring the static water contact angle made on the surface of the membrane. Lower contact angle means more hydrophilicity of the membrane (Yi et al. 2010). Figure 21.6 shows that the contact angle of the membrane diminishes with increase in wt% of additive. The membrane M3 (with 3 wt% of modifier) have lowest water contact angle and M0 (without additive) have highest water contact angle. Thus, the hydrophilicity of the membranes was amplified with increase in concentration of additive in the membranes. This could be explained by the following reason that the PEGME has –CH3 group at one end, in order that CH_3 has superior attraction towards base polymer as a result of hydrophobic behaviour of base polymer, accordingly it will remain together with polymer matrix for longer time. Thus, PEGME is not only functioning as a pore constructor but also working as a hydrophilic modifier for the PSF membrane.

21.3.3 Porosity and Equilibrium Water Content Analysis

Membrane porosity represents the porous nature as well as performance of the membrane. If porosity increases, pure water flux and permeability are expected to be increased due large pore area of the membrane. From Fig. 21.7 it was found that with increase in wt% of additive, porosity of the membranes increases. The porosity of

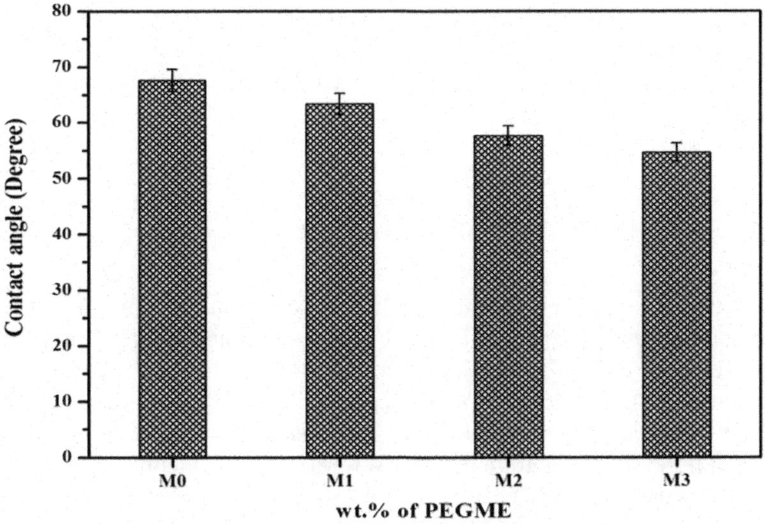

Fig. 21.6 Effect of additive dose on hydrophilic character of the membranes

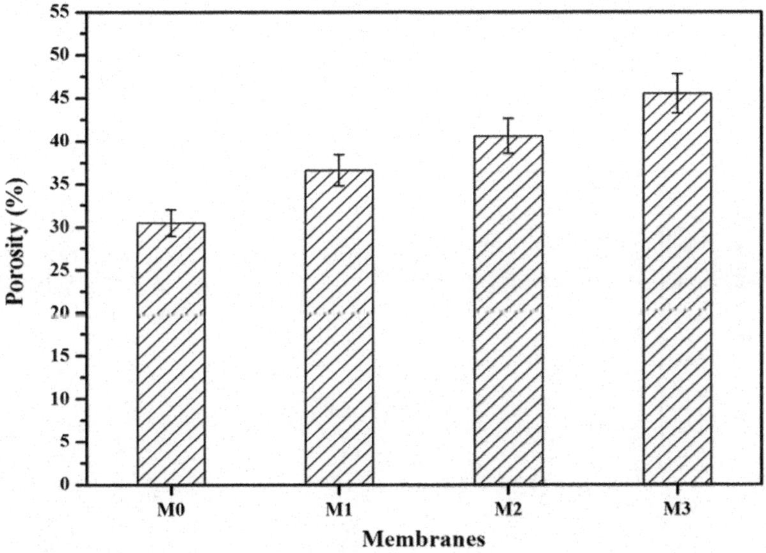

Fig. 21.7 Effect of additive dose on porosity of the membranes

the membranes was improved from 30 to 46% for plain (M0) to blended membrane (M3) due to addition of additive. The equilibrium water content is mainly linked to the porosity of the membrane and varies with the porosity of the membrane. In this study, the trend of equilibrium water content was analogous to the porosity of the membrane. The minimum equilibrium water content was found for M0 membrane whereas, the maximum equilibrium water content was observed for M3 membrane. These results disclose that blending of PEGME5000 additive improved the porosity and equilibrium water content of membrane upto 3 wt%.

21.3.4 Pure Water Flux Measurement

The pure water flux (PWF) of the membranes was examined through batch filtration setup from the collected quantity of permeated water. Initially the PWF of all membranes was very high and after 90 min of filtration process, it become constant due to compaction or densification of membrane. Figure 21.8 shows the PWF flow of all the membranes for 240 min. The steady state PWF values were noted after 240 min of pure water penetration. The PWF was increased with additive concentration and M3 have higher PWF than other membranes. The value of PWF of M3 is 20 times than the flux of M0 membrane. Thus the order of PWF for all membrane was M0 < M1 < M2 < M3. As a consequence, increase in PWF with addition of additive represents the rise in permeability of the membrane. The values of steady state PWF for plain and blended membranes were 15.5 and 151.8 L/m^2h, respectively. These findings undoubtedly propose that blending of PEGME5000 in different concentration with PSF polymer is capable to control the permeation performance of the polymeric membrane.

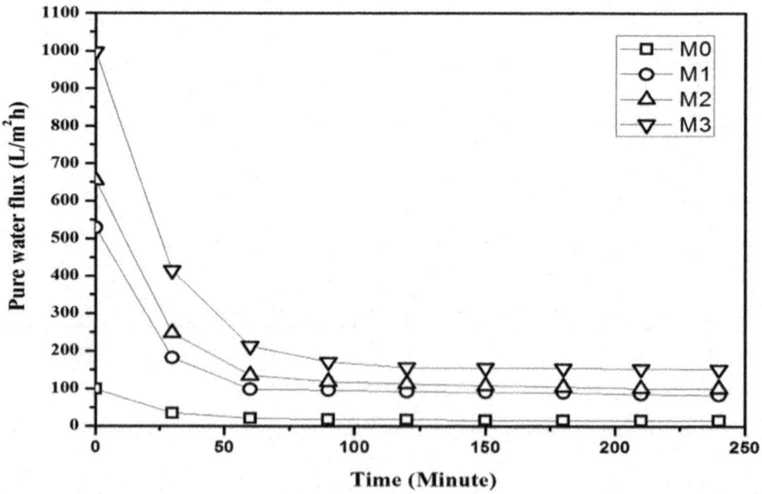

Fig. 21.8 Pure water flux measurement with respect to time

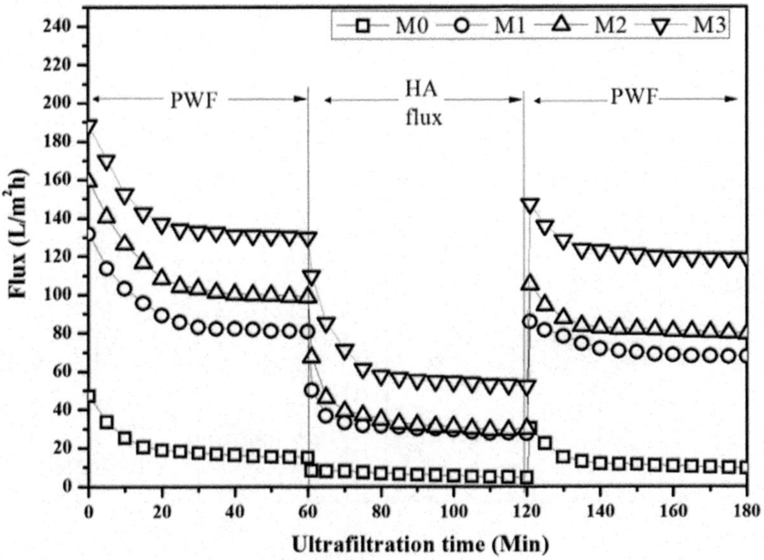

Fig. 21.9 Humic acid ultrafiltration experiment

21.3.5 Humic Acid Removal Analysis

Ultrafiltration experiments had been conducted using humic acid solution and the HA flux profile was shown in Fig. 21.9. The HA flux trend was similar with the trend of PWF however the value of HA flux was relatively less than the PWF because of adsorption and/or deposition of HA molecules on the surface or inside the pores of membrane. It was also found that the membrane M3 have grater HA flux than M0 membrane.

Figure 21.10 shows the fouling ratios which were determined from the HA ultrafiltration experiments. Irreversible fouling and total fouling was decreased with additive concentration whereas reversible fouling was increased. Decrease in irreversible fouling and total fouling indicated the increase in HA flux. The HA removal percentage was also observed to be increased with additive dose due to development of mesh type gel film on the surface of the membrane during HA ultrafiltration process. The HA removal percentage was near about 98% through the M3 membrane (Fig. 21.11).

Figure 21.12 represent the flux recovery ratio in HA ultrafiltration experiments. It was observed that the membrane M3 have higher flux recovery ratio than the M0, M1 and M2 membranes. The value of flux recovery ratio was improved from 0.63 to 0.92 for M0–M3 membranes.

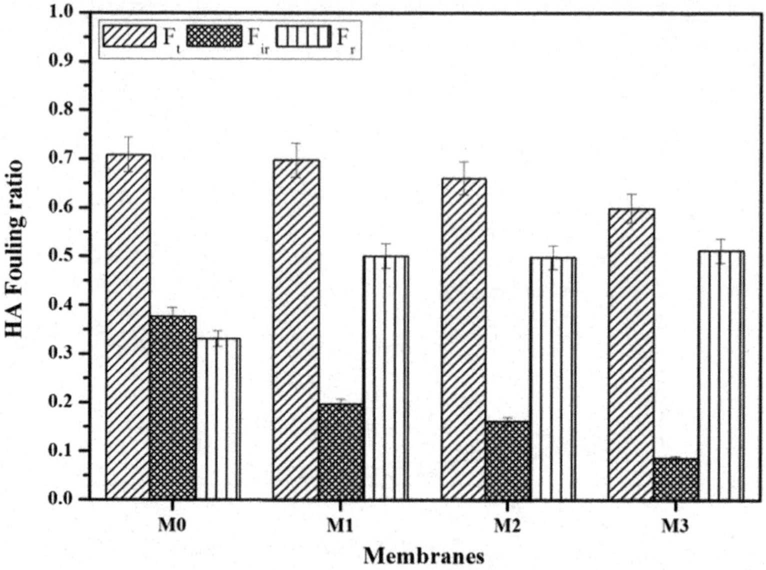

Fig. 21.10 Membrane fouling in humic acid ultrafiltration process

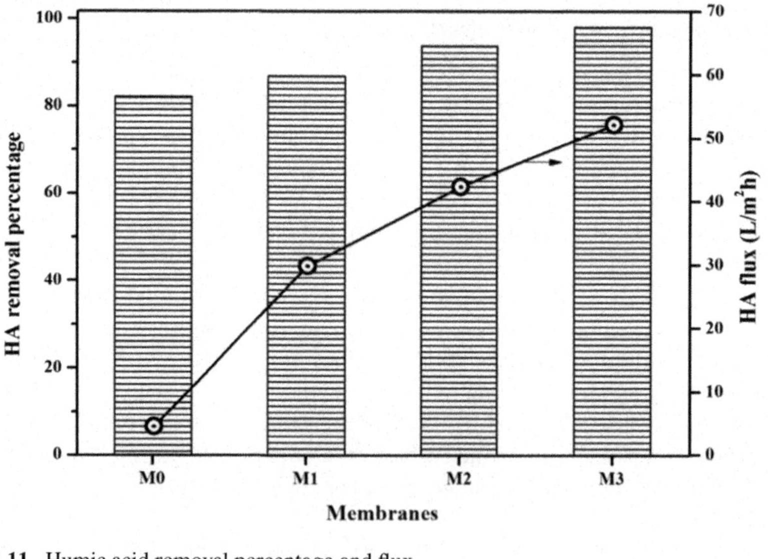

Fig. 21.11 Humic acid removal percentage and flux

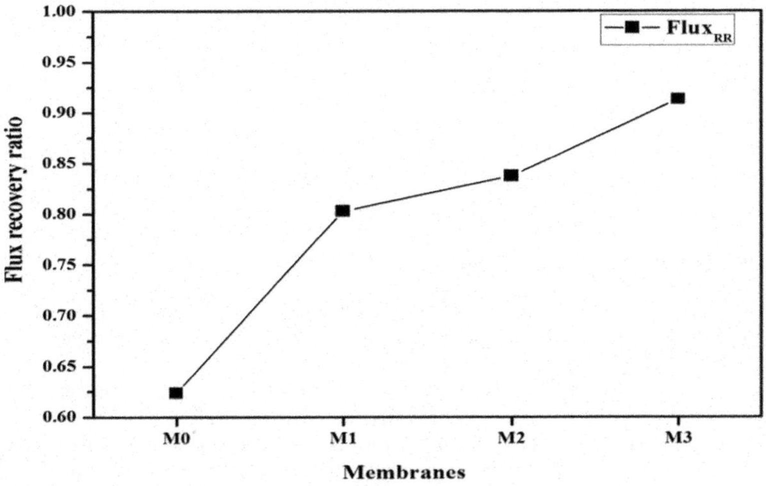

Fig. 21.12 Flux recovery in humic acid ultrafiltration

21.4 Conclusion

A set of four membranes was synthesized with different compositions of hydrophilic PEGME of 5000 molecular weight additives. The effect of additive on PWF, hydrophilicity, structure, morphology, HA removal percentage and porosity was investigated. Morphology of the fabricated membranes was found to be in ultrafiltration range based on pore size of the membranes. The existence of expected chemical groups was identified by FTIR spectra. The contact angle trend was M0 > M1 > M2 > M3 which shows the increase in hydrophilicity. The porosity and PWF were increased with additive concentration. Due to rise in hydrophilicity of the membranes, total and irreversible fouling was reduced. It was also found that the blending of additive in PSF membrane is suitable for the removal of HA. The maximum HA removal was observed with M3 membrane. Hence, addition of polymeric hydrophilic additive not only improves the fouling resistance characteristic of the membrane, but it is also effective for the removal of HA from water.

References

Beril Melbiah JS, Nithya D, Mohan D (2017) Surface modification of polyacrylonitrile ultrafiltration membranes using amphiphilic Pluronic F127/$CaCO_3$ nanoparticles for oil/water emulsion separation. Colloids Surf A516:147–160

Ghaemi N, Madaeni SS, Alizadeh A, Daraei P, Badieh MMS, Falsafi M, Vatanpour V (2012) Fabrication and modification of polysulfone nanofiltration membrane using organic acids: morphology, characterization and performance in removal of xenobiotics. Sep Purif Technol 96:214–228

Huang X, Zhang J, Wang W, Liu Y, Zhang Z, Li L, Fan W (2015) Effects of PVDF/SiO_2 hybrid ultrafiltration membranes by Sol-Gel method for the concentration of fennel oil in herbal water extract. RSC Adv 5(24):18258–18266

Hwang L, Chen J, Wey M (2013) The properties and filtration efficiency of activated carbon polymer composite membranes for the removal of humic acid. Desalination 313:166–175

Kumar M, Gholamvand Z, Morrissey A, Nolan K, Ulbricht M, Lawler J (2016) Preparation and characterization of low fouling novel hybrid ultrafiltration membranes based on the blends of GO−TiO_2 nanocomposite and polysulfone for humic acid removal. J Memb Sci 30061–30068.

Lee J, Jeong S, Ye Y, Chen V, Vigneswaran S, Leiknes T, Liu Z (2013) Protein fouling in carbon nanotubes enhanced ultrafiltration membrane: fouling mechanism as a function of pH and ionic strength. Sep Purif Technol 118:226–233

Li JH, Xu YY, Zhu LP, Wang JH, Du CH (2009) Fabrication and characterization of a novel TiO_2 nanoparticle self-assembly membrane with improved fouling resistance. J Memb Sci 326(2):659–666

McCloskey BD, Park HB, Ju H, Rowe BW, Miller DJ, Freeman BD (2012) A bioinspired fouling-resistant surface modification for water purification. J Membr Sci 413:82–90

Padaki M, Isloor AM, Ismail AF, Abdullah MS (2012) Synthesis, characterization and desalination study of novel PSAB and mPSAB blend membranes with Polysulfone (PSf). Desalination 295:35–42

Pakizeh M, Moghadam AN, Omidkhah MR, Namvar-Mahboub M (2013) Preparation and characterization of dimethyldichlorosilane modified SiO_2/PSf nanocomposite membrane. Korean J Chem Eng 30:751–760

Panda SR, Mukherjee M, De S (2015) Preparation, characterization and humic acid removal capacity of chitosan coated iron oxide polyacrylonitrile mixed matrix membrane. J Water Process Engg 6:93–104

Panda SR, De S (2014) Preparation, characterization and performance of $ZnCl_2$ incorporated polysulfone (PSF)/polyethylene glycol (PEG) blend low pressure nanofiltration membranes. Desalination 347:52–65

Qiu S, Wu L, Pan X, Zhang L, Chen H, Gao C (2009) Preparation and properties of functionalized carbon nanotube/PSF blend ultrafiltration membranes. J. Memb. Sci. 342:165–172

Saini B, Khuntia S, Sinha MK (2019) Incorporation of cross-linked poly(AA-co-ACMO) copolymer with pH responsive and hydrophilic properties to polysulfone ultrafiltration membrane for the mitigation of fouling behaviour. J Memb Sci 572:184–197

Saini B, Sinha MK (2018) Effect of hydrophilic poly (ethylene glycol) methyl ether additive on the structure, morphology, and performance of polysulfone flat sheet ultrafiltration membrane. J Appl Polym Sci 47163:1–14

Saini B, Sinha MK, Dash SK (2018) Mitigation of HA, BSA and oil/water emulsion fouling of PVDF ultrafiltration membranes by SiO_2-g-PEGMA nanoparticles. J Water Process Eng 30:100603

Saraswathi MSA, Kausalya R, Kaleekkal NJ, Nagendran RA (2017) BSA and humic acid separation from aqueous stream using polydopamine coated PVDF ultrafiltration membranes. J Env Chem Engg S2213-3437:30242–30247

Shen X, Xie T, Wang J, Liu P, Wang F (2017) Nanoparticles for efficient oil/water separation. RSC Adv 5262–5271

Sinah MK, Purkait MK (2013) Increase in hydrophilicity of polysulfone membrane using polyethylene glycol methyl ether. J Memb Sci 437:7–16

Yi Z, Zhu L, Xu Y, Zhao Y, Ma X, Zhu B (2010) Polysulfone-based amphiphilic polymer for hydrophilicity and fouling-resistant modification of polyethersulfone membranes. J Memb Sci 365:25–33

Zhao S, Wang Z, Wei X, Zhao B, Wang J, Yang S, Wang S (2011) Performance improvement of polysulfone ultrafiltration membrane using PANiEB as both pore forming agent and hydrophilic modifier. J Memb Sci 385–386(1):251–262

Zhao S, Yan W, Shi M, Wang Z, Wang J, Wang S (2014) Polyethersulfone/ZnO ultrafiltration membrane with enhanced permeability and fouling resistance. J Memb Sci S0376-7388(15):00015

Chapter 22
Comparison of GCM Derived Rainfall for Bharathapuzha River Basin

Lini R. Chandran and P. G. Jairaj

Abstract General Circulation Models (GCMs) enable one to understand how climate changes on a regional scale and are of great importance in planning for our habitat. Statistical downscaling is the commonly used downscaling method mainly due to its computational efficiency. Even though studies utilizing GCM outputs for watershed level studies for certain basins in India are available, only very few are reported for peninsular region of India. The present study is aimed at estimating the better performing GCM for a river basin of Kerala State, India, which is bounded by the orographic features on one side and sea on the other side. Such type of studies is of increasing importance presently as the rainfall patterns and climate are seen to have a striking change in the recent past in the state. Multivariate Linear Regression with Principal Component Analysis was performed in this study for statistical downscaling of rainfall in the study region. Two GCMs namely CGCM3 and BCM2 were used for development of the downscaling models. The model was calibrated and tested with 50 years of NCEP reanalysis data and was used for projecting the precipitation in the study region. The performance of these GCMs was compared for arriving at a better model for water resources planning in the region. It was seen that BCM2 performed better in projecting the rainfall in the area.

22.1 Introduction

Assessment of the water resources potential in a region is very essential for the preparation of many developmental plans in the region. There are many methods for estimating the available water resources in an area at a future period. But the General Circulation Models (GCMs) which mathematically represents the atmospheric and ocean dynamics and earth systems are generally used for projecting the hydrological variables for a future period. This is because climate is an important factor influencing the hydrological variables in a region. Many researchers have used various methods to establish the relationship between climatic variables simulated by the

L. R. Chandran (✉) · P. G. Jairaj
Department of Civil Engineering, College of Engineering Trivandrum, Thiruvananthapuram, India
e-mail: linisam@gmail.com

A. Pandey et al. (eds.), *Climate Impacts on Water Resources in India*, Water Science and Technology Library 95, https://doi.org/10.1007/978-3-030-51427-3_22

GCMs and the hydrological variables of interest. These models have incorporated many complex global processes and perform well in simulating variables at large spatial resolution (Reichler and Kim 2008). But the sub-grid scale features are less accurately represented by these models (Wilby et al. 2002). The regional hydrological studies involve small scale processes and this necessitates the need for downscaling of the GCM variables usually at catchment level.

The statistical downscaling is a simple and practical method for establishing the relationship between the atmospheric variables which are better simulated by GCMs and the hydrological variables at catchment level. They transform the coarse scale atmospheric variables to point scale information which are essential for impact studies (Wilby and Dawson 2013). The quantitative relations are usually derived in the form of transfer function (Hewitson and Crane 1996). The local hydrological variables at a future period are projected by means of the developed transfer function using the climate variables simulated by GCMs for those period. It is assumed that the relationship between predictors and predictand remains same at changing scenarios or that stationarity exists in the process (Mondal and Mujumdar 2015).

At present there are numerous GCMs developed by various modeling agencies. The uncertainties in the models due to different mathematical modeling procedures and parameterizations results in the variation of model simulation of future climate. There may be many predictors available which are supposed to influence the predictand of interest. But suitable predictors that are highly correlated to the predictand has to be determined so as to simplify and effectively compute the transfer function. Selection of appropriate predictors is the most vital step in any downscaling process (Cavazos and Hewitson 2005). Multivariate linear regression (MVLR) is a simple way of estimating the relation between the predictors and predictand. In this study a MVLR model is developed for a river basin in Kerala for estimating the precipitation in the catchment using the predictors from two GCMs CCCMA-CGCM3 and BCCR-BCM2.0. These models are found to perform better in Indian region, particularly in South India (Srinivasa Raju and Nagesh Kumar 2014). The better performing GCM is suggested based on the ability of the model to closely simulate the observed precipitation in the study area.

22.2 Study Region and Data Used

22.2.1 Catchment

The River basin of Bharathapuzha River is used for the study (Fig. 22.1). The extent of the study region is from 10°15′ to 10°40′ N latitudes and 76°00′ to 76°35′ E longitudes and extends over Malappuram, Thrissur and Palakkad districts of Kerala and Coimbatore district in Tamil Nadu. Bharathapuzha is the second longest West flowing perennial river in Kerala State. It originates in the Western Ghats and flows west draining to the Arabian Sea near Ponnani town. The catchment area is spread

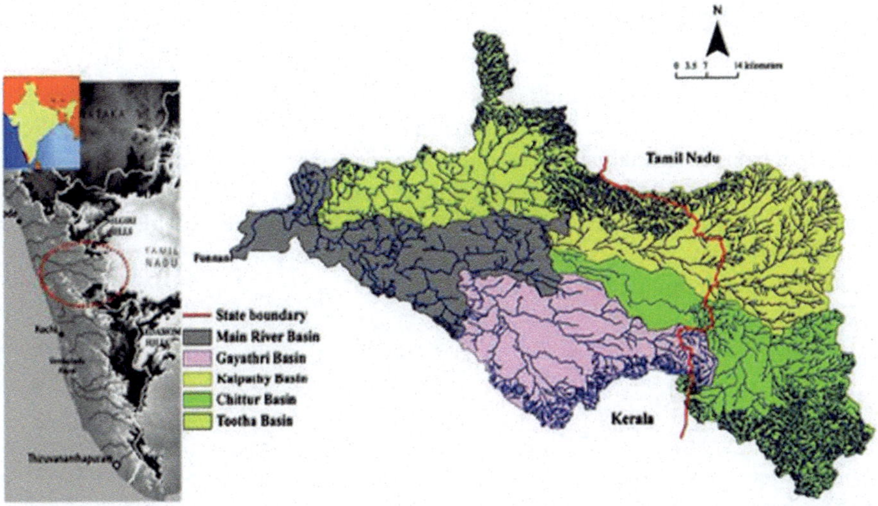

Fig. 22.1 Bharathapuzha River basin

along the two states with around 70% falling in Kerala and the rest in Tamil Nadu. The total length of the river from its origin to out fall is about 209 km and drains a total area of 6186 km^2. The basin is experiencing water shortage recently and large scale land use changes are seen there. The river flows through varying topography from highlands, midlands and to lowlands before draining to the sea (India-WRIS website 2018).

22.2.2 *Data Used*

The predictor data used for the study were extracted from the reanalysis data prepared by National Center for Environmental Prediction (NCEP/NCAR Reanalysis data) (Kalnay et al. 1996). The data has a resolution of 2.5° × 2.5°. The predictor data over the 9 grid points ranging from 7.5 N to 12.5 N latitudes and 72.5 E to 77.5 E longitudes were extracted. Based on previous studies, 27 predictor variables which were seen to influence the precipitation of a region were selected for the study. The predictor data for all the 9 grid points for the time period from January 1951 to December 2000 was extracted. The predictand was monthly mean precipitation over the basin. Gridded precipitation data from National Climate Centre, Indian Meteorological Department with a grid spacing of 0.25° × 0.25° was used for the model development. The GCM simulated predictor variables from the Canadian Centre for Climate Modelling and Analysis (CCCMA) CGCM3 and Bjerknes Centre for Climate Research (BCCR) BCM2.0 for B2 scenario were used for projecting the precipitation in the region.

22.3 Methodology

The selection of suitable predictors forms the most important part in any downscaling study (Hewitson and Crane 1996; Cavazos and Hewitson 2005). Hence the predictors extracted from the reanalysis data were analysed and the optimum predictors selected for performing regression. The suitability of using the selected GCM predictors to project the predictand was ensured. The downscaling model was developed using multivariate linear regression and their performance compared by projecting the predictand for future period.

22.3.1 Predictor Selection

The predictors were extracted for NCEP reanalysis data for the period January 1951 to December 2000. The spatial extent of influence of predictors were verified and a 3×3 grid region of 9 grid points spanning from 7.5 N to 12.5 N latitudes and 72.5 E to 77.5 E longitudes were selected as the zone of influence. 27 predictor variables were extracted and their suitability to reliably simulate the predictand were confirmed. The predictors analysed were the pressure level variables namely zonal component of wind or u-wind, meridional component of wind or v-wind, specific humidity, omega which is indicative of vertical velocity, geopotential height and air temperature at various pressure levels namely 1000, 925, 850, 700 mb and the surface variables such as max temperature, min temperature and mean sea level pressure. Scatterplots and cross-correlations are generally used for ascertaining the suitability of a predictor (Dibike and Coulibaly 2006). The product moment correlation, Spearman's rank correlation and Kendall's tau for all the predictors were calculated and those predictors with maximum correlation with predictand were selected for downscaling.

22.3.2 Model Development

The dependency of GCM variables to project the predictand using the model developed with the selected predictors was ascertained prior to actual model development. Scatter plots were generated between the NCEP predictors and GCM predictors for all the selected predictor variables to confirm this dependency (Anandhi et al. 2009). The resolution of NCEP and GCM variables were different. The GCM variables were interpolated to NCEP grid points before performing the earlier step. After ensuring the dependency, a multivariate linear regression model was developed. The predictors were standardized. Principal Component Analysis (PCA) was performed to reduce the variables to a set of linearly uncorrelated variables called Principal Components (PCs). PCs which captured more than 95% of the variance were used for forming the feature vectors. Calibration of the model was done using predictor data for the period January 1951 to December 1980. The model was tested for the period January 1981 to December 2000.

22.4 Results and Discussion

22.4.1 Analysis and Selection of Probable Predictor Variables

From earlier studies, it is seen that many pressure level and surface variables influence the precipitation in a region. Hence the pressure level variables namely zonal component of wind or u-wind, meridional component of wind or v-wind, specific humidity (Sp. Hum), omega which is indicative of vertical velocity, geopotential height (GPH) and air temperature at various pressure levels of 1000, 925, 850 and 700 mb and the surface variables such as maximum temperature, minimum temperature and mean sea level pressure (MSLP) were used in the analysis. Cross correlations were determined for all the 27 reanalysis predictor variables selected and is tabulated in Table 22.1 P, S and K in the table represent Product moment correlation, Spearman's rank correlation and Kendall's tau respectively. Seven predictors namely v-wind (1000 mb), u-wind (850 mb), specific humidity (925 mb), omega (700 mb), geopotential height (850 mb), maximum temperature and mean sea level pressure with largest correlation were selected as the most suitable predictors. However, all these predictors could not be used for model development as data at few grid points for some of these variables were missing for the GCM simulations used. Hence three predictors namely mean sea level pressure, specific humidity (925 mb) and geopotential height (850 mb) were used for model development.

22.4.2 Correlation Analysis of GCM and Reanalysis Predictors

The selected predictor variables were used for the model development. For projecting the precipitation accurately using the developed model, there should be dependency between the NCEP predictors used for model development and the GCM predictors used for the projection. Hence this dependency was checked by plotting scatterplots between them for all the predictors used for downscaling. The scatterplots clearly indicated that these predictor variables are realistically simulated by the GCMs. The scatterplot showing the dependency between NCEP predictors and CGCM3 predictors is shown in Fig. 22.2.

Table 22.1 Cross correlations estimated for predictors

Sl. No	Predictors	Correlation coefficients		
		P	S	K
1	Mslp	−0.64	−0.71	−0.50
2	v-wind (1000 mb)	0.63	0.64	0.55
3	v-wind (925 mb)	0.24	0.13	0.08
4	v-wind (850 mb)	−0.47	−0.54	−0.36
5	v-wind (700 mb)	−0.19	−0.22	−0.15
6	u-wind (1000 mb)	0.76	0.80	0.60
7	u-wind (925 mb)	0.79	0.83	0.63
8	u-wind (850 mb)	0.83	0.86	0.66
9	u-wind (700 mb)	0.77	0.74	0.53
10	Sp. Hum (1000 mb)	0.60	0.67	0.47
11	Sp. Hum (925 mb)	0.77	0.88	0.68
12	Sp. Hum (850 mb)	0.77	0.86	0.66
13	Sp. Hum (700 mb)	0.73	0.82	0.62
14	Omega (1000 mb)	−0.31	−0.27	−0.19
15	Omega (925 mb)	−0.25	−0.23	−0.17
16	Omega (850 mb)	−0.13	−0.16	−0.11
17	Omega (700 mb)	−0.75	−0.85	−0.65
18	Minimum temperature	0.26	0.39	0.26
19	Maximum temperature	−0.58	−0.64	−0.42
20	GPH (1000 mb)	−0.64	−0.71	−0.51
21	GPH (925 mb)	−0.68	−0.74	−0.54
22	GPH (850 mb)	−0.70	−0.76	−0.55
23	GPH (700 mb)	−0.60	−0.63	−0.44
24	Air Temp (1000 mb)	−0.19	−0.16	−0.10
25	Air Temp (925 mb)	−0.30	−0.29	−0.18
26	Air Temp (850 mb)	−0.05	0.03	0.03
27	Air Temp (700 mb)	0.28	0.35	0.23

22.4.3 Development of MVLR Model

The extracted predictor variables were standardized. 3 predictor variables over the 9 grid points were extracted which formed 27 variables for performing regression. The monthly average precipitation for the study region was computed from the observed rainfall data. The calibration of the model was done using monthly average predictor values taken from reanalysis data.

Multivariate Linear Regression was performed to establish the relationship between the predictors and predictand. These predictor variables may be mutually

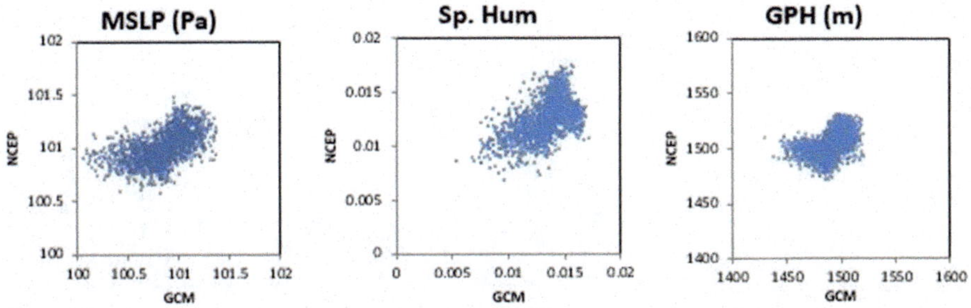

Fig. 22.2 Scatterplot revealing dependency of CGCM3 and NCEP predictors

correlated and in order to reduce the dimensionality of the data Principal component analysis was performed on these variables. The feature vector was formed that accounted to 95% variance. Using this feature vector and observed data the model was calibrated. The developed model was tested using the reanalysis data extracted for the testing period. Comparison of the observed and simulated precipitation during the testing period is given in Fig. 22.3. The predictor values of the GCM simulated variables for the projection period were also extracted. The GCMs used for this study were the CGCM3 model developed by Canadian Centre for Climate Modelling and Analysis and BCM2.0 model developed by the agency Bjerknes Centre for Climate Research (BCCR). These GCMs were specifically selected as their performance for simulating precipitation in Indian region have earlier been ascertained. The location of NCEP reanalysis predictors and GCM simulated variables are different because of the difference in their resolutions. Hence the predictor variables extracted from the GCMs were re-gridded to NCEP variable locations and were standardized. The principal components were calculated along the principal directions extracted earlier and the future precipitation over the study area was downscaled using the model.

The performance of the developed regression model to simulate the precipitation on the study region using the predictors from the two GCMs was compared with

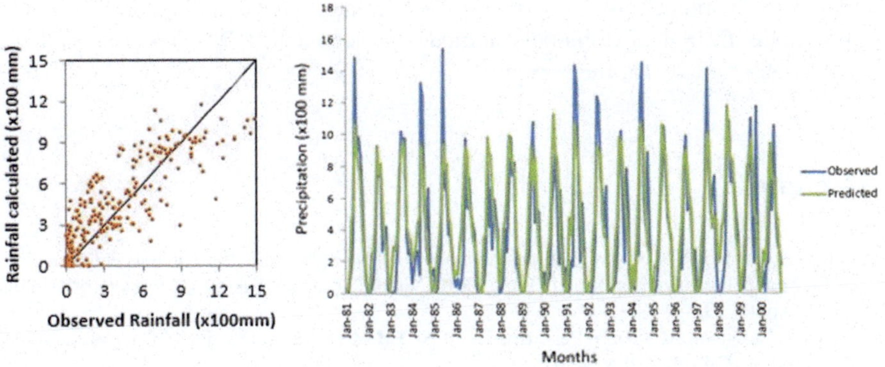

Fig. 22.3 Comparison of observed and simulated precipitation during the testing period

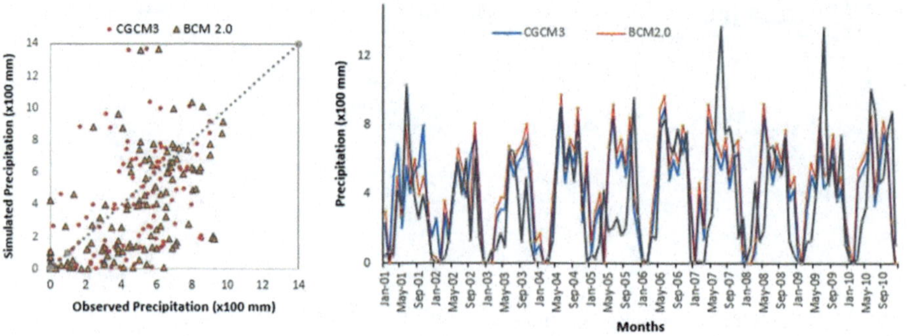

Fig. 22.4 Comparison of mean monthly observed and GCMs simulated precipitation for the time slice 2001–2010

the observed precipitation for a time slice from Jan 2001 to Dec 2010 as shown in Fig. 22.4. It is observed that in both cases the extreme values were simulated with less accuracy. The performance of the GCMs were assessed by calculating the error statistics RSQ and RMSE. The RSQ value for CGCM3 was 0.72 and that for BCM 2.0 was 0.78. The RMSE values for both cases were 2.76 and 1.6 respectively. Comparing these it is concluded that BCM2.0 could simulate the precipitation in the region better than the other GCM.

22.5 Summary and Conclusion

The precipitation in a study region in peninsular India was downscaled from the climate models using multivariate regression analysis. The predictors for downscaling were arrived at by finding the extent of correlation of these variables with the predictand. Two GCMs were used in the study to ascertain the better performing GCM in the area. Principal Component analysis was done for reducing the dimensionality of the predictors. Necessary error estimates were computed to arrive at the better model from the two models studied. It is seen that BCM2.0 performed better than the other model for the region.

References

Anandhi A, Srinivas VV, Nagesh Kumar D, Nanjundiah S (2009) Role of predictors in downscaling surface temperature to river basin in India for IPCC SRES scenarios using support vector machine. Int J Climatol 29:583–603

Cavazos T, Hewitson BC (2005) Performance of NCEP variables in statistical downscaling of daily precipitation. Clim Res 28:95–107

Dibike YB, Coulibaly P (2006) Temporal neural networks for downscaling climate variability and extremes. Neural Netw 19(2):135–144
Hewitson BC, Crane RG (1996) Climate downscaling: techniques and application. Climate Res 7:85–95
India-WRIS website (2018) https://www.india-wris.nrsc.gov.in/wrpinfo. Last accessed 01 Jan 2018
Kalnay E et al (1996) The NCEP/NCAR 40-year reanalysis project. Bull Am Meteor Soc 77(3):437–471
Mondal A, Mujumdar PP (2015) "Regional hydrological impacts of climate change: implications for water management in India" Hydrological sciences and water security: past, present and future. In: Proceeding of Eleventh Kovacs Colloquium, Paris, France, IAHS Publ., p 366
Reichler T, Kim J (2008) How well do coupled models simulate today's climate? Bull Am Meteorol Soc 89:303–311
Srinivasa Raju K, Nagesh Kumar D (2014) Ranking of global climate models for India using multicriterion analysis. Clim Res 60:103–117
Wilby RL, Dawson CW (2013) The statistical downscaling model: insights from one decade of application. Int J Clim 33(7):1707–1719
Wilby RL, Dawson CW, Barrow EM (2002) SDSM—a decision support tool for the assessment of regional climate change impacts Environ Model Softw 17:147–159

Chapter 23
Long-Term Historic Changes in Temperature and Potential Evapotranspiration Over Betwa River Basin

Ashish Pandey, Deen Dayal, S. S. Palmate, S. K. Mishra, S. K. Himanshu, and R. P. Pandey

Abstract In this study, trend in minimum temperature, maximum temperature, diurnal temperature range (DTR) and potential evapotranspiration (PET) has been assessed for the period of 1901–2013 over 21 stations in Betwa river basin, part of Central India. Modified Mann–Kendall (MMK) test has been employed for detecting the long-term trend in time series of variables. Further, Sen's slope estimator has been employed for estimation of magnitude of the trend. The results revealed increasing trend for minimum and maximum temperature at seasonal and annual bases barring monsoon season. DTR trends were found to be insignificant for all stations at winter, post-monsoon season and annual bases although a decreasing trend over a very few stations in pre-monsoon and monsoon seasons was observed. In case of PET, most stations exhibited insignificant trends, and only a few stations showed an increasing trend in pre-monsoon and winter seasons as well as on annual basis. This study may be useful in understanding the climatic conditions in the context of water resources planning and management.

A. Pandey (✉) · D. Dayal · S. S. Palmate · S. K. Mishra · S. K. Himanshu
Department of Water Resources Development and Management, Indian Institute of Technology Roorkee, Roorkee, India
e-mail: ashish.pandey@wr.iitr.ac.in

D. Dayal
e-mail: deemishra26@gmail.com

S. S. Palmate
e-mail: santoshpalmate@gmail.com

S. K. Mishra
e-mail: skm61fwt01@gmail.com

S. K. Himanshu
e-mail: sushilkumarhimanshu@gmail.com

R. P. Pandey
National Institute of Hydrology, Roorkee, India
e-mail: rpp.nihr@gov.in

A. Pandey et al. (eds.), *Climate Impacts on Water Resources in India*, Water Science and Technology Library 95, https://doi.org/10.1007/978-3-030-51427-3_23

23.1 Introduction

Climate change has gained significant global attention over the past few decades due to its deleterious impacts on a wide range of sectors. Alterations in hydro-climatology may adversely affect the fresh water availability, which has risen as a critical issue for the present generation (Gleick 1989; Singh et al. 2014). The reports of Intergovernmental Panel on Climate Change (IPCC) predicts that the Indian sub-continent will be subjected to frequent climatic aberrations, rise in temperature, substantial rainfall deficit and consequential water stress (Subash et al. 2011). The spatial distribution and magnitude of trends in temperature, as well as potential evapotranspiration (PET), would be extremely important and beneficial for natural resource management (Subash et al. 2011). The trends in PET are of great significance for agricultural water resources management (Huo et al. 2013).

There are many methods to analyze the time-series but it is essential to be aware of the methods which are valid and useful to meet the required test assumptions. It is also essential to examine and determine historic changes in the hydro-climatic parameters (Tabari et al. 2011). From the statistical point of view, trend analysis often refers to a technique for extracting an underlying pattern of behavior in a time series. Generally, statistical non-parametric methods are preferred to carry out trend analysis as they evade the problems aroused due to data skew (Yu et al. 1993; Yue and Pilon 2004a). The parametric tests applied on hydro-climatic time series are based on numerous assumptions, which may lead to erroneous results in practice. For nonparametric approaches, fewer assumptions are needed. Among nonparametric methods Mann–Kendall (MK) test (Mann 1945; Kendall 1975) is the most popular, which has been extensively used in hydrology literature (Pujol et al. 2007; Kumar et al. 2010; Shadmani et al. 2012; Gocic and Trajkovic 2013; Goyal 2014; Sayemuzzaman and Jha 2014; Tosunoglu and Kisi 2017; Swain et al. 2017; Dayal et al. 2018). However, it ignores the serial correlation in a time series (Hamed and Rao 1998; Yue and Wang 2002; Yue et al. 2003). It is observed that hydro-climatic time series is either serially correlated or auto-correlated (Yue et al. 2003; Yue and Wang 2004b; Tabari and Aghajanloo 2013), which leads to misinterpretation of the results. Therefore, modified MK test has been applied in this study to deal with the problem of auto-correlation.

The studies on trends of meteorological parameters at river basin level are only a few (Wang et al. 2007; Irmak et al. 2012; Krysanova et al. 2015; Dayal et al. 2018). Most of these studies are only focused on the trend of precipitation whereas temperature is also an important factor and should be considered for proper planning of underdeveloped basins. Lack of water resources infrastructure is one of the reasons behind underdevelopment of the Betwa river basin. Thus, it is required to study the temporal pattern of minimum and maximum temperature, diurnal temperature range (DTR), and Potential Evapotranspiration (PET) over the Betwa river basin.

23.2 Study Area

Betwa river basin is positioned in the central part of India. It covers five districts of southern Uttar Pradesh (UP) and ten districts of Madhya Pradesh (MP). It extends from 22° 52′ N to 26° 03′ N latitude and 77° 6′ E to 80° 14′ E longitude as shown in Fig. 23.1. The basin covers an area of 43,930 km^2 out of which 69% comes under MP and 31% comes under UP. The river traverses a length of 590 km in total (358 km in UP and 232 km in MP). Most of the basin area is under cultivation and has diverse land covers (topography and vegetation) in composite pattern (Dayal et al. 2019). The mean rainfall over the basin is 1140 mm per year. The area-averaged daily temperature varies from a minimum of 8.1 °C to a maximum of 42.3 °C (Suryavanshi et al. 2014; Palmate et al. 2017; Dayal et al. 2018).

23.3 Meteorological Data

For trend analysis, the monthly data for minimum and maximum temperature, DTR, and PET were obtained for 113 years (1901–2013) from India Water Portal (IWP) and India Meteorological Department (IMD). There were 21 stations over the Betwa basin as shown in Fig. 23.1. For seasonal analysis, the monthly data was rearranged into four seasons, i.e., monsoon, autumn, winter and summer corresponding to June–September, October–November, December-February and March–May respectively. In Indian context, summer and autumn seasons are also regarded as pre-monsoon and post-monsoon, respectively.

23.4 Methodology

In this study, two non-parametric approaches i.e. Modified Mann–Kendall (MMK) and Sen's slope tests, are employed to detect the trends in the time series. The MMK test has the advantage over the conventional Mann–Kendall (MK) test in terms of accounting the autocorrelation for all lags among the data points (Hamed and Rao 1998). The MK test is a rank based non-parametric test, which is widely used for detecting trend as well as its significance. However, it assumes the autocorrelation among the data points to be insignificant, which is not always true in practice. The autocorrelation among the data points may lead to misinterpretation of the MK test results (Hamed and Rao 1998). The MMK test basically modifies the computation of variance of the MK test when the data possess serial correlation. The detailed methodology of MK and MMK test can be obtained from hydro-climatology literature (Hirsch et al. 1982; Ehsanzadeh et al. 2011; Dayal et al. 2018).

Sen's slope estimator is also a non-parametric test (Theil 1950; Sen 1968) to estimate the slope of dataset. This is having an advantage of limited influence of

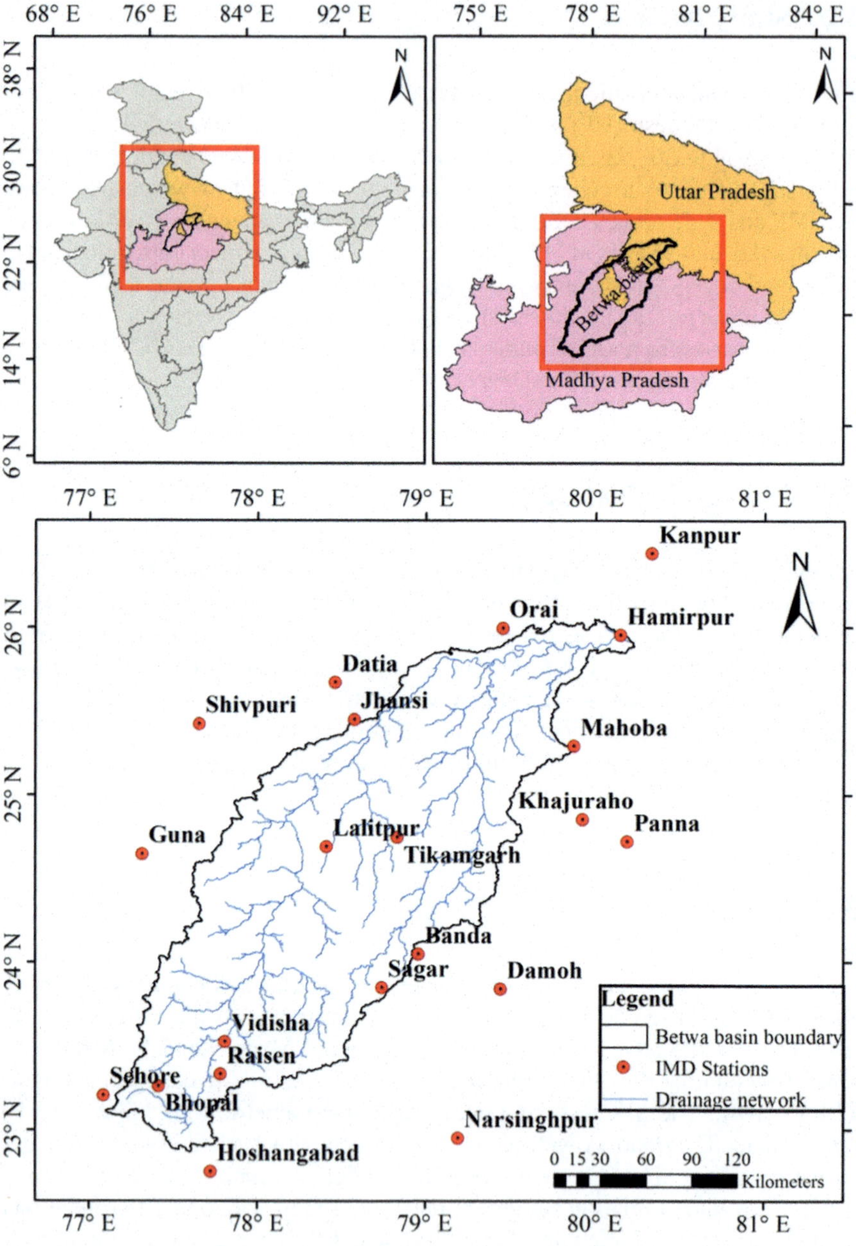

Fig. 23.1 Location map of the Betwa basin and meteorological stations

the outliers over the slopes, in contrast to the parametric tests e.g. linear regression, where the outliers majorly affect the trend results. Further, Sen's slope test is also used to compute the magnitude of trend in terms of relative change (RC) using the median of all the slopes, number of data points and mean value of all the data. The details of the Sen's slope test is available in numerous recent literature (Shadmani et al. 2012; Gocic and Trajkovic 2013; Suryavanshi et al. 2014; Krysanova et al. 2015; Dayal et al. 2018).

23.5 Results and Discussion

The results of trend analysis (using MMK and Sen's slope tests) for maximum and minimum temperature, DTR and PET over 21 stations in Betwa basin are obtained and discussed in subsequent sub-sections.

The trends in these parameters at 10, 5, and 1% significance level are identified from the Z_{mk} value. The results of Sen's slope estimator test (slope, intercept and RC) are reported only for the stations which possess significant Z_{mk} trends (10% or better significance level).

23.5.1 Minimum and Maximum Temperature

The results of MMK test (Z_{mk} values) at seasonal and annual scales and their corresponding significance level for both the maximum and minimum temperature are presented in Table 23.1. The Sen's slope test results for minimum temperature for the stations with significant Z_{mk} value are presented in Table 23.2. Results show that, during summer season, five stations (Kanpur, Lalitpur, Mahoba, Raisen, and Shivpuri) showed increasing trend at 10% significance level, five stations (Damoh, Datia, Guna, Sehore and Tikamgarh) showed increasing trend at 5% significance level and remaining stations except Panna showed increasing trend at 1% significance level. The slope of the increasing trend during summer season ranged from 0.004 to 0.012. During monsoon season, four stations (Datia, Kanpur, Shivpuri, and Orai) exhibited decreasing trend, and only one station (Hoshangabad) showed increasing trend with slope ranging from 0.003 to 0.008. Remarkably rising trend (at 1% significance level) has been observed for autumn season over all the stations with slope value ranging from 0.009 to 0.019. Similarly, increasing trend at 5% significance level for Damoh and increasing trend at 1% significance level for remaining stations has been observed during the winter season with slope varying from 0.005 to 0.017. In case of annual mean minimum temperature, five stations (Datia, Hamirpur, Jhansi, Panna, and Shivpuri) showed positive trend at 5% significance level whereas, the remaining stations showed positive trend at 1% significance level with Sen's slope varying between 0.003 and 0.009. Figure 23.2 presents a spatial view of the temporal variation of the minimum temperature at seasonal and annual scales over Betwa basin.

Table 23.1 Results of trend analysis (Z_{mk} values) for mean minimum and maximum temperature

Station	Summer		Monsoon		Autumn		Winter		Annual	
	T_{min}	T_{max}	T_{min}	T_{max}	T_{min}	T_{max}	T_{min}	T_{max}	T_{min}	T_{max}
Banda	3.81[a]	3.20[a]	−0.45	0.29	4.86[a]	5.72[a]	5.23[a]	6.27[a]	3.44[a]	4.11[a]
Bhopal	3.03[a]	2.52[b]	0.97	−0.62	5.31[a]	4.66[a]	8.61[a]	3.90[a]	3.91[a]	2.54[b]
Damoh	2.45[b]	3.94[a]	0.28	1.53	3.45[a]	4.53[a]	2.41[b]	5.89[a]	2.96[a]	3.97[a]
Datia	2.22[b]	3.47[a]	−2.05[e]	−1.00	3.96[a]	3.68[a]	5.87[a]	4.82[a]	2.52[b]	3.72[a]
Guna	2.39[b]	3.80[a]	1.07	0.81	3.19[a]	3.75[a]	11.98[a]	5.20[a]	3.40[a]	4.82[a]
Hamirpur	2.93[a]	3.17[a]	−0.96	−0.77	4.01[a]	3.63[a]	4.74[a]	2.19[b]	2.49[b]	2.83[a]
Hoshangabad	3.03[a]	4.25[a]	2.03[b]	2.21[b]	3.75[a]	4.68[a]	7.36[a]	3.91[a]	4.58[a]	4.97[a]
Jhansi	2.89[a]	4.46[a]	−0.46	−0.81	3.59[a]	4.02[a]	4.18[a]	4.38[a]	2.44[b]	3.89[a]
Kanpur	1.78[c]	1.22	−5.33[d]	−3.97[d]	4.42[a]	1.93[c]	6.58[a]	2.44[b]	2.79[a]	0.41
Khajuraho	2.81[a]	3.50[a]	1.01	1.11	4.17[a]	4.94[a]	5.01[a]	6.22[a]	4.75[a]	3.21[a]
Lalitpur	1.78[c]	2.14[b]	−0.97	−0.93	2.91[a]	3.15[a]	4.55[a]	5.98[a]	3.48[a]	3.32[a]
Mahoba	1.78[c]	1.96[c]	−1.56	−1.67[f]	3.35[a]	4.41[a]	6.87[a]	4.60[a]	4.37[a]	3.06[a]
Narsinghpur	3.05[a]	3.37[a]	1.59	1.51	3.80[a]	3.53[a]	13.89[a]	6.54[a]	4.21[a]	3.28[a]
Orai	3.27[a]	3.35[a]	−1.88[f]	−2.24[e]	4.81[a]	2.86[a]	7.33[a]	3.23[a]	3.55[a]	2.38[b]
Panna	1.41	3.70[a]	−0.88	0.32	4.34[a]	4.36[a]	4.68[a]	3.86[a]	2.03[b]	3.24[a]
Raisen	1.95[c]	2.06[b]	0.94	0.48	3.46[a]	4.10[a]	4.00[a]	4.12[a]	3.32[a]	3.77[a]
Sagar	3.22[a]	3.40[a]	0.82	−0.23	3.73[a]	4.03[a]	10.17[a]	7.64[a]	4.51[a]	4.61[a]
Sehore	2.56[b]	2.85[a]	0.39	0.38	3.67[a]	4.43[a]	6.51[a]	6.49[a]	3.47[a]	3.72[a]
Shivpuri	1.65[c]	2.38[b]	−1.70[f]	−0.85	3.36[a]	2.29[b]	4.95[a]	5.27[a]	2.37[b]	3.55[a]
Tikamgarh	2.31[b]	3.34[a]	0.00	0.23	2.94[a]	3.89[a]	8.91[a]	4.70[a]	3.57[a]	4.13[a]

(continued)

Table 23.1 (continued)

Station	Summer		Monsoon		Autumn		Winter		Annual	
	T_{min}	T_{max}	T_{min}	T_{max}	T_{min}	T_{max}	T_{min}	T_{max}	T_{min}	T_{max}
Vidisha	3.20[a]	3.79[a]	1.45	0.65	3.48[a]	4.06[a]	6.24[a]	7.03[a]	4.35[a]	4.79[a]

Note [a,b,c] stands for significant positive trend, whereas [d,e,f] for significant negative trend, at 1%, 5% and 10% significance level respectively

Table 23.2 Value of slope, intercept and relative change for mean minimum temperature

Station	Summer			Monsoon			Autumn			Winter			Annual		
	Slope	RC	Intercept	Slope	RC	Intercept	Slope	RC	Intercept	Slope	RC	Intercept	Slope	RC	Intercept
Banda	0.011	6.12	20.26	–	–	–	0.017	12.57	14.37	0.015	19.40	8.00	0.008	4.78	18.00
Bhopal	0.007	3.79	21.73	–	–	–	0.016	10.92	15.30	0.011	11.26	10.55	0.008	4.67	18.47
Damoh	0.005	2.80	21.78	–	–	–	0.011	7.68	15.78	0.005	4.90	11.04	0.005	3.01	18.75
Datia	0.007	3.58	21.22	−0.005	−2.09	25.17	0.010	7.10	15.20	0.010	12.87	8.35	0.004	2.66	18.40
Guna	0.006	3.22	21.66	–	–	–	0.012	7.96	15.73	0.009	10.19	9.99	0.006	3.82	18.54
Hamirpur	0.012	6.72	20.23	–	–	–	0.019	13.63	14.32	0.017	22.14	7.66	0.008	5.14	17.98
Hoshangabad	0.006	2.97	22.23	0.006	2.72	22.95	0.017	11.86	15.27	0.010	9.61	11.41	0.009	5.15	18.63
Jhansi	0.011	5.74	21.00	–	–	–	0.015	10.31	15.00	0.012	14.45	8.51	0.006	3.51	18.29
Kanpur	0.005	2.67	20.77	– 0.008	– 3.43	26.64	0.009	6.68	15.12	0.012	15.79	7.92	0.003	2.00	18.57
Khajuraho	0.007	3.95	20.95	–	–	–	0.012	8.59	15.23	0.007	7.86	9.50	0.008	4.58	18.23
Lalitpur	0.006	2.80	21.63	–	–	–	0.013	7.97	15.72	0.011	10.25	10.06	0.006	3.15	18.57
Mahoba	0.005	2.32	20.56	–	–	–	0.013	8.65	14.63	0.010	12.08	8.28	0.005	2.86	17.99
Narsinghpur	0.007	3.56	21.42	–	–	–	0.018	13.01	14.57	0.009	9.33	10.63	0.008	5.12	17.89
Orai	0.009	4.79	20.74	−0.005	−2.23	25.93	0.014	10.31	14.78	0.014	18.89	7.74	0.006	3.80	18.26
Panna	–	–	–	–	–	–	0.015	10.60	15.02	0.008	9.49	9.58	0.005	2.98	18.33
Raisen	0.004	2.24	22.15	–	–	–	0.016	11.21	15.43	0.008	7.70	11.33	0.006	3.69	18.81
Sagar	0.009	4.59	21.82	–	–	–	0.018	12.02	15.57	0.013	12.98	10.73	0.009	5.20	18.56
Sehore	0.007	3.28	22.23	–	–	–	0.015	9.46	15.60	0.011	9.62	11.16	0.007	3.96	18.85
Shivpuri	0.005	2.45	21.58	−0.003	−1.53	24.57	0.010	7.05	15.68	0.009	10.17	9.28	0.004	2.49	18.55
Tikamgarh	0.007	3.59	21.42	–	–	–	0.010	7.20	15.62	0.008	9.15	9.59	0.006	3.80	18.50

(continued)

Table 23.2 (continued)

Station	Summer			Monsoon			Autumn			Winter			Annual		
	Slope	RC	Intercept	Slope	RC	Intercept	Slope	RC	Intercept	Slope	RC	Intercept	Slope	RC	Intercept
Vidisha	0.009	4.68	21.44	–	–	–	0.017	11.95	15.25	0.012	12.86	10.23	0.009	5.44	18.25

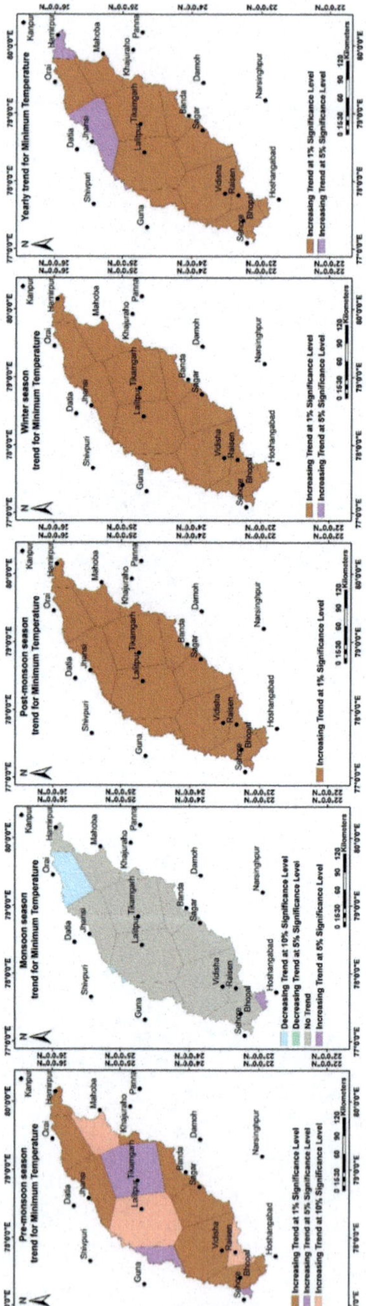

Fig. 23.2 Spatial trend (at various significance levels) of mean minimum temperature over the study area

For maximum temperature also, the data at monthly scale was obtained and used to compute mean value for seasonal and annual scale, which was assessed for autocorrelation and trend detection. As mentioned above, the Mann–Kendall Z values (Z_{mk}) are presented in Table 23.1. For the stations possessing significance in trend, Sen's slope estimator results are given in Table 23.3. It is observed that, during summer season, all the stations exhibited positive trend. However, only one station (Mahoba) and four stations (Bhopal, Lalitpur, Raisen, and Shivpuri) showed positive trend at 10 and 5% significance level respectively. Kanpur is the only station to have a non-significant positive trend. The remaining fifteen stations showed a rising trend of maximum temperature at 1% significance level. The Sen's slope ranged from 0.004 to 0.012 over different stations. During monsoon season, three stations i.e. Kanpur, Mahoba and Orai exhibited decreasing trend at 1, 10 and 5% significance level respectively, and only Hoshangabad station showed increasing trend at 5% significance level. Increasing trend at 10% significance level in Kanpur, at 5% significance level in Shivpuri and at 1% significance level in the remaining stations were observed for the autumn season with Sen's slope varying from 0.006 to 0.019. Positive trend at 5% significance level in two stations (Hamirpur and Kanpur) and at 1% significance level in remaining stations has been observed during the winter season with Sen's slope varying from 0.005 to 0.012. In case of annual mean maximum temperature, two stations (Bhopal and Orai) showed rising trend at 5% significance level whereas, the remaining stations except Kanpur showed rising trend at 1% significance level. The Sen's slope ranged between 0.003 and 0.009. The Kanpur station possessed a significant decreasing trend for monsoon season, significant increasing trend for autumn and winter seasons, and non-significant increasing trend for summer season, which resulted in a no significant increasing or decreasing trend at annual basis. The spatial view of the temporal variation of the mean maximum temperature at seasonal and annual scales over Betwa basin is presented in Fig. 23.3.

23.5.2 Diurnal Temperature Range (DTR)

Monthly mean DTR data was used to estimate the trend and its significance using Mann–Kendall statistic (Z_{mk}). The Sen's slope test results are presented in Table 23.4. The results show that, during summer season, only two stations (Lalitpur and Sagar) showed decreasing trend at 5% significance level. For monsoon season, three stations (Hoshangabad, Mahoba, and Raisen) showed negative trend at 10% significance level whereas, two stations (Sagar and Vidisha) showed negative trend at 5% significance level. However, an increasing trend has been observed in Narsinghpur at 10% significance level during the autumn season. No significant trend has been observed during the winter season. In case of annual mean DTR, only Banda station

Table 23.3 Value of slope, intercept and relative change for mean maximum temperature

Station	Summer			Monsoon			Autumn			Winter			Annual		
	Slope	RC	Intercept	Slope	RC	Intercept	Slope	RC	Intercept	Slope	RC	Intercept	Slope	RC	Intercept
Banda	0.010	3.06	37.03	–	–	–	0.018	6.78	29.42	0.011	4.85	24.18	0.008	2.90	31.62
Bhopal	0.005	1.51	37.86	–	–	–	0.012	4.50	30.19	0.007	2.81	26.77	0.005	1.67	31.95
Damoh	0.009	2.85	37.19	–	–	–	0.019	7.14	29.18	0.009	3.71	25.75	0.009	3.14	31.26
Datia	0.011	3.38	37.06	–	–	–	0.013	4.70	30.15	0.010	4.48	23.80	0.007	2.34	31.76
Guna	0.010	3.13	37.05	–	–	–	0.015	5.38	30.22	0.012	5.41	25.18	0.008	2.92	31.57
Hamirpur	0.008	2.25	37.43	–	–	–	0.011	4.02	30.17	0.006	2.83	24.32	0.004	1.55	32.12
Hoshangabad	0.009	2.65	37.77	0.004	1.49	31.39	0.018	6.63	29.63	0.006	2.46	27.35	0.009	3.05	31.71
Jhansi	0.011	3.24	37.16	–	–	–	0.014	5.18	29.99	0.009	3.91	24.20	0.006	2.21	31.75
Kanpur	–	–	–	−0.008	– 2.52	35.82	0.006	2.24	30.81	0.005	2.48	24.18	–	–	–
Khajuraho	0.012	3.69	36.97	–	–	–	0.018	6.51	29.54	0.010	4.23	24.87	0.009	3.15	31.40
Lalitpur	0.006	1.65	37.15	–	–	–	0.013	4.36	29.87	0.010	4.05	25.02	0.006	1.78	31.55
Mahoba	0.005	1.27	37.16	−0.004	– 1.22	33.93	0.014	4.52	29.79	0.009	3.79	24.30	0.005	1.56	31.70
Narsinghpur	0.012	3.57	36.18	–	–	–	0.017	6.48	28.11	0.010	4.34	25.36	0.008	3.06	30.20
Orai	0.007	2.12	37.43	−0.005	– 1.46	35.22	0.010	3.57	30.46	0.007	3.17	23.96	0.003	1.21	32.20
Panna	0.007	2.15	37.15	–	–	–	0.015	5.61	29.27	0.008	3.49	25.17	0.007	2.35	31.44
Raisen	0.004	1.29	37.89	–	–	–	0.016	5.84	29.76	0.006	2.70	26.75	0.006	2.10	31.82
Sagar	0.007	2.18	37.18	–	–	–	0.016	6.07	29.48	0.010	4.52	25.61	0.007	2.45	31.31
Sehore	0.007	1.90	38.52	–	–	–	0.015	4.91	30.76	0.011	3.91	27.82	0.008	2.38	32.53
Shivpuri	0.007	2.11	36.99	–	–	–	0.008	2.93	30.44	0.009	4.02	24.43	0.005	1.73	31.68
Tikamgarh	0.008	2.44	37.17	–	–	–	0.013	4.87	29.93	0.010	4.61	24.75	0.007	2.33	31.62
Vidisha	0.009	2.85	37.06	–	–	–	0.016	6.05	29.69	0.012	5.10	25.57	0.008	2.79	31.34

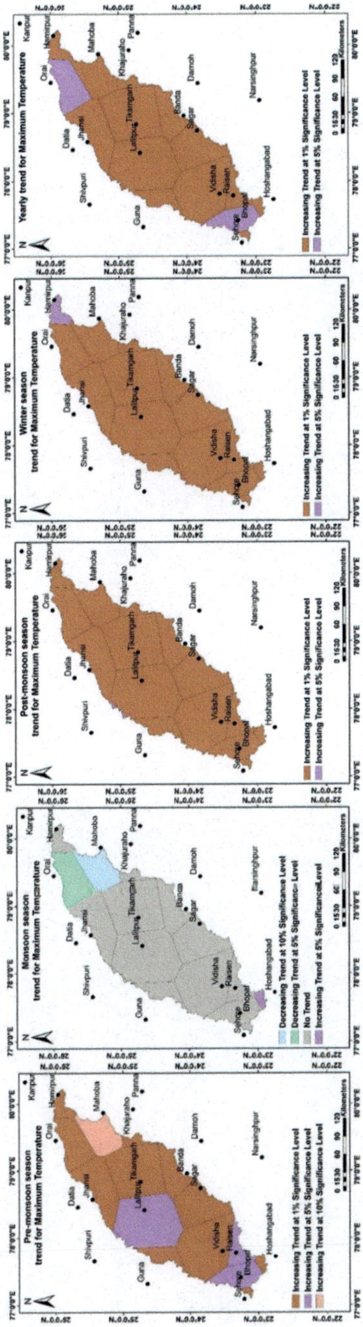

Fig. 23.3 Spatial trend (at various significance levels) of mean maximum temperature over the study area

Table 23.4 Value of slope, intercept and relative change for DTR

Station	Summer			Monsoon			Autumn			Winter			Annual		
	Slope ($\times 10^{-5}$)	RC	Intercept	Slope ($\times 10^{-5}$)	RC	Intercept	Slope ($\times 10^{-5}$)	RC	Intercept	Slope ($\times 10^{-5}$)	RC	Intercept	Slope ($\times 10^{-5}$)	RC	Intercept
Banda	–	–	–	–	–	–	–	–	–	–	–		−57.5	−0.476	13.67
Hoshangabad	–	–	–	−20.8	−0.279	8.43	–	–	–	–	–	–	–	–	–
Lalitpur	−27.3	−0.179	15.52	–	–	–	–	–	–	–	–	–	–	–	–
Mahoba	–	–	–	−11.1	−0.126	9.04	–	–	–	–	–	–	–	–	–
Narsinghpur	–	–	–	–	–	–	18.6	0.155	13.51	–	–	–	–	–	–
Raisen	–	–	–	−38.3	−0.509	8.50	–	–	–	–	–	–	–	–	–
Sagar	−38.9	−0.287	15.35	−30.8	−0.412	8.48	–	–	–	–	–	–	–	–	–
Vidisha	–	–	–	−26.4	−0.345	8.69	–	–	–	–	–	–	–	–	–

showed negative trend at 10% significance level. Figure 23.4 show the spatial distribution of seasonal and annual mean DTR trends across the 21 stations over Betwa river basin. Since both the maximum and minimum temperature trends were positive over most of the stations, the DTR might not have undergone any significant temporal changes, which can be observed from the negligible values of Sen's slope for almost all the stations.

23.5.3 Potential Evapotranspiration (PET)

In this analysis, monthly PET data has been used to estimate the Mann–Kendall Z values and the corresponding significance levels at 10, 5 and 1%. The results of Sen's slope estimator test are outlined in Table 23.5. The result shows that, during summer (or pre-monsoon) season, all the stations showed non-significant trends except Sehore station, which showed an increasing trend at 10% significance level with a slope value of 0.04. Similarly, for autumn as well as monsoon season, no station exhibited significant trend. A significant rising trend (at 1% significance level) in three stations (Lalitpur, Mahoba, and Sehore) have been observed during the winter season with Sen's slope varying from 0.046 to 0.062. In case of annual PET, only one station (Sehore) and four stations (Guna, Hoshangabad, Raisen, and Vidisha) showed positive trends at 10% and 5% significance level respectively. The Sen's slope over these stations ranged from 0.105 to 0.223. Figure 23.5 gives a pictorial view of the trends and significance levels of PET at seasonal and annual scales over the Betwa river basin.

23.6 Conclusion

In this study, the trend of various meteorological variables was analyzed using the modified MK test and Sen's slope estimator method. The study reveals that the minimum, maximum and average temperatures have increased for all the stations during summer, autumn and winter seasons at three significance levels. The diurnal temperature range is decreasing at some stations during monsoon and summer seasons. A decreasing trend of the DTR was found, which may be due to more increase in minimum temperature as compared to increase in maximum temperature. This may be due to local influences i.e. irrigation, urban growth, land use variations etc. The PET parameter doesn't show any trend in summer and monsoon seasons except at one station (which is showing increasing trend during summer).

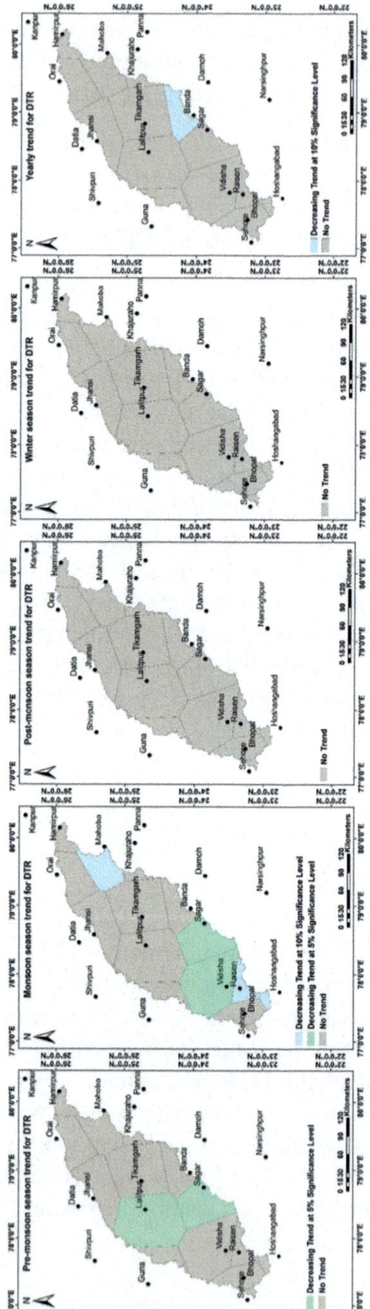

Fig. 23.4 Spatial trend (at various significance levels) of DTR over the study area

Table 23.5 Value of slope, intercept and relative change for PET

Station	Summer			Monsoon			Autumn			Winter			Annual		
	Slope	RC	Intercept	Slope	RC	Intercept	Slope	RC	Intercept	Slope	RC	Intercept	Slope	RC	Intercept
Guna	–	–	–	–		–	–		–	–	–	–	0.219	1.03	2386.07
Hoshangabad	–	–	–	–	–		–		–	–	–	–	0.223	1.05	2386.65
Lalitpur	–	–	–	–	–		–		–	0.048	1.01	483.10	–	–	–
Mahoba	–	–	–	–	–	–	–	–	–	0.046	0.97	479.11	–	–	–
Raisen	–	–	–	–	–	–	–	–	–	–	–	–	0.165	0.78	2384.38
Sehore	0.040	0.53	768.84	–	–	–	–	–	–	0.062	1.19	532.44	0.105	0.43	2450.99
Vidisha	–	–	–	–	–	–	–	–	–	–	–	–	0.220	1.04	2376.42

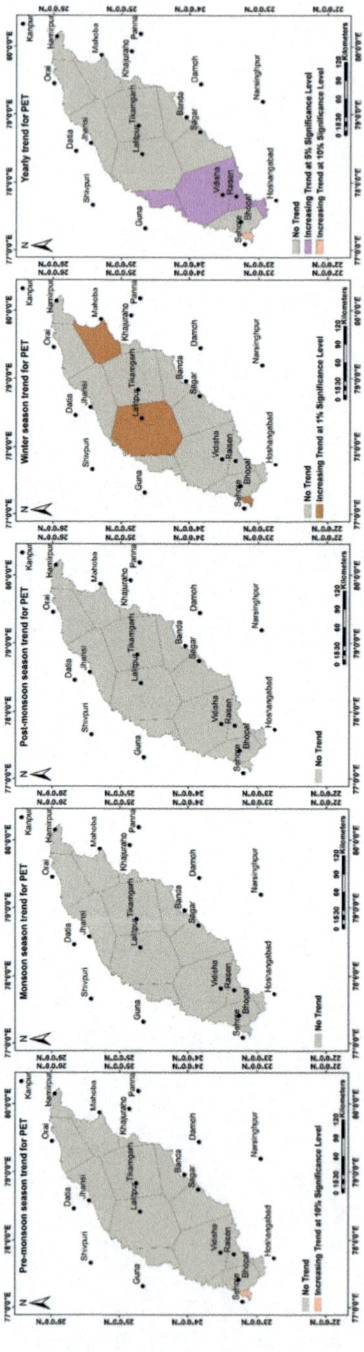

Fig. 23.5 Spatial trend (at various significance levels) of PET over the study area

However, during the winter season, PET shows increasing trend at three stations. In case of annual PET, five stations show increasing trend. This study reveals the long-term changes in climatic variables and their spatiotemporal distribution pattern, which will be helpful for effective water management in the Betwa river basin.

References

Dayal D, Pandey A, Himanshu SK, Palmate SS (2018) Long term historic changes of precipitation and aridity index over an Indian River Basin. In: World environmental and water resources congress 2018: groundwater, sustainability, and hydro-climate/climate change, pp 262–272

Dayal D, Swain S, Gautam AK, Palmate SS, Pandey A, Mishra SK (2019) Development of ARIMA model for monthly rainfall forecasting over an Indian River Basin. In: World environmental and water resources congress 2019: watershed management, irrigation and drainage, and water resources planning and management, pp 264–271

Ehsanzadeh E, Ouarda TB, Saley HM (2011) A simultaneous analysis of gradual and abrupt changes in Canadian low streamflows. Hydrol Process 25(5):727–739

Gleick PH (1989) Climate change, hydrology, and water resources. Rev Geophys 27(3):329–344

Gocic M, Trajkovic S (2013) Analysis of changes in meteorological variables using Mann-Kendall and Sen's slope estimator statistical tests in Serbia. Global Planet Change 100:172–182

Goyal MK (2014) Statistical analysis of long term trends of rainfall during 1901–2002 at Assam India. Water Res Manage 28(6):1501–1515

Hamed KH, Rao AR (1998) A modified Mann-Kendall trend test for autocorrelated data. J Hydrol 204(1–4):182–196

Hirsch RM, Slack JR, Smith RA (1982) Techniques of trend analysis for monthly water quality data. Water Resour Res 18(1):107–121

Huo Z, Dai X, Feng S, Kang S, Huang G (2013) Effect of climate change on reference evapotranspiration and aridity index in arid region of China. J Hydrol 492:24–34

Irmak S, Kabenge I, Skaggs KE, Mutiibwa D (2012) Trend and magnitude of changes in climate variables and reference evapotranspiration over 116-yr period in the Platte River Basin, central Nebraska–USA. J Hydrol 420:228–244

Kendall MG (1975) Rank correlation methods. Charles Griffin, London

Krysanova V, Wortmann M, Bolch T, Merz B, Duethmann D, Walter J, Kundzewicz ZW (2015) Analysis of current trends in climate parameters, river discharge and glaciers in the Aksu River basin (Central Asia). Hydrol Sci J 60(4):566–590

Kumar V, Jain SK, Singh Y (2010) Analysis of long-term rainfall trends in India. Hydrol Sci J 55(4):484–496

Mann HB (1945) Non-parametric tests against trend. Econometrica 13:245–259

Palmate SS, Pandey A, Kumar D, Pandey RP, Mishra SK (2017) Climate change impact on forest cover and vegetation in Betwa Basin India. Appl Water Sci 7(1):103–114

Pujol N, Neppel LUC, Sabatier R (2007) Regional tests for trend detection in maximum precipitation series in the French Mediterranean region. Hydrol Sci J 52(5):956–973

Sayemuzzaman M, Jha MK (2014) Seasonal and annual precipitation time series trend analysis in North Carolina, United States. Atmos Res 137:183–194

Sen PK (1968) Estimates of the regression coefficient based on Kendall's tau. J American Statistical Assoc 63(324):1379–1389

Shadmani M, Marofi S, Roknian M (2012) Trend analysis in reference evapotranspiration using Mann-Kendall and Spearman's Rho tests in arid regions of Iran. Water Resour Manage 26(1):211–224

Singh VP, Khedun CP, Mishra AK (2014) Water, environment, energy, and population growth: implications for water sustainability under climate change. J Hydrol Eng 19(4):667–673

Subash N, Sikka AK, Ram Mohan HS (2011) An investigation into observational characteristics of rainfall and temperature in Central Northeast India—a historical perspective 1889–2008. Theoret Appl Climatol 103:305–319

Suryavanshi S, Pandey A, Chaube UC, Joshi N (2014) Long-term historic changes in climatic variables of Betwa Basin India. Theoret Appl Climatol 117(3–4):403–418

Swain S, Patel P, Nandi S (2017) Application of SPI, EDI and PNPI using MSWEP precipitation data over Marathwada, India. In: 2017 IEEE International geoscience and remote sensing symposium (IGARSS), IEEE, pp 5505–5507 IEEE.

Tabari H, Aghajanloo MB (2013) Temporal pattern of aridity index in Iran with considering precipitation and evapotranspiration trends. Int J Climatol 33(2):396–409

Tabari H, Somee BS, Zadeh MR (2011) Testing for long-term trends in climatic variables in Iran. Atmos Res 100:132–140

Theil H (1950) A rank-invariant method of linear and polynomial regression analysis. K Ned Akad Wet 53:386–392

Tosunoglu F, Kisi O (2017) Trend analysis of maximum hydrologic drought variables using Mann-Kendall and Şen's innovative trend method. River Res Appl 33(4):597–610

Wang Y, Jiang T, Bothe O, Fraedrich K (2007) Changes of pan evaporation and reference evapotranspiration in the Yangtze River basin. Theoret Appl Climatol 90(1–2):13–23

Yu YS, Zou S, Whittemore D (1993) Non-parametric trend analysis of water quality data of rivers in Kansas. J Hydrol 150(1):61–80

Yue S, Pilon P (2004a) A comparison of the power of the t test, Mann-Kendall and bootstrap tests for trend detection. Hydrol Sci J 49(1):21–37

Yue S, Wang C (2004b) The Mann-Kendall test modified by effective sample size to detect trend in serially correlated hydrological series. Water Res Manage 18(3):201–218

Yue S, Wang CY (2002) Applicability of prewhitening to eliminate the influence of serial correlation on the Mann-Kendall test. Water Res Res 38(6):1–4

Yue S, Pilon P, Phinney BOB (2003) Canadian streamflow trend detection: impacts of serial and cross-correlation. Hydrol Sci J 48(1):51–63

Chapter 24
Climate Change Detection in Upper Ganga River Basin

Chetan Sharma and C. S. P. Ojha

Abstract Upper Ganga river basin is composed of complex topography and more likely to be affected by climate change. Mann–Whitney-Pettit method was used to detect the significant shift/change in precipitation in Upper Ganga river basin and to find the time of change. Fine resolution precipitation data provided by Climatic research unit (CRU) was adopted as reference data. Significant decreasing trends at all locations was found at 5% significance level using Mann–Kendall test (MK). Most of the locations indicated the year 1992 as change point at 5% significance level. Higher decreasing trends were observed in the lower reaches while upper reaches exhibited earlier change point.

24.1 Introduction

The fifth assessment report (AR5) prepared by Working Group I of Intergovernmental Panel on climate change (IPCC) confirmed that climate change is happening due to increase in GHGs and changes in Land use (IPCC 2013). The climate change is found to be affecting different hydro-meteorological variables, i.e., precipitation, temperature, etc. Alexander et al. (2006) found that there is decrease in number of cold nights and increase in warm nights globally. The intensity of extreme rainfall events is also found to be increased after mid-twentieth century. Kumar et al. (2011) found that there is overall increase in temperature in Indian subcontinent. Jain and Kumar (2012) studied the trends of annual rainfall in major Indian river basin and found that 6 river basins are showing increasing trends while 15 river basin showing decrease in rainfall. Sharma and Ojha (2019) found that there have been increase in the annual precipitation in majority far latitude regions while mid latitude areas indicated decrease in precipitation. Significant decrease in the precipitation and wet day frequency in

C. Sharma (✉)
Department of Civil Engineering, MITS, Gwalior, India
e-mail: chetan.cvl@gmail.com

C. S. P. Ojha
Department of Civil Engineering, Indian Institute of Technology Roorkee, Roorkee, India
e-mail: cspojha@gmail.com

A. Pandey et al. (eds.), *Climate Impacts on Water Resources in India*, Water Science and Technology Library 95, https://doi.org/10.1007/978-3-030-51427-3_24

the Ganga basin was also observed in last century. Mondal and Mujumdar (2012) presented effect of climate change in annual rainfall and streamflow patterns in Mahanadi river basin, India at basin scale. Significant decreasing trends in precipitation was observed and the location of change point was detected from the year 1974–1999. Different studies are also available showing decreasing trends in rainfall patterns in Uttarakhand, India where river Ganga originates (Pranuthi et al 2014; Sharma et al. 2015; Sharma and Ojha 2018).

It is evident that climate change in affecting Indian river basins. In this study an effort has been made to find if climate change is affecting precipitation patterns in the Upper Ganga river basin. The exact location of climate change is also found out in this study.

24.2 Study Area

Upper Ganga river basin, encompassed between latitude 29°45′ N–31°25′ N and longitude 77°05′ E–80°06′ E up to district Haridwar, is considered as study area. It is characterized by complex topography composed of snowclad mountains having elevation upto ~6400 m to plains near Haridwar (Sharma et al. 2019). Mighty river Ganga originates from Gangotri in Upper Ganga river basin and travel ~2600 km to meet Bay of Bengal. Study area is shown in Fig. 24.1.

High spatial resolution monthly total precipitation available at grid interval of 0.5° latitude × 0.5° longitude provided by Climatic Research Unit (CRU) (Harris et al. 2014) for year 1901–2016 was used as reference data in this study. The precipitation data was further re-gridded by using spline interpolation to get the precipitation at the grid interval of 0.25° × 0.25°.

24.3 Methodology

Monthly precipitation data was used to develop annual total precipitation at each grid point. These annual values were used to compute magnitude of trends, significance of trends using Mann–Kendall test (Kendall 1975) and to find the location of change point.

Least square linear slopes were computed to find trend in annual precipitation at each grid point. Mann-Kendal (MK) test is a popular method to find significance of trends in a time series (Sharma and Ojha 2020). The significance of the trends was checked at 5% significance level.

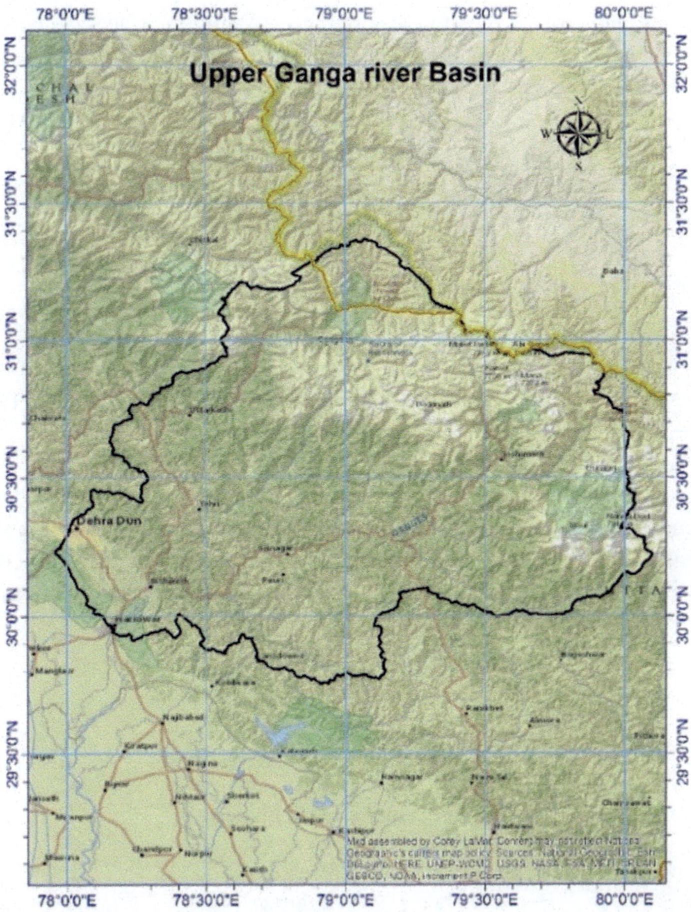

Fig. 24.1 Study area: upper Ganga river basin

24.3.1 Mann-Kendal Test (MK)

The test static 'S_{MK}' of the MK test is calculated as

$$S_{MK} = \sum_{i=1}^{n-1} \sum_{j=i+1}^{n} \operatorname{sgn}(a_j - a_i) \tag{24.1}$$

here $a_1, a_2, \ldots, a_n$ are n data points in a time series.

$$\operatorname{sgn}(a_j - a_i) = \begin{cases} 1\, \mathit{if}\, (a_j - a_i) > 0 \\ 0\, \mathit{if}\, (a_j - a_i) = 0 \\ -1\, \mathit{if}\, (a_j - a_i) < 0 \end{cases} \tag{24.2}$$

A positive value of 'S_{MK}' indicates rising trend and vice-versa. The significance of 'S_{MK}' is checked by computing standardized test static 'Z_{MK}' as

$$Z_{MK} = \begin{cases} \frac{S_{MK}-1}{\sqrt{VAR(S_{MK})}} \text{ if } S_{MK} > 0 \\ 0 \text{ if } S_{MK} = 0 \\ \frac{S_{MK}+1}{\sqrt{VAR(S_{MK})}} \text{ if } S_{MK} < 0 \end{cases} \quad (24.3)$$

$$VAR(S_{MK}) = \frac{n(n-1)(2n+5) - \sum_{k=1}^{p} t_k(t_k - 1))(2t_k + 5)}{18} \quad (24.4)$$

where p is number of groups with equal value and t_k is number of data points in the kth group.

If $-Z\alpha_{/2} \leq Z_{MK} \leq Z\alpha_{/2}$, (Z is standard normal deviate) indicate that trend is not significant at α significance level.

The change point detection i.e., to find the location of change point in a time series, was calculated using non-parametric change point detection method, Mann Whitney-Pettit (MWP) (Pettitt 1979). It is a non-parametric test of change detection. The MWP test is explained in the following paragraph.

24.3.2 Mann–Whitney-Pettit Test

For a time series $A(t)$ of length N, index U_t is given by

$$U_t = \sum_{i=1}^{t} \sum_{j=t+1}^{N} \text{sgn}(a_j - a_i) \quad (24.5)$$

where,

$$\text{sgn}(a_j - a_i) = \begin{cases} 1, for, a_j > a_i \\ 0, for, a_j = a_i \\ -1, for, a_j < a_i \end{cases} \quad (24.6)$$

A plot of $|U_t|$ versus t increases up to a peak value where change become significant and start decreasing after that. The location of Peak value of $|U_t|$ indicate possible change point

$$(K_t)_{MWP} = \max(U_t) \quad (24.7)$$

Change point is considered as significant at defined significance level α if approximate value of significance probability of change point p_t exceeds 1-α. p_t is given by

$$p_t = 1 - \exp\left(\frac{-6(K_t)^2_{MWP}}{N^3 + N^2} \right) \tag{24.8}$$

24.4 Results and Discussion

Least square linear slopes were used to find the trends of annual precipitation and presented in Fig. 24.2a. Most of these trends were found significant at 5% significance level using MK test. It can be seen that majority of locations are showing net negative trends which indicates that overall annual rainfall has decreased in the basin. Higher decreasing trends were observed in the southern regions. It can also be observed that magnitude of decreasing trends is relatively higher in lower reaches of river Ganga than upper reaches. All of these decreasing trends were found significant using MK test.

Spatial patterns of the location of change point in Upper Ganga river basin is shown in Fig. 24.2b. It can be noted that the majority of locations are showing the year 1992 as change point. Change is detected earlier in some Northern and Western locations in Upper Ganga river basin.

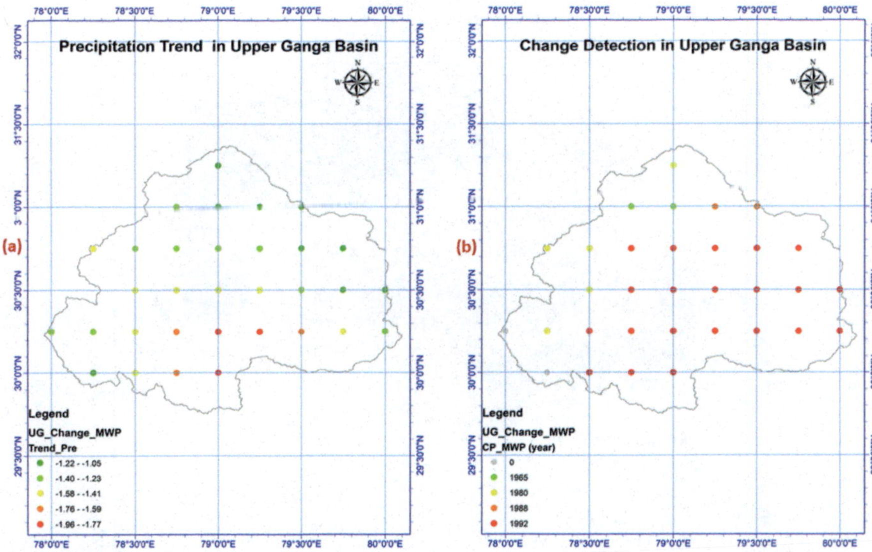

Fig. 24.2 Upper Ganga river basin (**a**) Trends (mm/ year) significant at 5% significance level. (**b**) Years of change detection using MWP test

To better understand these spatial patterns of trends and change point, a digital elevation model (DEM) representing spatial patterns of terrain in upper Ganga river basin was prepared as provided in Fig. 24.3. It can be seen that southern region of the study area is at lower elevation than northern parts of Upper Ganga river basin. Northern region is dominated by snow cladded mountains and topographically complex. The spatial patterns of trends show that annual rainfall in the lower reaches of the basin have significantly deceased than the region of higher elevations. It may be due to the fact that lower reaches are becoming highly populated and anthropogenic activities are causing more harm to the natural climate (Zhang et al. 2007). Lower catchment is the home of much larger population than upper reaches and highly dependent on precipitation for meeting agricultural and domestic needs. So, higher decrease in relatively plain areas might have severely affected water availability and affected local population.

Change detection studies suggested that the change is detected earlier in high altitude regions and late in lower reaches. It shows that even though, the magnitude

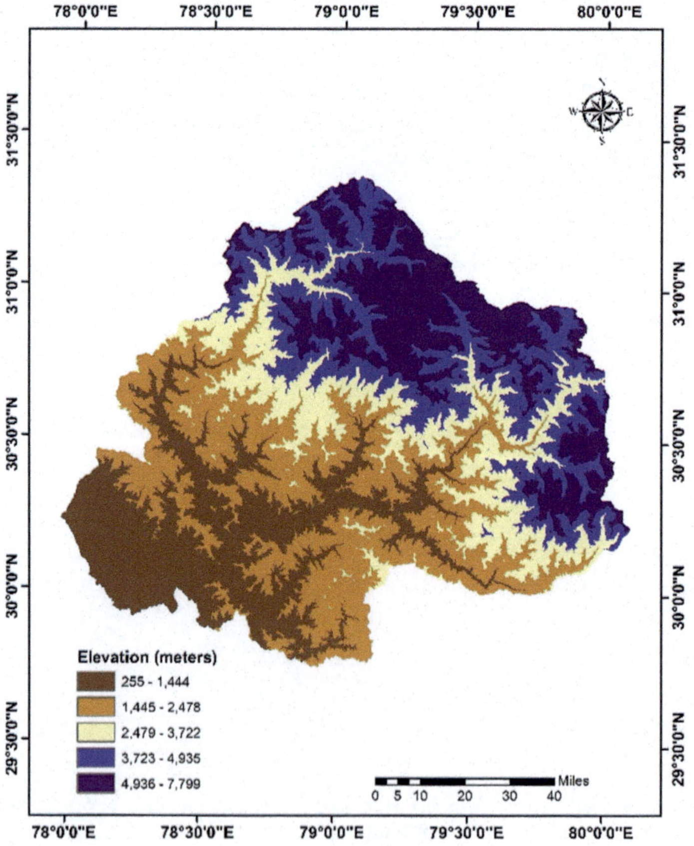

Fig. 24.3 Digital elevation map of upper Ganga river basin

of decreasing trends were lower in the upper reaches, but higher reaches of the basin are getting affected by the climate change earlier. A possible reason to explain the early change point in the upper reaches might be linked to relatively lower annual total precipitation in these regions than lower reaches. Due to comparatively lower precipitation, even a lower decreasing trend of precipitation could be large enough for the falling trends to become significant earlier than the regions with higher annual precipitation.

24.5 Summary and Conclusions

Trend estimation of annual rainfall at different locations in Upper Ganga river basin (up to Haridwar) were presented in this study. It was found that all the locations were showing significant decreasing trends at 5% significance level. Comparatively higher decreasing trends were found in plain areas than high altitude regions. Change point detection using MWP test indicated that majority of regions show change in the year 1992. Earlier change was detected in the mountainous regions of Upper Ganga river basin. It shows that high altitude regions in Upper Ganga river basin gets affected earlier due to climate change. This study shows that both upper and lower reaches of Upper Ganga river basin are highly prone to climate change.

References

Alexander LV, Zhang X, Peterson TC et al (2006) Global observed changes in daily climate extremes of temperature and precipitation. J Geophys Res 111:D05109. https://doi.org/10.1029/2005JD006290

Harris I, Jones PD, Osborn TJ, Lister DH (2014) Updated high-resolution grids of monthly climatic observations-the CRU TS3.10 Dataset. Int J Climatol 34:623–642. https://doi.org/10.1002/joc.3711

IPCC (2013) Summary for policymakers. In: Climate change 2013: the physical science basis. Contribution of Working Group I to the Fifth Assessment Report of the Intergovernmental Panel on Climate Change [Stocker, T.F., D. Qin, G.-K. Plattner, M. Tignor, S.K. Allen, J. Boschung, A. Nauels, Y. Xia,. Cambridge, United Kingdom and New York, NY, USA, Cambridge, United Kingdom and New York, NY, USA].

Jain SK, Kumar V (2012) Trend analysis of rainfall and temperature data for India. Curr Sci 102:37–49. https://doi.org/10.2307/24080385

Kendall MG (1975) Rank Correlation Methods, 4th edn. Charless Griffin, London

Kumar KK, Kamala K, Rajagopalan B et al (2011) The once and future pulse of Indian monsoonal climate. Clim Dyn 36:2159–2170. https://doi.org/10.1007/s00382-010-0974-0

Mondal A, Mujumdar PP (2012) On the basin-scale detection and attribution of human-induced climate change in monsoon precipitation and streamflow. Water Resour Res 48:1–18. https://doi.org/10.1029/2011WR011468

Pettitt AN (1979) A Non-Parametric Approach to the Change-Point Problem. J R Stat Soc Ser C Applied Stat 28:126–135.

Pranuthi G, Dubey SK, Tripathi SK, Chandniha SK (2014) Trend and change point detection of precipitation in urbanizing districts of Uttarakhand in India. Indian J Sci Technol 7:1573–1582. https://doi.org/10.17485/IJST/2014/V7I10/49365
Sharma C, Arora H, Ojha CSP (2015) Assessment of the effect of climate change on historical and future rainfall in Uttarakhand. In: Hydro-2015 international, Roorkee
Sharma C, Ojha CSP (2018) Spatio-temporal variability of snow cover of Yamunotri catchment, India. In: IGARSS 2018–2018 IEEE international geoscience and remote sensing symposium. IEEE, pp 5192–5194
Sharma C, OjhaCSP, Shukla AK et al. (2019) Modified approach to reduce GCM bias in downscaled precipitation: a study in Ganga River Basin. Water (Switzerland) 11. 10.3390/w11102097
Sharma C, OjhaCSP (2019) Changes of annual precipitation and probability distributions for different climate types of the world. Water (Switzerland) 11. 10.3390/w11102092
Sharma C, Ojha CSP (2020) Modified Signal to Noise Ratio Method for Early Detection of Climate Change. J Hydrol Eng (accpeted). https://doi.org/10.1061/(ASCE)HE.1943-5584.0001943
Zhang X, Zwiers FW, Hegerl GC et al. (2007) Detection of human influence on twentieth-century precipitation trends. Nature 448: 461–465. 10.1038/nature06025

Chapter 25
Rainfall Variability Assessment—A Case Study of Rokel-Seli River Basin in Sierra Leone

Saramadie Thorlu-Bangura, Mitthan Lal Kansal, and Surendra Kumar Chandniha

Abstract Rainfall is a most important factor for climate study. It is depending on various factors of hydrological cycle. Actual projection of rainfall pattern is quite difficult due to involvement of uncertainty with respect to space and time. Hence, it is very difficult to access actual occurrence of rainfall in daily basis. For climate study various agencies are developed number of climate data base as per long term historical hydro-meteorological data sets. Using these climatic data base, one can project the rainfall patters under particular uncertainty bands and understand the rainfall variability at particular space. Understanding the trend and variability of rainfall is also important to determine the supplemental water requirements for crops as well as water resources planning during their critical water deficit periods. The aim of this study is to investigate the past variation of rainfall and to identify the trend over Rokel-Seli river basin in Sierra Leone for a period of 45 years (1961–2005) of rainfall data. Rokel-Seli river basin is importance to the country's economy as it supplies water to the Bumbuna hydroelectric power scheme as well as water for the agriculture, fisheries, mining and transportation and for ecological purposes. The long-term trend has been detected using the Mann Kendell (MK) and Modified Mann Kendall (MKK) test(s) for historical time series in terms of monthly, seasonal and annual basis. Further, shift change point has been detected for break point identification using SNHT and MWP test(s). Moreover, rainfall has been projected till

S. Thorlu-Bangura
Ministry of Water Resources, Freetown, Sierra Leone
e-mail: thorlubangura@gmail.com

S. Thorlu-Bangura · M. L. Kansal (✉)
Water Resources Development and Management, Indian Institute of Technology Roorkee, Roorkee, India
e-mail: mlk@wr.iitr.ac.in

S. K. Chandniha
Department of Soil and Water Engineering, Indira Gandhi Krishi Vishwavidyalaya, Raipur, Chhattisgarh, India
e-mail: chandniha.surendra@gmail.com

A. Pandey et al. (eds.), *Climate Impacts on Water Resources in India*, Water Science and Technology Library 95, https://doi.org/10.1007/978-3-030-51427-3_25

2050s under different climate scenarios with various CMIP5 emission conditions, i.e., RCP-2.6, RCP-4.5 and RCP-8.5. Project rainfall and its trends can be useful for future prospective of agriculture and water resources planning and mitigations under consideration of Climate change.

25.1 Introduction

Rainfall variability assessment is quite important for determining the water availability and requirement for a catchment which can influence water use and allocation. Precipitation is most important factor which influences the living beings directly and indirectly. However, precipitation play key role in hydrological cycle (Prabhakar et al. 2019). The overall hydrology is initiated with precipitation and is regulate hydrology, vegetation and water bodies, and it is also significant for agricultural production (King et al. 2014). The occurrence and variability of precipitation influences to a large extent which crops can grow in different areas/regions throughout the world (Silberstein et al. 2012). A good knowledge of rainfall variability is relevant to help optimize farm production in a sustainable manner (Gajbhiye et al. 2015).

The Rokel-Seli river basin is of critical importance to the country's economy as it supplies water to the Bumbuna hydroelectric power scheme as well as water for agriculture, fisheries, mining and transportation and for ecological purposes. As per future prospective, quantum of water availability from precipitation is also quite important (Zhao et al. 2008). Hence, allocation of water for irrigation, industrial and domestic use is also very important and interlinked with hydrological cycle. Understanding the trend and variability of precipitation is also necessary to determine the supplemental water requirements of crops during their critical growth periods (Liu and Lin 2004). The response of hydrologic circulation to climate and land use changes is important in studying the historical, present, and future evolution of aquatic ecosystems.

The present study highlights the rainfall variability over the Rokel-Seli River Basin in Sierra Leone. The adopted analysis provides key information of basin's water availability and hence this work offers benchmark information that can be used to increase the capacity of long-range water resource planning and management, land use planning, agricultural water development and conservation, and industrial water use over the next several decades at basin level. Climate change impacts the rainfall distribution all around the world. The variation in rainfall distribution would alter the storage, recharge surface runoff and soil moisture (Masafu et al. 2016). Moreover, the rainfall variation can increase or decrease the discharge and water availability to a river basin. Increased variation in the intensity and frequency of precipitation is one of the major impacts of climate change (Anandhi et al. 2008).

25.2 Description of Study Area

The Rokel-Seli River, which is the largest river in Sierra Leone is 356 km (221 miles) in length and has a width varying from 6.4 to 16.1 km (4–10 miles). It stretches across the entire northern region before joining the Atlantic Ocean. It has a basin area of 10,699 km^2 which infringes four major districts (Koinadugu, Bombali, Tonkolili and Port Loko districts) having 31% (2,159,119) of the total country population (7,075,641) of Sierra Leone. Location map of study area is shown in Fig. 25.1.

This basin is characterized by a heterogeneous forest-savanna mosaic and experiences a humid tropical climate with annual rainfall averaging 2435 mm and mean monthly temperature of 27.78 °C. Elevation difference is about 956 m and varies from 19 to 975 m. There are two main seasons: rainy season (May–October) and dry season (November–April). There are several small traditional villages in the area with rice cultivation in wet depressions and harvesting of non-timber forest products such as oil palm nuts. Between the upstream and downstream, lots of mining and agricultural activities are taking place such that the rainfall pattern within this area varies considerably.

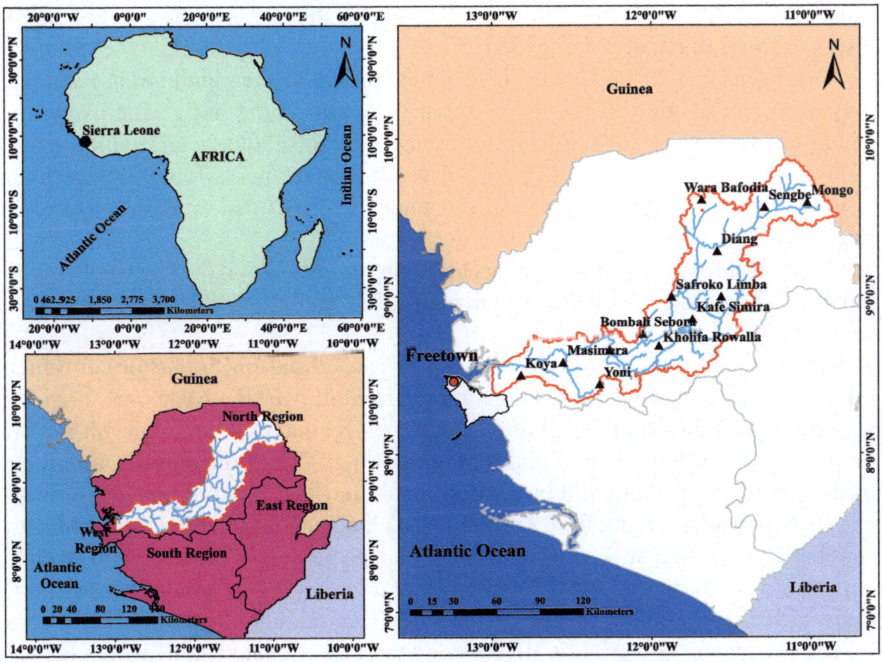

Fig. 25.1 Location map of study area

25.3 Methodology

25.3.1 Assessment of Rainfall Variability

In other to assess the rainfall variability, the rainfall data used for the thirteen (13) stations in the Rokel-Seli river basin from 1961 to 2005 (45 years) was obtained from the Sierra Leone water security website (https://www.salonewatersecurity.com/data) which was created to serve as a repository for hydrological (rainfall, surface water and groundwater) data. This was achieved by the Ministry of Water Resources, the lead government institution of Sierra Leone responsible for monitoring water resources in collaboration with multiple and diverse organizations re-established for hydrological monitoring activities. The extracted monthly, annual and seasonal rainfall data for the entire basin (station-wise) can be shown in Table 25.1.

The average annual rainfall for the basin was estimated as weighted rainfall. The area covered by each rain gauge station with their corresponding annual and seasonal rainfall values were computed to give the mean annual rainfall of RSRB as shown in Table 25.2.

The mean annual and seasonal depth of rainfall of Rokel-Seli river basin was estimated to be 2435, and 2190 and 245 mm in the wet season and dry Season respectively as illustrated in Fig. 25.2.

This means that 90% of the average annual rainfall occurs in the wet season with only 10% contribution of rainfall in the dry season on the basin. The assessment of rainfall variability is of crucial importance for stakeholders and policy makers to provide information for an improved water management. Table 25.3 gives the statistics relating to the variability of the annual and seasonal rainfall of Rokel-Seli river basin.

Considerable aerial variation exists in the annual rainfall on the basin with highest rainfall of magnitude 3589 mm annually and, 3336 and 715 in the wet and dry season respectively.

Based on the rainfall variability analysis carried out on the historical rainfall, it has been detected that the average annual rainfall over the basin was 2435 mm, varying from minimum rainfall of 895 mm to 3589 maximum rainfall. Most of the rainfall occurs during the wet season, contributing 90% of the annual rainfall over the basin ranging from minimum rainfall of 844 mm to 3336 mm maximum rainfall with average annual rainfall of 2190 mm which occurs during the months of May–October. The dry season which occurs between the periods of November to April has only a maximum rainfall of 715 mm contribute only 10% of annual rainfall on the basin with a minimum of 11 mm with mean annual rainfall of 245 mm (Table 25.3).

The coefficient of variation of the annual and seasonal rainfall varies between 14 and 52 with an average value of about 25. The coefficient of variation is least at stations of high rainfall and largest in regions of scanty rainfall as indicated in Fig. 25.3 at station Seng be (upstream) and station Yoni (downstream) for both annually and seasonally.

Table 25.1 Historical average monthly, annual and seasonal rainfall (mm) of RSRB for the period of 1961–2005

Station	J	F	M	A	M	J	J	A	S	O	N	D	Annual	Wet season	Dry season
Bombali	10	15	47	115	300	361	374	483	435	265	91	13	2507	2217	290
Diang	12	17	31	91	276	394	493	545	480	278	126	28	2772	2467	305
Kafe Simera	6	11	18	74	251	360	460	599	508	318	106	18	2729	2496	233
Kholifa Rowala	14	14	25	97	279	323	337	487	427	338	125	20	2486	2191	295
Koya	3	9	25	74	188	300	355	421	362	279	80	10	2106	1905	202
Malal Mara	5	7	17	72	310	361	449	550	483	268	91	14	2627	2421	206
Masimera	10	11	31	97	246	306	312	443	418	339	109	11	2333	2064	269
Mongo	9	8	21	76	203	284	312	406	363	221	62	9	1972	1787	185
SafrokoLimba	9	13	25	104	307	351	426	518	455	281	104	13	2605	2338	268
Sambaia	9	11	16	66	223	365	417	508	439	269	84	17	2425	2220	204
Sengbe	7	6	20	83	241	317	341	419	348	203	61	9	2054	1868	186
Warra Bafodia	7	9	20	92	257	310	356	454	437	364	150	17	2474	2179	295
Yoni	13	10	24	120	333	399	445	519	442	235	68	11	2618	2373	245
Average Rainfall	**9**	**11**	**25**	**89**	**263**	**341**	**390**	**488**	**431**	**281**	**97**	**15**	**2435**	**2194**	**245**
Standard Deviation	**3**	**3**	**8**	**17**	**43**	**37**	**61**	**58**	**49**	**48**	**27**	**5**	**256**	**232**	**45**

Table 25.2 Mean annual and seasonal weighted station rainfall (mm) of RSRB for the period of 1961–2005

Rain gauge station	Station area (km^2)	Weightage factor	Station reading of average rainfall			Weighted station rainfall		
			Annual	Wet season	Dry season	Annual	Wet season	Dry season
Bombali	499	0.043	2507	2217	290	109	96	13
Diang	1449	0.126	2772	2467	305	350	311	38
Kafe Simera	923	0.08	2729	2496	233	219	201	19
Kholifa Row	1427	0.124	2486	2191	295	309	272	37
Koya	843	0.073	2106	1905	202	155	140	15
Malal Mara	606	0.053	2627	2421	206	139	128	11
Masimera	763	0.066	2333	2064	269	155	137	18
Mongo	1141	0.099	1972	1787	185	196	178	18
SafrokoLimba	201	0.018	2605	2338	268	46	41	5
Sambaia	1494	0.13	2425	2220	204	315	289	27
Sengbe	710	0.062	2054	1868	186	127	115	12
Warra Bafodia	764	0.067	2474	2179	295	165	145	20
Yoni	661	0.058	2618	2373	245	151	137	14
Total	**11,481**	**1**				**2435**	**2190**	**245**

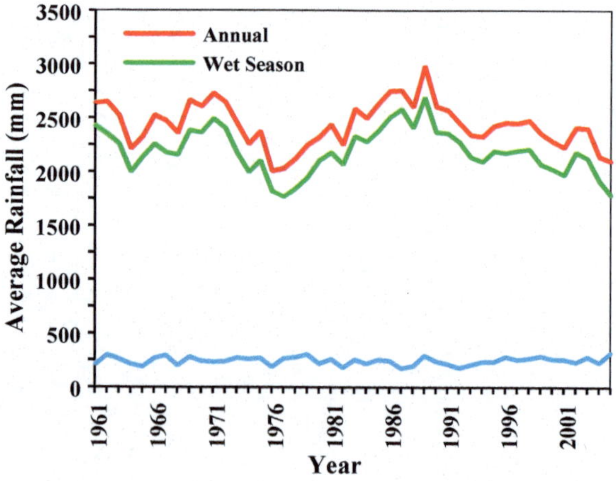

Fig. 25.2 Average annual and seasonal rainfall distribution over RSRB during 1961–2005

This analysis can be useful to detect the changes in the precipitation-streamflow relationship and quantifies the impact of precipitation to runoff (i.e. response of streamflow to climate change) in the basin. It was observed that two stations; one at the extreme upstream and the other at downstream gives highest values of coefficient of variation ranging between 30 and 52% for both annual and seasonal. This means,

Table 25.3 Average annual and seasonal rainfall Statistics of RSRB for the period of 1961–2005

Station	Annual					Wet season					Dry season				
	Min.	Max.	Mean	SD	%CV	Min.	Max.	Mean	SD	%CV	Min.	Max.	Mean	SD	%CV
Bombali Sebora	887	3855	2507	468	19	857	3245	2217	402	18	30	3245	290	142	49
Diang	1099	4082	2772	560	20	1075	3747	2467	557	23	24	512	305	94	31
Kafe Simira	849	3498	2729	444	16	835	3397	2496	448	18	14	564	233	104	45
Kholifa Rowalla	892	3274	2486	439	18	883	3042	2191	450	21	9	548	295	121	41
Koya	799	3447	2106	468	22	799	3369	1905	463	24	0	521	202	96	48
Malal Mara	940	3299	2627	365	14	927	3009	2421	332	14	13	509	206	91	44
Masimera	976	3354	2333	350	15	976	3076	2064	372	18	0	562	269	124	46
Mongo	745	3261	1972	431	22	741	3193	1787	424	24	4	450	185	95	52
SafrokoLimba	992	3426	2605	391	15	975	3090	2338	351	15	18	531	268	105	39
Sambaia	936	3514	2425	504	21	918	3163	2220	489	22	18	419	204	75	37
Sengbe	638	3579	2054	552	27	636	3422	1868	544	29	2	379	186	88	47
Warra Bafodia	940	3496	2474	464	19	928	3248	2179	434	20	12	568	295	118	40
Yoni	941	4569	2618	758	29	939	4369	2373	741	31	3	493	245	111	45
Rokel-Seli river basin	**895**	**3589**	**2435**	**256**	**11**	**884**	**3336**	**2190**	**232**	**11**	**11**	**715**	**245**	**45**	**18**

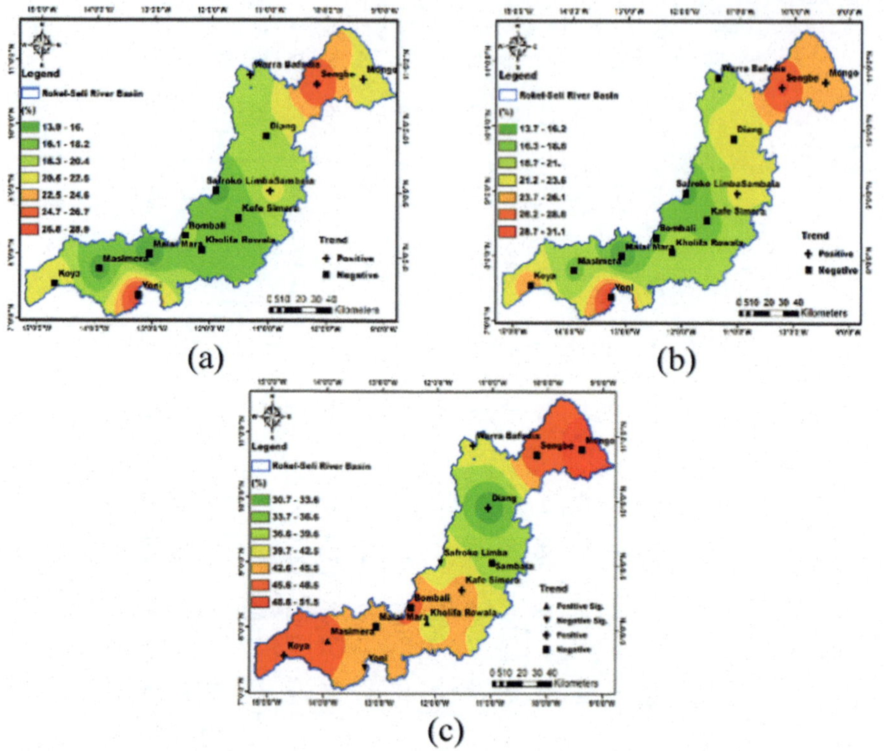

Fig. 25.3 Inter-annual variability of rainfall (%CV) for **a** annual, **b** wet and **c** dry season over RSRB during the period of 1961–2005

less rainfall occurs at those points on the basin during the period of 1961–2005. There is less variation in most part of the basin considering the middle part ranging between 13 and 16% noting that, much of the rainfall is being received at this point. The absolute variability of rainfall distribution was 256 and 232 annually and wet season for the entire basin (Table 25.3); noting that there is significant magnitude of deviation from their mean values of rainfall. However, the standard deviation for the dry season was found to be 45 depicting that, there is low dispersion or variability from the mean rainfall. Hence, there is little or no significant amount of rainfall in the dry season.

25.3.2 Rainfall Trend Analysis

Trend analyses of precipitation are useful to explore the impact of climate change on water resources. Thus, trend analyses of precipitation are required to facilitate more sustainable water planning and management. In this study, the long-term trend has been detected using the Mann–Kendall (MK) test for historical time series in terms of monthly, seasonally and annually.

25.3.2.1 Mann–Kendall Test (MKT)

The non-parametric Mann–Kendall test is commonly employed to detect monotonic trends in series of environmental data, climate data or hydrological data (Hamed and Rao 1998). The null hypothesis, H_0, is that the data come from a population with independent realizations and are identically distributed. The alternative hypothesis, H_A, is that the data follow a monotonic trend (Pohlert 2016). In this study the Mann–Kendall test (MKT) and the Modified Mann–Kendall test (MMKT) has been used to detect the rainfall trend during the period of 1961–2005 over RSRB. The Mann–Kendall test statistic S is calculated using the formula as follows:

$$S = \sum_{i=1}^{N=1} \cdot \sum_{j=i+1}^{N=1} \operatorname{sgn}(x_j - x_i) \tag{25.1}$$

where x_j and x_i are the annual values in years j and i, j > i respectively, and N is the number of data points. The value of sgn $(x_j - x_i)$ is computed as follows:

$$\operatorname{sgn}(x_j - x_i) = \begin{cases} 1 \ \textit{if} \ (x_j - x_i > 0) \\ 0 \ \textit{if} \ (x_j - x_i) = 0 \\ -1 \ \textit{if} \ (x_j - x_i) < 0 \end{cases} \tag{25.2}$$

These statistics represents the number of positive differences minus the number of negative differences for all the differences considered. For large samples ($N > 10$), the test is conducted using a normal approximation (Z statistics) with the mean and the variance as follows:

$$E[S] = 0 \tag{25.3}$$

$$Var(S) = \frac{1}{18}\left[\begin{array}{l} N(N-1)(2N+5) - \sum_{p=1}^{q} \cdot t_p(t_p - 1) \\ (2t_p + 5) \end{array}\right] \tag{25.4}$$

Here q is the number of tied (zero difference between compared values) groups, and t_p is the number of data.

values in the pth group. The values of S and VAR(S) are used to compute the test statistic Z as:

$$Z = \begin{cases} \frac{S-1}{\sqrt{Var(S)}} & if \quad S > 0 \\ 0 & if \quad S = 0 \\ \frac{S+1}{\sqrt{Var(S)}} & if \; S(S) < 0 \end{cases} \tag{25.5}$$

The presence of a statistically significant trend is evaluated using the Z value. A positive value of Z indicates an upward trend and its negative value a downward trend.

25.3.2.2 Modify Mann–Kendall Test (MMKT)

The Modified Mann–Kendall test has been used for trend detection of autocorrelation series. Therefore, in this analysis the autocorrelation between the ranks of the observations 'pk' has been estimated after subtracting the non-parametric Sen's median slope from the slope.

$$\frac{n}{n^*} = 1 + \frac{2}{n(n-1)(n-2)} \times \sum_{p}^{q} (n-k)(n-k-1)$$

$$(n-k-2)p \tag{25.6}$$

Significant values of 'pk' have only been used for calculating the variance correction factor n/n* and it was calculated from the equation proposed by Hamed and Rao (1998).

where:

n represents the actual number of observations,

n* is represented as effective number of observations to account for the autocorrelation in the data and pk is the autocorrelation function for the ranks of the observations.

The corrected variance is then given as (Hamed and Rao 1998).

$$Var^*(S) = Var(S) \times \frac{n}{n^*} \tag{25.7}$$

where $Var(S)$ is from Eq. (25.4).

25.3.3 Magnitude of Rainfall Trend (Theil Sen's Slope)

Sen (1968) developed a non-parametric method to estimate the magnitude (slope) of the trend in a time series (Sen 1968). This method assumes a linear trend in the time series. In this method, the slope Qi of all data value pairs are calculated according to:

$$Qi = \frac{x_j - x_k}{j - k} \tag{25.8}$$

where j > k. If there are n values x_j in the time series we get as many as $N = \frac{n(n-1)}{2}$ slope estimates Qi. The Sen's estimator of the slope is the median of these N values of Qi. The N values of Qi are ranked from the smallest to the largest and the Sen's estimator as follows:

$$Q = \left(Q_{\left[\frac{n+1}{2}\right]}\right), \quad \text{if N is odd} \tag{25.9}$$

Or

$$Q = \frac{1}{2}\left(Q_{\frac{n}{2}} + Q_{\left[\frac{n+2}{2}\right]}\right), \quad \text{if N is even} \tag{25.10}$$

The two-sided test is carried out at 100 (1 – a) % of the confidence interval to obtain the true slope for the non-parametric test in the series. The positive or negative slope Q_i is obtained as upward (increasing) or downward (decreasing) trend. In the present study, the test was carried out at 5% significance level, therefore when Z value exceeds ±1.96 null hypotheses is rejected and show the existence of trend in the series as in Table 25.4.

The Z-statistics value was analyzed (Chandniha et al. 2017) as follows:

- $-1.96 < Z < 1.96$ = No Trend (Not significant)
- $Z > 1.96$ = Increase in trend (i.e. positively significant)
- $Z < -1.96$ = Decrease in trend (i.e. negatively significant).

The values of +Z and −Z indicates upward and downward trend respectively. The Z values of Mann–Kendall test accept the null hypotheses of no trend when;

$$\pm Z \leq Z_1 - x/2$$

where x is the level of significance at two tailed trend tests.

The Theil Sen's slope estimator is a robust method of robustly fitting a line of sample points in the plane by choosing the medians of the slope of all line through pairs of the points. It has been viewed as the most popular nonparametric analysis for determining a linear trend and therefore this can be illustrated by box plot in Fig. 25.4.

Table 25.4 The sen slope and Z-statistic values of annual and seasonal rainfall using MK and MMK for RSRB during 1961–2005

Station	MK			MMK			Sen Slope		
	Annual	Wet Season	Dry Season	Annual	Wet Season	Dry Season	Annual	Wet Season	Dry Season
Bombali	−0.323	−0.245	−0.440	−0.254	−0.245	−0.532	−2.034	−1.203	−0.530
Diang	−0.479	−0.773	0.851	−0.479	−0.773	0.851	−2.890	−4.057	0.736
Kafe Simera	−1.947	−2.299	0.382	−1.947	**−2.299**	0.448	−9.758	−11.090	0.457
Kholifa Rowala	−1.105	−1.575	2.827	−1.105	−1.873	*2.827*	−5.961	−9.484	4.009
Koya	−0.088	−0.303	1.458	−0.068	−0.269	1.458	−0.430	−1.367	1.550
Malal Mara	−0.303	−0.284	−0.225	−0.339	−0.284	−0.225	−0.990	−1.247	−0.272
Masimera	−0.538	−1.712	3.238	−0.628	−1.622	*3.533*	−2.032	−6.709	4.789
Mongo	0.264	0.812	−0.147	0.208	0.631	−0.180	1.666	3.715	−0.132
SafrokoLimba	−1.438	−1.145	−2.358	−1.252	−0.915	**−2.358**	−7.537	−4.661	−2.979
Sambaia	0.665	1.017	−1.800	0.842	1.336	−1.800	4.920	7.126	−1.647
Sengbe	1.301	1.301	−0.695	1.371	1.347	−0.779	6.591	6.928	−0.844
Warra Bafodia	0.127	−0.470	1.272	0.164	−0.372	1.272	0.267	−1.836	1.267
Yoni	−1.849	−1.477	−3.042	−1.399	−0.939	**−3.042**	−17.354	−12.920	−3.917

Italic values: Indicates the significant increasing trend during 1961–2005
Bold values: Indicates the significant decreasing trend during 1961–2005

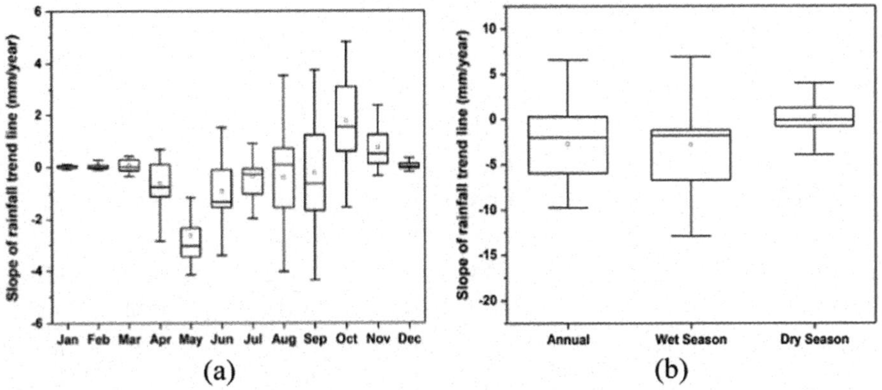

Fig. 25.4 **a** Box plot of the Theil-Sen slopes for annual and seasonal rainfall time series of RSRB. **b** Box plot of the Theil-Sen slopes for monthly rainfall time series

25.3.4 Homogeneity Test in the Times Series (1961–2005)

Application of homogenization on climatic time series preserve the climatic signal and reduce the impact of non-climatic factors in the time series. Therefore, it is important to address these factors in order to develop homogenized records for studying climate change. Change-point analysis examines climate data discontinuities and it directly addresses the question of where the change in the mean value of the observations is likely to have occurred.

Homogeneity in trends was tested using the Standard Normal Homogeneity Test (SNHT) to obtain the homogeneous and heterogeneous trend in the data time series with significance level of 5%. Change point (*P* values) has been computed using 10,000 Monte Carlo simulations base on Mann–Whitney-Pettit (MWP) Test and SNHT (Alexanderson 1986; Alexandersson and Moberg 1997).

The test interpretation, H_0: stands for the homogeneous series and Ha: there is a date at which there is change in the data as shown in Table 25.5. As the computed p-value is greater than the significance level alpha = 0.05, one cannot reject the null hypothesis H_0.

Trend analysis was done applying MK and MMK tests and Sen's slope on monthly, annual and seasonal rainfall data for all stations in the basin. The MK and MMK statistics at 5% significance level are shown in Table 25.4. Among all the 13 stations in the basin, 3 stations show a negatively significant trend in wet and dry season and just 2 stations show positively significant trend in the dry season. From the Z-statistics value, there is not any significant annual trend in all stations, although 9 stations show negative trend and 4 shows positive trend annually during the period of 1961–2005.

Monthly and annual trends are shown in Fig. 25.4 using the box plot of Theil-Sen's slope method for the Rokel-Seli River basin. The box central line represents median, upper and lower lines represent the 75th and 25th percentile, respectively. Also, the

Table 25.5 The MWP test and SNHT test for RSRB (1961–2005)

Station	Pettitt's test			SNHT test		
	P-value	Year	Trend	*P*-value	Year	Trend
Bombali	0.594	1972	Ho	0.736	1972	Ho
Diang	0.875	1998	Ho	0.740	2003	Ho
Kafe Simera	0.118	1973	Ho	0.232	2002	Ho
Kholifa Rowala	0.448	1993	Ho	0.589	1998	Ho
Koya	0.342	1973	Ho	0.159	1972	Ho
Malal Mara	0.667	1980	Ho	0.913	1980	Ho
Masimera	0.178	1991	Ho	0.644	1991	Ho
Mongo	0.267	1972	Ho	0.157	1972	Ho
SafrokoLimba	0.136	1967	Ho	0.044	1998	Ho
Sambaia	0.189	1967	Ho	0.060	1967	Ho
Sengbe	0.059	1982	Ho	0.253	1973	Ho
Warra Bafodia	0.546	1973	Ho	0.599	1973	Ho
Yoni	0.063	1992	Ho	0.098	1992	Ho

upper and lower lines represent the maximum and minimum values of rainfall slopes. Based on the analysis, the dry season shows a positive slope and similarly all the months found in the dry season (November–April) also shows positive slope in the monthly box plot. This denotes that, during periods of high rainfall the Theil-Sen's slope will show negative trend and vis-à-vis during low rainfall. Using MWP and SNHT tests in identifying homogeneity in the data series (Prabhakar et al. 2018), results in Table 25.5 shows that there was no shift change point detected in the data series for the period of 1961 and 2005. Hence all P-values shows Ho (Null Hypothesis) trend (i.e. all data in the series are homogenous during the time series) (Yusof and Kane 2013).

25.3.5 Rainfall Projections

Rainfall projections are necessary in determining the water balance and future water allocation in the basin. Thus, evaluating the future variation of hydrologic cycle and water resources has special significance for regional planning and water resources management. It helps in the assessment of the future impact of climate change over the basin which affects changes in the hydrologic cycle of the basin. Global Climate Models (GCMs) are the fundamental tools that provide future projections of climate variables in the changing ecosystem. Therefore, studies dealing with the climate change impact assessment at catchment scale require downscaling of GCM projections to an appropriate scale to represent the catchment heterogeneity (Silberstein et al. 2012). Various statistical and dynamic downscaling methods have been adopted in the past to downscale large scale atmospheric variables from the GCMs to a regional scale or to a finer scale representative of a catchment (Silberstein et al. 2012; Anandhi et al. 2008). For the purpose of this study, statistical downscaling method has been applied.

25.3.5.1 Statistical Downscaling Method (SDSM)

Statistical downscaling method was applied for downscaling the monthly rainfall of RSRB for future rainfall projections using traditional downscaling regression-based approach (Chandniha and Kansal 2016); Multiple-Linear Regression (MLR). The general formula of MLR is written as:

$$Y^{MLR} = \alpha + \sum_{i=1}^{n} \beta_i X_i + \varepsilon \qquad (25.11)$$

where, Y^{MLR} is the estimated predictand (rainfall), α is the intercept; β is the regression coefficients, X_i is the predictor (26 parameters and ε is the error term. Regression coefficients at 95% confidence level were estimated with Durbin-Watson technique for residuals estimation. SPSS (Ver. 24) was applied for model fit, correlation values, coefficient of determination, descriptive statistics etc.

The following steps were used to carry out the methodology for rainfall projections:

1. Perform consistency check of observed monthly rainfall (Predictand) using Hydrognomon (Ver. 4) for the period of 1961–2005
2. Transfer National Centre of Environmental Predictions (NCEP)and Global Circulation Models (GCM) predictors for the study area from Canadian Centre for Climate Modeling and Analysis (CanESM2) https://www.climate-scenarios.canada.cac corresponding to Representative Concentration Pathways (RCP2.6, RCP4.5 and RCP8.5) emission scenarios and then convert the predictors from daily to monthly basis taking average of each predictor over the month.
3. Identify calibration period (from 1961 to 1990) and validation period (from 1991 to 2005) for the calibration and validation of the model
4. Develop the empirical relationship between historical rainfall (predictand) and the 26 predictors using MLR technique on calibration period data
5. Using expert opinion based on scatter plot, partial correlation, correlation etc. the most suitable predictors were identified
6. Rainfall (predictand) were estimated during the validation period using the selected predictors and compared with the observed values
7. The probable error in the observed and the estimated values were calculated and bias correction was applied to correct the predicated values
8. The important indicators of the goodness of regression were checked by the following parameter; Nash-Sutcliffe Error Estimate (NS-EE), Coefficient of

Correlation (CC), Normalized Mean Square Error (NMSE), and the Root Mean Square Error (RMSE)
9. The suggested series of NCEP Corresponding to RCP2.6, RCP4.5 and RCP8.5 scenarios and the selected predictors were used to generate the future series of the predictand for the periods of 2020s and 2050s.

In applying the above methodology for the future projection of rainfall over Rokel-Seli river Basin the daily observed predictor data of atmospheric variables derived from NCEP 2.80° (latitude) × 2.80° (longitude) grid-scale for 45 years (1961–2005) were obtained from CanESM2. The data was extracted between latitude 8.90° N to 9.30° N and longitude −11.98° W to −11.5° W (BOX_125X_36Y). The Representative Concentration Pathways (RCP2.6, 4.5 8.5) emission scenarios were also downloaded from CanESM2. Full descriptions of NCEP variables (predictors) are elaborated in the Table 25.6.

The Geostrophic air flow velocity, Vorticity, Zonal velocity component, Meridional velocity, Divergence and Wind direction are variables derived using the geostrophic approximation at different atmospheric levels.

The vorticity measures the rotation of the air, Zonal velocity component is the velocity component along a line of latitude (i.e. east–west), Meridional velocity component is the velocity component along a line of longitude (i.e. north–south), Divergence relates to the stretching and outflow of air from the base of an anti-cyclone. Wind direction variable is the only variable which is not normalized. The same parameters are considered in comparing the results based on MLR method.

The MLR equations derived for each rainfall station can be given as follows:

Bombali: $207.1 + 24.3X_{15} - 9.8X_{19} + 30.3X_{20} + 169.3X_{22}$

Diang: $253.6 + 45X_7 + 68.4X_{11} + 103.3X_{22} + 39.5X_{23} + 45.4X_{24} - 45.4X_{25}$

Kafe Simera: $236.5 + 29.4X_{15} - 0.5X_{19} + 18.4X_{20} + 215.7X_{22}$

Kholifa Rowala: $226.4 + 29.4X_7 + 41.8X_{11} + 69.1X_{22} + 41.5X_{23} + 30.2X_{25}$

Koya: $182.1 + 13.8X_{15} - 0.3X_{19} + 18.7X_{20} + 164.2X_{22}$

Malal Mara: $210.5 + 42.8X_{15} + 12.9X_{19} + 35.3X_{20} + 183.1X_{22}$

Masimera: $213.2 + 22.8X_7 + 40X_{11} + 64.7X_{22} + 39.9X_{23} + 32.8X_{25}$

Mongo: $162.1 + 25X_{15} - 1.5X_{19} + 24.4X_{20} + 142.3X_{22}$

SafrokoLimba: $244.6 + 68.8X_{11} + 56.8X_{22} + 52.3X_{23} + 25.1X_{24} + 29X_{25}$

Sambaia: $223.8 + 25.6X_7 + 22.1X_{11} + 184.9X_{22} - 0.9X_{23} - 13.7X_{25}$

Sengbe: $162.2 + 25.1X_{15} + 2.9X_{19} + 36.8X_{20} + 137.9X_{22}$

Warra Bafodia: $227.1 + 16.5X_7 + 25.6X_{11} + 100.6X_{22} + 26.9X_{23} + 37X_{25}$

Yoni: $199.4 + 56.4X_{15} - 13.6X_{19} + 41.8X_{20} + 158.3X_{22}$

Table 25.6 Variables and description of NCEP and GCM predictors (26 variables)

S. No.	Variable	Description	Unit
1	ncepmslpgl	Mean sea level pressure	Pa
2	ncepp1_fgl	Geostrophic air flow velocity	m/s
3	ncepp1_ugl	ZonaL velocity component	m/s
4	ncepp1_vgl	Meridional velocity component	m/s
5	ncepp1_zgl	Vorticity	m/s
6	ncepp1thgl	Wind direction	m/s
7	ncepp1zhgl	Divergence	m/s
8	ncepp5_fgl	Geostrophic air flow velocity	m/s
9	ncepp5_ugl	ZonaL velocity component	m/s
10	ncepp5_vgl	Meridional velocity component	m/s
11	ncepp5_zgl	Vorticity	m/s
12	ncepp5thgl	Wind direction	m/s
13	ncepp5zhgl	Divergence	m/s
14	ncepp8_fgl	Geostrophic air flow velocity	m/s
15	ncepp8_ugl	ZonaL velocity component	m/s
16	ncepp8_vgl	Meridional velocity component	m/s
17	ncepp8_zgl	Vorticity	m/s
18	ncepp8thgl	Wind direction	
19	ncepp8zhgl	Divergence	m/s
20	ncepp500gl	500 hPa geopotential height	m
21	ncepp850gl	850 hPa geopotential height	m
22	nceprcpgl	Near surface relative humidity	%
23	nceps500gl	Specific humidity at 500 hPa height	kg/kg
24	nceps850gl	Specific humidity at 850 hPa height	kg/kg
25	ncepshumgl	Near surface specific humidity	kg/kg
26	nceptempgl	Mean temperature at 2 m	K

Where, X_7 = Divergence; X_{11} = vorticity; X_{15} = Zonal velocity component; X_{19} = Divergence; X_{20} = 850 hPa geopotential height; X_{22} = near surface relative humidity; X_{23} = Specific humidity at 500 hPa height; X_{24} = specific humidity at 850 hPa height; X_{25} = near surface specific humidity.

25.3.6 Calibration and Validation

The daily observed predictor data obtained from NCEP reanalysis, normalized over the period of 1961–1990 and hence this period was selected for calibration, while the validation period was selected from 1991 to 2005 to normalize the calibration and validation model. Observed and estimated rainfall during calibration and validation period is shown in Fig. 25.5 and scatter plot of both cases are represented in Fig. 25.6. The important parameters or indicators of the goodness of the regression for each of the stations during the calibration and validation periods are also shown in Table 25.7.

During the calibration and validation periods the average monthly values were calculated for both observed and the estimated rainfall as in Table 25.8.

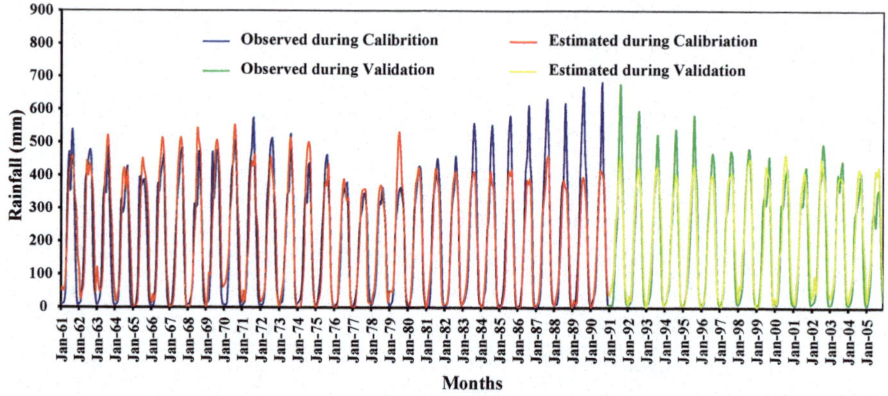

Fig. 25.5 Observed and estimated monthly rainfall during calibration and validation periods

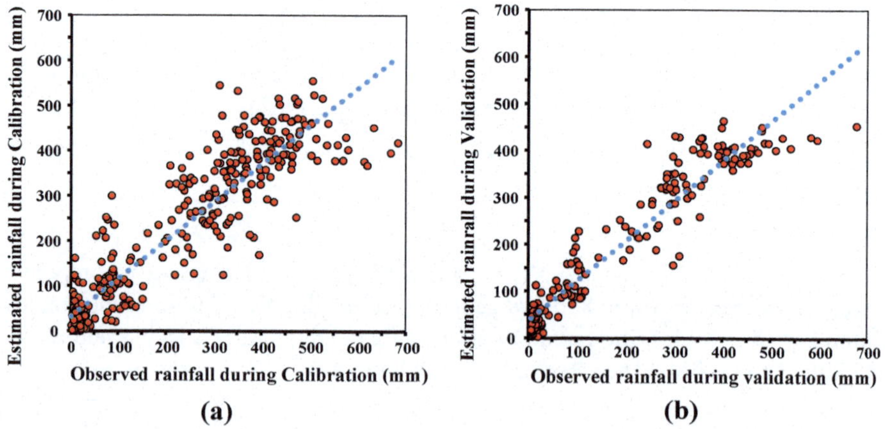

Fig. 25.6 Scattered plot between observed and estimated rainfall during **a** calibration and **b** validation

Table 25.7 Important indicators of accuracy of results during calibration and validation for rainfall time series at the various stations of Rokel-Seli River Basin

Station	NCEP	Calibration/Validation	RMSE (mm)	NMSE	NASH	CC
Bombali	1961–1990	Calibration	115.43	0.16	0.65	0.81
	1991–2005	Validation	104.45	0.24	0.72	0.86
Diang	1961–1990	Calibration	136.27	0.13	0.71	0.81
	1991–2005	Validation	129.56	0.30	0.70	0.84
Kafe Simera	1961–1990	Calibration	130.53	0.19	0.67	0.84
	1991–2005	Validation	126.55	0.32	0.71	0.85
Kholifa Rowalla	1961–1990	Calibration	114.72	0.15	0.69	0.82
	1991–2005	Validation	88.25	0.20	0.74	0.86
Koya	1961–1990	Calibration	103.99	0.10	0.75	0.82
	1991–2005	Validation	91.14	0.27	0.71	0.85
Malal Mara	1961–1990	Calibration	120.20	0.16	0.70	0.84
	1991–2005	Validation	127.17	0.33	0.70	0.85
Masimera	1961–1990	Calibration	104.99	0.14	0.70	0.83
	1991–2005	Validation	76.45	0.17	0.77	0.88
Mongo	1961–1990	Calibration	96.72	0.10	0.76	0.82
	1991–2005	Validation	93.99	0.31	0.70	0.84
SafrokoLimba	1961–1990	Calibration	114.40	0.11	0.76	0.83
	1991–2005	Validation	89.71	0.19	0.79	0.90
Sambaia	1961–1990	Calibration	112.68	0.11	0.79	0.84
	1991–2005	Validation	98.33	0.25	0.76	0.87
Sengbe	1961–1990	Calibration	120.60	0.35	0.58	0.77
	1991–2005	Validation	93.93	0.30	0.65	0.82
Wara Bafodia	1961–1990	Calibration	114.83	0.18	0.60	0.81
	1991–2005	Validation	87.93	0.18	0.79	0.89
Yoni	1961–1990	Calibration	159.08	0.32	0.58	0.78
	1991–2005	Validation	114.49	0.31	0.60	0.82

The equations derived after SDSM base on MLR approach during calibration and validation periods are applied for future projections under Couple Model Intercomparison phase-5(CMIP5) emission scenarios; RCP 2.6 RCP 4.5 and RCP 8.5 which are used in this study.

For the purpose of this work, the projected rainfall has been categorized in various time step as; past (1961–2005), 2020s (2011–2040) and 2050s (2041–2070) scenarios as tabulated in Table 25.9.

Further, climate change scenarios also help the future planning of various activities which are associated with water and dependent with rainfall over the catchment.

Table 25.8 Calibration and validation values of monthly average rainfall of Rokel-Seli River Basin

Rainfall (mm)	(OBS/NCEP)	Monthly												R^2	Annual			
		J	F	M	A	M	J	J	A	S	O	N	D		Min	Max	Average	SD
Calibration	Observed	8	11	24	93	277	354	401	496	431	275	93	15	0.848	1962	2820	2477	271
	Estimated	20	21	34	87	210	352	428	433	414	296	143	36		1966	2926	2474	289
Validation	Observed	11	12	26	81	233	315	370	473	430	293	104	15	0.905	1893	2798	2363	267
	Estimated	17	28	48	104	211	325	404	420	395	304	145	42		2048	2852	2444	255

Table 25.9 Past and future rainfall statistics under RCP2.6, RCP4.5 and RCP8.5 of RSRB

Projection scenarios		Past and predicted monthly average rainfall (mm)											
		J	F	M	A	M	J	J	A	S	O	N	D
RCP-2.6	Past	9	11	25	89	263	341	390	488	431	281	97	15
	2020s	23	16	36	143	284	373	433	458	435	233	131	20
	2050s	15	18	30	144	288	385	451	460	339	251	143	20
RCP-4.5	Past	9	11	25	89	263	341	390	488	431	281	97	15
	2020s	16	19	37	122	269	362	420	454	389	279	115	26
	2050s	29	23	45	145	292	379	443	434	410	309	118	17
RCP-8.5	Past	9	11	25	89	263	341	390	488	431	281	97	15
	2020s	30	25	31	145	287	372	439	461	408	306	108	18
	2050s	29	26	30	135	292	379	443	434	410	309	114	21

Projection scenarios		Annual Rainfall (mm)			
		Min.	Max.	Average	SD
RCP-2.6	Past	1972	2772	2435	182
	2020s	1967	3250	2744	173
	2050s	1928	3270	2545	172
RCP-4.5	Past	1972	2772	2435	182
	2020s	2725	3907	2508	171
	2050s	2602	4568	2645	174
RCP-8.5	Past	1972	2772	2435	182
	2020s	2450	5946	2630	177
	2050s	2524	5175	2623	175

The forecasted rainfall is expected to help the policy makers and the stakeholders for making effective water resources planning.

It can also be observed that, the MLR model fits between the observed and estimated monthly average rainfall during the calibration and validation periods with a coefficient of determination of 0.848 and 0.905 respectively as indicated in Table 25.8. Applying the recommended MLR model, average annual rainfall estimated for the period of 1961–1990 was 2474 mm as compared to the observed rainfall of 2477 mm. Similarly, the estimated average rainfall during 1990–2005 was 2444 mm compared to 2363 mm as observed rainfall. The projected mean annual rainfall for the periods of 2020s and 2050s according to CMIP5 emission scenarios are; 2744 and 2545 mm (RCP 2.6), 2508 and 2645 mm (RCP 4.5) and 2630 and 2623 mm (RCP8.5) as against observed rainfall of 2435 mm. The overall forecasted rainfall (taking the average of all 3 scenarios) for 2020s is 2627 mm and for 2050s is 2604 mm as compared to the observed value of 2435 mm. As projected for the average rainfall over Rokel-Seli River Basin, the expected rainfall to occur in 2020s and 2050s is about 7–8% higher than the observed rainfall during the periods of 1961–2005.

25.3.7 Dependable Annual Rainfall of Rokel-Seli River Basin

It is very important to know the annual rainfall dependability in planning water resources over the basin in order to obtain the relationship between the magnitude of the event and its probability of exceedance. Hence the 75% or 95% dependable annual rainfall are of utmost concern to identify the minimum water availability.

Therefore, in this study, the dependable annual rainfall has been estimated for the past (1961–2005) and future time series of 2020s (2011–2041) and 2050s (2041–2070) as shown in Table 25.10 and Fig. 25.7.

The 95% dependable rainfall for the past, 2020s and 2050s were estimated as 2034 mm, 2320 mm and 2462 mm respectively. This justifies the values of minimum average rainfall that occurred in the basin and which could be expected in 2020s and 2050s. Hence these values are very useful for future water resources development and management on the basin.

Table 25.10 Dependable annual rainfall of RSRB for past (1961–2005), 2020s (2011–2040) and 2050s (2041–2070)

Percentage dependability	Dependable rainfall (mm)		
	Past	2020s	2050s
5	2749	2751	2815
50	2451	2551	2588
75	2284	2507	2529
90	2125	2412	2499
95	2034	2320	2462

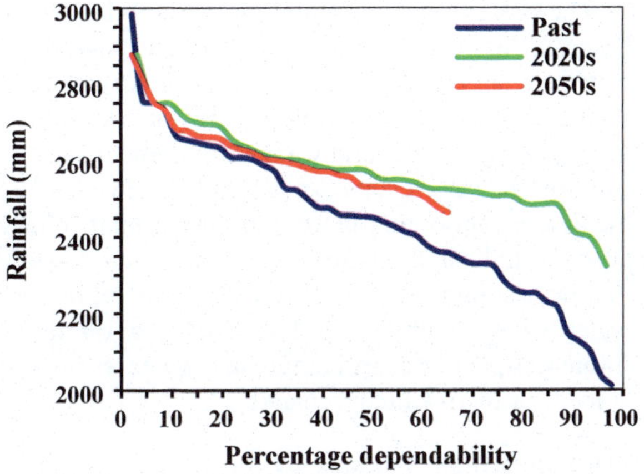

Fig. 25.7 Rainfall dependable curves for past (1961–2005), 2020s (2011–2040) and 2050s (2041–2070)

25.4 Conclusions

The rainfall over Rokel-Seli river basin varies considerably from the upstream and downstream showing high value of coefficient of variation annually and seasonally. Thus, the high values of CV at these stations shows that less rainfall occurs at these parts of the basin and high rainfall occurring most part at mid of the basin with average rainfall of 2435 mm. The non-parametric (MK and MMK) tests were used to detect rainfall trends over the basin during the period of (1961–2005). The accuracy of MMK test in terms of significance level was more precise than MK test at the same level of significance, showing only 3 stations having negatively significant trend and two stations with positively significant trends in the wet and dry seasons. No significant trends were identified annually.

The Sen's slope magnitude varies between −17.354 and 6.951 mm/year annually in the basin (1961–2005), therefore the seasonal slopes were mostly negative for the wet season and positive for the dry season. The Mann–Whitney–Pettitt and SNHT tests were used to identify possible break points in precipitation during the 45 year period. However, from the results of the study, it can be concluded that there was no shift change point in the data series. Hence the rainfall data was homogenous for all throughout the time series for the period of 1961–2005.

Rainfall projections has been carried out in the basin for the periods of 2020s and 2050s. The average annual rainfall projected in the 2020s (2011–2040) is 2627 mm and for 2050s (2041–2017) is 2604 mm as compared to the observed value of 2435 mm. This means that there would be increase in rainfall of about 7–8% unto 2050s due to climate change. The observed dependable annual rainfall at 95% is 2034 mm and the projected dependable rainfall at 95% dependability in 2020s and 2050s is 2320 mm and 2462 mm respectively. It is very important to know the dependable rainfall in planning water resources over the basin to identify the minimum water available for various water requirement, also to determine the projected minimum rainfall which could be used for future water resources management.

The present study highlights the rainfall variability over the Rokel-Seli River Basin in Sierra Leone. The adopted analysis provides key information of basin's water availability and hence this work offers benchmark information that can be used to increase the capacity of long-range water resource planning and management, land use planning, agricultural water development and conservation, and industrial water use over the next several decades at basin level. The results of the study also help in the assessment of the future impact of climate change over the basin which affects changes in the hydrologic cycle of the basin. Also, this study can be used as a guide to simulate rainfall variability over other basins in Sierra Leone to determine the water balance within those basins. Project rainfall and its trends can be useful for future prospective of agriculture and water resources planning and mitigations under consideration of Climate change.

Acknowledgement This work is supported by the Department of Water Resources Development and Management, India Institute of Technology-Roorkee (IIT-R), India. The authors are thankful to the Ministry of Water Resources, Government of Sierra Leone for providing the rainfall data for the analysis of the study area. We also grateful to National Institute of Hydrology (NIH), Roorkee, India for their technical support.

References

Alexandersson H (1986) A homogeneity test applied to precipitation data. J Climatol 6(6):661–675

Alexandersson H, Moberg A (1997) Homogenization of Swedish temperature data. Part I: homogeneity test for linear trends. Int J Climatol J Roy Meteorol Soc 17(1):25–34

Anandhi A, Srinivas VV, Nanjundiah RS, Nagesh Kumar D (2008) Downscaling precipitation to river basin in India for IPCC SRES scenarios using support vector machine. Int J Climatol 28:401–420. https://doi.org/10.1002/joc.1529

Chandniha SK, Kansal ML (2016) Rainfall estimation using multiple linear regression based statistical downscaling for Piperiya watershed in Chhattisgarh. J Agrometeorol 18(1):106

Chandniha SK, Meshram SG, Adamowski JF, Meshram C (2017) Trend analysis of precipitation in Jharkhand State, India. Theor Appl Climatol 130(1–2):261–274

Gajbhiye S, Meshram C, Mirabbasi R, Sharma SK (2015) Trend analysis of rainfall time series for Sindh river basin in India. https://doi.org/10.1007/s00704-015-1529-4

Hamed KH, Rao AR (1998) A modified Mann-Kendall trend test for autocorrelated data. J Hydrol 204(1–4):182–196

King KW, Fausey NR, Williams MR (2014) Effect of subsurface drainage on streamflow in an agricultural headwater watershed. J Hydrol 519:438–445

Liu XY, Lin ED (2004) Impact of climate change on water requirement of main crops in North China. J Hydraul Eng 2:77–82

Masafu CK, Trigg MA, Carter R, Howden NJ (2016) Water availability and agricultural demand: an assessment framework using global datasets in a data scarce catchment, Rokel-Seli River, Sierra Leone. J Hydrol Reg Stud 8:222–234

Prabhakar AK, Singh KK, Lohani AK (2018) Regional level long-term rainfall variability assessment using Mann-Kendall test over the Odisha state of India. J Agrometeorol 20(2):164–165

Prabhakar AK, Singh KK, Lohani AK, Chandniha SK (2019) Assessment of regional-level long-term gridded rainfall variability over the Odisha State of India. Appl Water Sci 9(4):93

Pohlert T (2016) Non-parametric trend tests and change-point detection. CC BY-ND, 4

Sen PK (1968) Estimates of the regression coefficient based on Kendall's tau. J Am Stat Assoc 63(324):1379–1389

Silberstein RP, Aryal SK, Durrant J, Pearcey M, Braccia M, Charles SP, Boniecka L, Hodgson GA, Bari MA, Viney NR, McFarlane DJ (2012) Climate change and runoff in South-Western Australia. J Hydrol 475:441–455

Yusof F, Kane IL (2013) Volatility modeling of rainfall time series. Theoret Appl Climatol 113(1–2):247–258

Zhao D, Kuenzer C, Fu C, Wagner W (2008) Evaluation of the ERS scatterometer-derived soil water index to monitor water availability and precipitation distribution at three different scales in China. J Hydrometeorol 9(3):549–562

Chapter 26
Removal of Fluoride from Drinking Water Supplies

Stephano M. Alphayo and M. P. Sharma

Abstract The presence of fluoride in excess of permissible limit in drinking water has been resulting into serious health problems in several countries. Studies have shown that the rift valley regions in Tanzania are the most affected areas, where only 18.4% of 190 water sources showed fluoride values <1.5 mg/L as per acceptable limit by WHO, 56.8% as per Tanzanian standards and 43.2% has extreme fluoride concentrations. In such case, the excess fluoride removal to make the water potable is the only remedy. Various defluoridation methods varying in terms of cost of treatment, materials, efficiency and technology are available. This paper discusses de-fluoridation technique using RGAC and CAC as the most suitable technique for rural communities in Tanzania. This method is not only economical but also simple to perform. The optimum fluoride removal conditions are 6 mL/min flow rate and 6 cm thickness of bed with 71.9% efficiency.

26.1 Introduction

Fluoride ion is highly reactive and is naturally found in two forms i.e., inorganic and organic chemical (Goswami et al. 2015). The organic form has no effect on dental caries but is used for the production of pesticides, nerve gas and other organic compounds. The inorganic form can be removed from the water by defluoridation process when found in excess. Excess fluoride in drinking water tends to cause health problems like pitting and loss of enamel, crippling skeletal fluorosis and non-skeletal effects like kidney, endocrine, thyroid and liver disorders. This has attracted the scientists to develop a suitable and cost-effective fluoride removal method.

S. M. Alphayo (✉)
Water Resources Department, Water Institute, Dar es Salaam, Tanzania
e-mail: stephen.alphayo@yahoo.com

M. P. Sharma
Hydro and Renewable Energy Department, Indian Institute of Technology Roorkee, Roorkee 247667, India
e-mail: mahendrapal.sharma@gmail.com

A. Pandey et al. (eds.), *Climate Impacts on Water Resources in India*, Water Science and Technology Library 95, https://doi.org/10.1007/978-3-030-51427-3_26

Chidambaram et al. (2003) removed fluoride using natural materials like red soil, charcoal, brick, fly-ash and serpentine. The study found red soil to have a good fluoride removal capacity followed by brick, fly-ash, serpentine and charcoal. Kaseva (2006) studied the optimization and application of the regenerated bone char media for the defluoridation of drinking water in Tanzania. The results indicated that the highest fluoride removal may be affected by adsorption. Kanyora et al. (2015) suggested the regenerated bone char as a viable option for reducing the excess fluoride in drinking water. Renuka and Pushpanji (2013) reviewed defluoridation techniques by considering the fact that fluorosis is an irreversible condition and can neither be cured nor prevented. Zafar and Ahmed (2015) also reviewed the fluoride releasing materials and the therapeutic effects of the released fluoride and found that these materials can be used as a rechargeable reservoir that can release recharged fluoride making its availability over a longer period of time. Waghmare and Arfin (2015) recommended that out of different techniques utilized for defluoridation of water, the adsorption procedure may be largely utilized as it can efficiently remove fluoride in terms of its simplicity, reproductivity and ease of operation. Goswami et al. (2015) found the suitability of inexpensive leaf adsorbents to effectively remediate fluoride-contaminated water. Sheetal et al. (2015) studied adsorption potential of granular activated carbon and found it suitable for fluoride removal. Chinoy et al. (2006) reported nalgonda technique as an effective household de-fluoridation technique but is very expensive, since the aluminium sulphate is not produced locally (Gumbo and Mkongo 1995).

In most of the fluoride-affected areas in Tanzania, alternate sources are either not available or implementation is restricted to techno-economic and social factors. In majority of cases, the excess fluoride removal to make the water potable is the only remedy. De-fluoridation techniques reported in literatures are suitable only for urban areas where skilled manpower and other resources are not limited. The present paper focuses on Regenerated Granular Activated Charcoal (RGAC) and Coconut Activated Charcoal (CAC) method as more suitable fluoride removal method for rural areas in Tanzania, where population is highly scattered and water supply infrastructures are inadequate. This method can efficiently remove fluoride from problem water.

26.2 Study Area

Arumeru and Arusha districts lies between (03° 22′ 21″ S and 36° 41′ 40″ E) latitude and longitude respectively. Arumeru and Arusha districts are one of the districts in the Arusha Region of Tanzania, which is bordered to the north, west, and southwest by Monduli district, to the southeast by Manyara Region, and to the east by the Kilimanjaro Region. As per census of 2012, the population of the Arumeru and Arusha districts were 591,342 and 416,442 respectively (National Bureau of Statistics (NBS) and Office of Chief Government Statistician (OCGS) 2013). Figure 26.1 shows the location of Arumeru and Arusha districts in Arusha region in Tanzania.

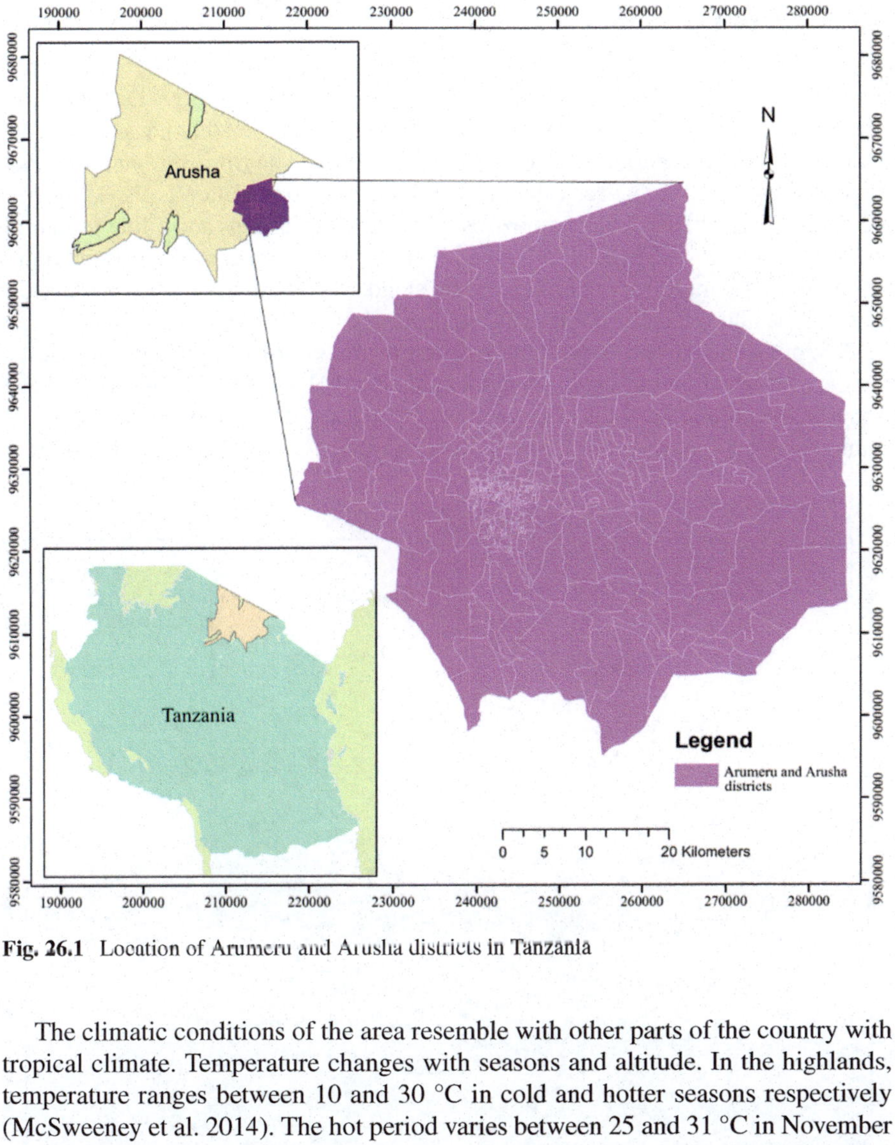

Fig. 26.1 Location of Arumeru and Arusha districts in Tanzania

The climatic conditions of the area resemble with other parts of the country with tropical climate. Temperature changes with seasons and altitude. In the highlands, temperature ranges between 10 and 30 °C in cold and hotter seasons respectively (McSweeney et al. 2014). The hot period varies between 25 and 31 °C in November and February respectively, while the coldest period ranges from 15 to 20 °C occurring between May and August. Lack of water in this area put the limits on average per capita daily water consumption from 8 L falling to 3–4 L during the dry season (Ghiglieri et al. 2010).

Arumeru district is distinguished as having the highest concentration of fluoride in their water bodies in the nation. The areas of natural waters have the contamination ranging between 5 and 30 mg/L (Hazla 2017). High fluoride concentration resulted from the alkaline volcanism, which is widespread throughout the East Africa Rift System extended from Syria through Jordan, Egypt, Libya, Algeria, Morocco and Sudan (Eirik 2013).

26.3 Data Collection

Fluoride data was obtained from Water quality officer, Ngurdoto Defluoridation Research Station, Pangani basin, Arusha Tanzania. This organization was established by Government of Tanzania as Defluoridation Centre. Data from 190 sampling locations was collected from year 1994 to 2015. However, some data was also taken from (Ghiglieri et al. 2010). Data was processed in different groups according to fluoride concentration i.e. <1.5 mg/L (WHO), < 4.0 mg/L (as per Tanzania fluoride maximum limit) and >4.0 mg/L (not complies to any standard). These groups are presented in green, blue and red color in Fig. 26.2 respectively.

Sampling and analysis were done at different times during the dry and wet seasons by the concerned agency using standard method (APHA 1998). The location of each sampling locations was plotted to view of distribution of all parameters. The 35/190 sampling locations showed fluoride concentration <1.5 mg/L (which is 18.4% of

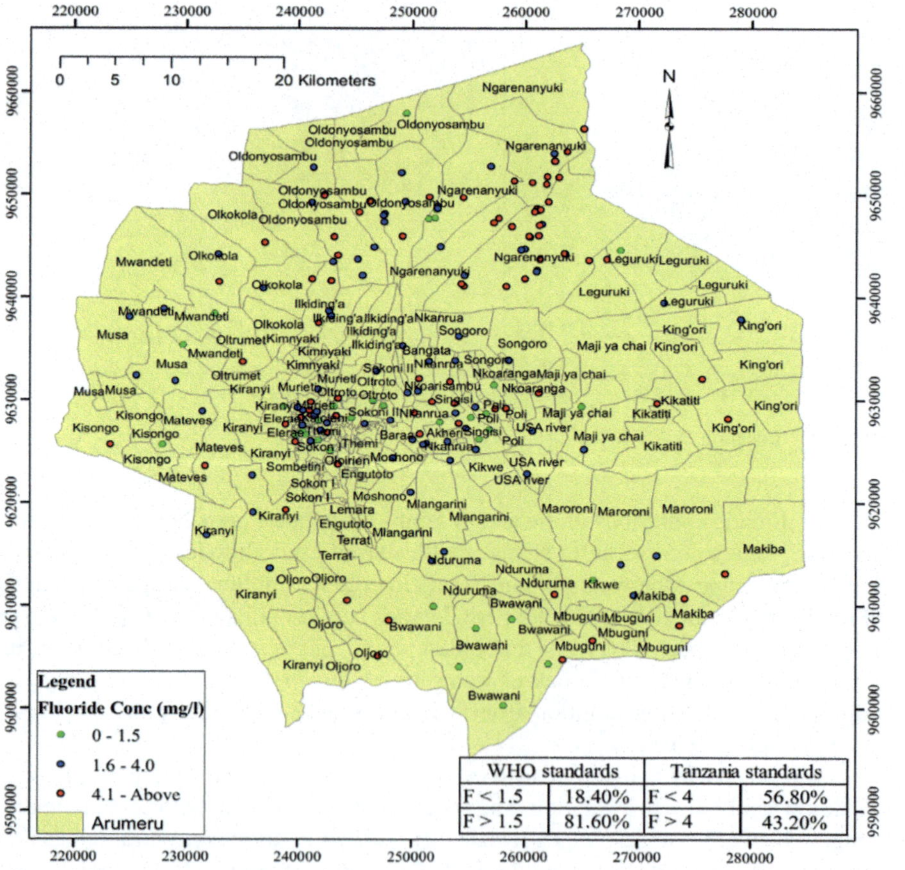

Fig. 26.2 Fluoride concentrations at different locations in Arumeru and Arusha districts in Tanzania

Table 26.1 Fluoride recommended values from different organizations/countries

Organization/Country	Fluoride (mg/L)	References
World Health Organization	1.0–1.5	WHO (2008)
India	1.0–1.5	Bureau of Indian Standards (1982)
Tanzania	1.0–4.0	Tanzania Bureau of Standards (2005)

the total data available). This range is acceptable by WHO. The 108/190 sampling locations showed fluoride concentration <4 mg/L (which is 56.8% of the total data available) and the range is acceptable by Tanzanian standards. The 82/190 sampling locations showed fluoride concentration >4.0 mg/L (which is 43.2% of the total data available) and the range is referred as extreme fluoride concentration areas not complying with any standard. The areas showing highest fluoride concentrations are Ngarenanyuki, Oldonyosambu, Maji ya chai, Makiba, Elerai, Olkokola, Bwawani, King'ori, Kikatiti and Kisongo. These areas seek a remedy for this exceeded fluoride concentration.

26.3.1 Fluoride Standards Available

Table 26.1 gives the limits of fluoride concentration as per WHO, India and Tanzania.

The range indicates that referred limits are the same except Tanzania where it is 1.0–4.0 mg/L.

26.3.2 Sources of Fluoride

The major source of fluoride is the ground water coming from fluoride containing rocks/aquifers (Brindha and Elango 2011). Other sources include air pollution, food, tea and manmade products. Coal burning for warmth and cooking contributes fluoride to the body through inhaled smokes. Industrial wastewater discharged directly into the water sources is another source of fluoride. Some dental products also contains very high concentrations of fluoride (Suneetha et al. 2015).

26.3.3 Health Effects of Fluoride

Excess Fluoride in human body have many health problems like dental and skeletal fluorosis, reduced immunity, effects on hormones, deformities of red blood cells, neurological manifestations, repeated abortion, gastrointestinal problems, muscle fiber degradation, urinary tract malfunction, male sterility. Table 26.2 shows the

Table 26.2 Health effects of fluoride

Health effects	Photos showing the effects of excess fluoride situation in the study area (www.thenasiotrust.org)	
Teeth (dental fluorosis)	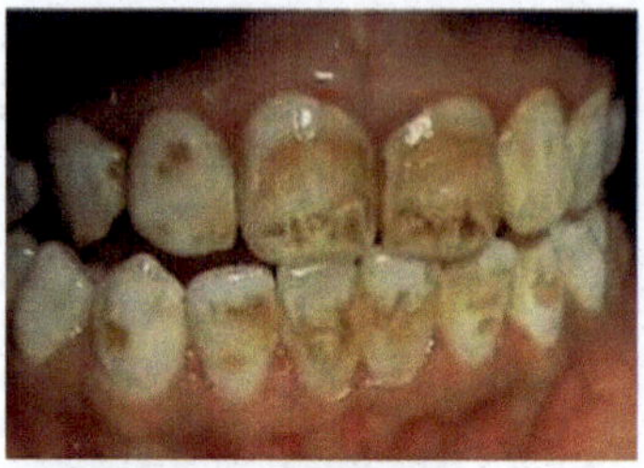	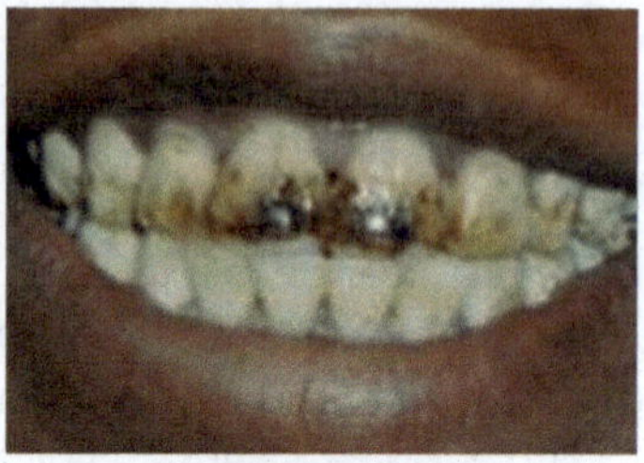
Skeleton fluorosis	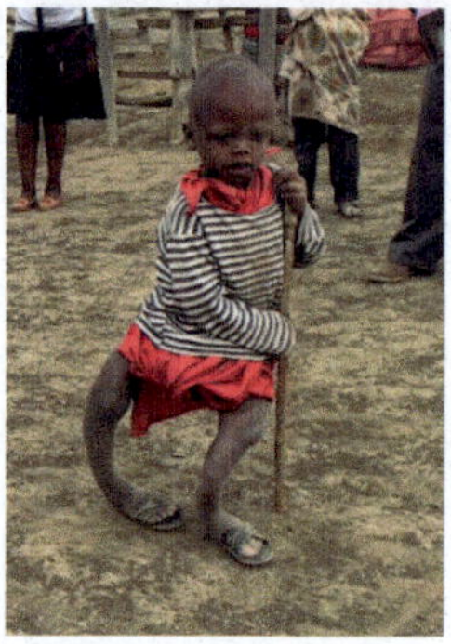	
Endocrine and neurologic systems		Das et al. (2006)
Immunological system		

pictorial views of health effects in study area.

The pitting of teeth and skeleton fluorosis is very common in Arumeru and Arusha, there is no cure for this disease. The only option is to prevent it from the coming generation.

26.3.4 Fluoride Removal Using Defluoridation Method

Defluoridation process can be carried out in large water supply systems under the supervision of qualified peoples. However, in developing countries like Tanzania, the cost of removal is the major constraints. Therefore, the decentralized defluoridation is the most recommended method for rural areas and is focused to fluoride removal for drinking and cooking water only. Different methods were compared to suggest a suitable method for rural areas in Tanzania. Merits and demerits discussed are summarized in the Table 26.3.

Table 26.3 Merits and demerits of some de-fluoridation methods

S. No.	Name of method	Merits	Demerits
1	Activated Alumina	• It requires minimum contact time for maximum defluoridation • It is indigenously available and cheap • Percentage of regeneration is considerably high	• Adsorption of fluoride is possible only at specific pH range • Frequent activation is needed • Regeneration cause disposal problems • Adsorption efficiency reduces with number of usage cycle
2	Nalgonda Technique	• Simplicity of design, construction, operation and maintenance • Low technology, adaptable at point of use • Beside fluoride turbidity, color, odor, pesticides and organic substance are also removed in this method	• The sludge disposal can cause a major environmental problem • Periodic analysis to calculate the correct dose of chemicals to be added • Careful control of the pH of treated water is required • The amount of alum required for the process may be large
3	Bone charcoal	• The cost of the technology used is low • The media used i.e. the bones required to make the bone char are locally available	• Bone char contains bacteria and is therefore unhygienic • Bone char may cause cultural and religious problems • The procedure is technique sensitive since the efficiency is a function of the charring process
4	RGAC and CAC	• No chemicals added to water • Low-cost and easy maintenance • Excellent for enhancing the taste and odor • Good at filtering out other chemicals as well as some microorganisms • Free from cultural and religion problems	• When filled with contaminants, it stops working and need to be replaced • The effectiveness of filtration is determined by factors such as flow rate and bed thickness • Not effective against some pathogenic bacteria and viruses, and can harbor bacteria

The cost comparison is arranged as follows; activated alumina > Nalgoda > Bone char > RGAC and CAC.

Currently, Ngurdoto Defluoridation Research Station in Tanzania have been dealing with fluoride removal by using bone char, which is a preferred method.

Table 26.4 Fluoride removal efficiency at different adsorbent depths/flow rates

S. No	Flow rate (mL/min)	Adsorbent depth (RGAC + CAC)		
		4 cm	6 cm	8 cm
		Removal efficiency (%)		
1	2	34.87	26.39	23.92
2	4	38.60	44.76	44.27
3	6	43.64	71.91	60.81
4	8	45.18	69.01	57.51

However, the social and economic reasons hinder the full participation of the community to use this method. This study recommends the use of RGAC and CAC as an alternative to bone charcoal.

26.3.4.1 Defluoridation Process Using RGAC and CAC

Charcoal (wood-fuel and coconut shells) is locally produced in rural areas in Arusha region in Tanzania. There can be collected and carbonization process is done at Ngurdoto Defluoridation Research Station. This is prepared at a temperature of 400° C in a furnace under controlled air. Charcoals obtained is activated by using 25% of NaOH, 25% of $CaCl_2$, 2M H_2SO_4 and 2M HNO_3 for 12 h. The charcoal obtained is washed by deionized water to remove residues of activating agents. It is finally dried and grinded to get granular activated carbon (Said and Machunda 2014).

The RGAC and CAC is filled in 4 glass columns in equal depth and results were recorded by varying flows and bed depth as shown in Table 26.4. The fluoride contaminated water is connected on the top and treated water is collected at the bottom of the column.

26.4 Results and Discussion

RGAC and CAC gives maximum removal at optimum bed depth of 6 cm and optimum flow rate of 6 mL/min. It is finally observed that the maximum removal efficiency of fluoride is 71.9% as shown in Table 26.4. This experiment was done several times in the agency Centre. The present data was obtained from (Sheetal et al. 2015).

Table 26.4 was obtained by using raw water at a constant concentration of 4 mg/L of fluoride. Flow rate and bed depth was varied. The results show that the optimum flow rates and bed depth required for maximum removal of Fluoride from the water is 6 mL/min and 6 cm respectively. This has a removal efficiency of 71.9%. Finally, RGAC and CAC can be regenerated by pyrolysis and by burning off of adsorbed organic substances at 800 °C under controlled air (Van Vliet 1991). This method is widely used and can regenerate carbon very well.

26.5 Conclusions

Fluorosis has contributed many public health problems in Tanzania. Drinking water coming from aquifers in rift valley regions is the main source of fluoride. Arumeru district in Tanzania showed highest values of fluoride in the country, where 81.6% of their water sources have fluoride concentration above 1.5 mg/L (refer limit by WHO). It can lead to chronic fluoride toxicity like dental fluorosis, skeletal fluorosis, crippling fluorosis and systemic fluorosis. Numerous defluoridation methods are worldwide used. Since majority of the methods are very expensive and are limiting only to urban areas. This paper discusses de-fluoridation technique using RGAC and CAC as the most suitable technique for Arumeru rural communities in Tanzania. The optimum fluoride removal conditions are 6 mL/min flow rate and 6 cm thickness of bed with 71.9% efficiency. This is recommended for rural communities in Tanzania.

References

APHA (1998) Standard Methods for the Examination of Water and Wastewater. Am Public Health Assoc/Am Water Works Assoc/Water Environ Fed 552:108–117

Brindha K, Elango L (2011) Fluoride in Groundwater: causes, implications and mitigation measures. In: Fluoride properties, applications and environmental management.

Bureau of Indian Standards (1982) Tolerance and classification. ISI-IS: 2296. New Delhi. https://wqaa.gov.in/WriteReadData/UserFiles/Documents/ToleranceandClassification.pdf

Chidambaram S, Ramanathan AL, Vasudevan S (2003) Fluoride removal studies in water using natural materials. Water SA 29(3):339–343. https://doi.org/10.4314/wsa.v29i3.4936

Chinoy NJ, Mehta D, Jhala DD (2006) Effects of fluoride ingestion with protein deficient or protein enriched diets on sperm function of mice. Fluoride 39(1):11–16

Das S, Maiti R, Ghosh D (2006) Fluoride-induced immunotoxicity in adult male albino rat: a correlative approach to oxidative stress. J Immunotoxicol 3(2):49–55. https://doi.org/10.1080/15476910600631587

Eirik Storøy Johansen (2013) The effects of fluoride on human health in Eastern Rift Valley, Northern Tanzania, p 35. https://www.duo.uio.no/

Ghiglieri G, Balia R, Oggiano G, Pittalis D (2010) Prospecting for safe (Low Fluoride) groundwater in the Eastern African Rift: The Arumeru District (Northern Tanzania). Hydrol Earth Syst Sci 14(6):1081–1091. https://doi.org/10.5194/hess-14-1081-2010

Goswami P, Sharma A, Sharma S, Verma S (2015) Defluoridation of water using low cost adsorbent. Int J Chem Stud 3(2):109–112

Gumbo FJ, Mkongo G (1995) Water defluoridation for rural fluoride affected communities in Tanzania. In: Proceedings of the 1st international workshop on fluorosis prevention and defluoridation of water, pp 109–114.

Hazla O (2017) Tanzania: fluoride polluted water haunts arusha. Tanzania Daily News, March

Kanyora A, Samuel K, Mark J (2015) Fluoride removal capacity of regenerated bone char in treatment of drinking water. Asian J Nat Appl Sci 4(1):30–36

Kaseva ME (2006) Optimization of regenerated bone char for fluoride removal in drinking water: a case study in Tanzania. J Water Health 4(1):139–147. https://doi.org/10.2166/wh.2005.062

McSweeney C, New M, Lizcano G (2014) UNDP Climate Change Country Profiles: Tanzania

National Bureau of Statistics (NBS), Office of Chief Government Statistician (OCGS) (2013) Population distribution by age and sex: The United Republic of Tanzania. Dar es Salaam. https://ihi.eprints.org/2169/1/Age_Sex_Distribution.pdf

Renuka P, Pushpanji K (2013) Review on defluoridation techniques of Water. Int J Eng Sci (IJES) 2(3):86–94

Said M, Machunda RL (2014) Defluoridation of water supplies using coconut shells activated carbon: batch studies. Int J Sci Res 3(7):2327–2331

Sheetal B, Mane SJJ, Tirthakar SNN (2015) Removal of fluoride from drinking water by using low cost adsorbent. Int J Res Eng Technol 4(4):349–351

Suneetha M, Syama Sundar B, Ravindhranath K (2015) Studies on defluoridation techniques: a critical review. Int J ChemTech Res 8(8):295–309

Tanzania Bureau of Standards (2005) National environmetal standards compendium. Dar es Salaam.

Van Vliet BM (1991) The regeneration of activated carbon. J South Afr Inst Min Metal 91(5):159–167

Waghmare SS, Arfin T (2015) Fluoride removal from water by various techniques—review. Int J Innov Sci Eng Technol 2(9):560–571.

WHO (2008) Guidelines for drinking-water quality. World Health Organization, Geneva. https://doi.org/10.1016/S1462-0758(00)00006-6

Zafar MS, Ahmed N (2015) Therapeutic roles of fluoride released from restorative dental materials. Fluoride 48(3):184–194

Chapter 27
Water Quality and Human Health

Rajesh Singh, Sujata Kashyap, and Ashish Pandey

Abstract Availability of fresh water has always been an important factor impacting human population and civilizations. However, with the growth of population and economy, the pollution from anthropogenic activities is degrading the quality of water. The degradation in water quality is affecting the health of mankind and ecosystem. It is benign to understand the water quality parameters impacting the human health for judicious management of water quality. This chapter discusses about the significance of water quality parameter impacting the drinking water from aesthetic, health, and operational point of view, methods of analysis, and treatment techniques for reducing the contaminants.

27.1 Introduction

"Water, water, everywhere, Nor any drop to drink", the lines from "the Rime of the ancient Mariner" written by Samuel Taylor Coleridge, has very well illustrated the importance of water quality long back in 1798, when the inland waters were not contaminated. Even the ancient literature, Rig Veda, talks about the quality of water for the well-being of mankind and the measures for improvising the quality of water in the verse 83.4

प्र वाता वन्ति पतयन्ति विद्युत उदोषधीर्जिहते पिंवते स्वः।
इरा विश्वम्भै भुवनाय जायते यत्पर्जन्यः पृथ्वी रेतसावति ।।

R. Singh (✉)
Environmental Hydrology Division, National Institute of Hydrology, Roorkee, India
e-mail: rsingh.nih@gmail.com; rsingh.nihr@gov.in

S. Kashyap
Axa Parenteral Limited, Roorkee, India
e-mail: sujata.iitr@gmail.com

A. Pandey
Department of Water Resources Development and Management, Indian Institute of Technology Roorkee, Roorkee, India
e-mail: ashish.pandey@wr.iitr.ac.in

A. Pandey et al. (eds.), *Climate Impacts on Water Resources in India*, Water Science and Technology Library 95, https://doi.org/10.1007/978-3-030-51427-3_27

The change in hydromorphology, rise in emerging pollutants, and spread of invasive species resulted in loss of pristine quality water bodies, due to which the water quality problems persist in developed and developing countries alike (WWAP 2019). Poor water quality increases water-related health risks. In developing countries, where only a small fraction of the wastewater is treated prior to release in the environment, several water-related diseases, including cholera and schistosomiasis prevail. WHO (2016) reported 4% of the population in low income economies suffered from diarrhea in 2015, among whom 60% were children below age of five. Access to potable water is essential for the development of a nation, since it improves human health, reduces health care costs, reduces mortality, and increases the working days, thereby increasing the wealth index of the citizens (WHO 2011a).

Approximately 820 million Indian people are facing high to extreme water stress situation (NITI 2019). Further, India's 70% water resources are contaminated with excess iron, fluoride, salinity, nitrate and arsenic, ranking it at 120th position out of 122 countries based on Water Quality Index (WQI). Therefore, National Institute for Transforming India (NITI) Aayog has framed a Composite Water Management Index for assessing and improving the efficient management of water resources of the country. In developing countries, consumption of unsafe water results in one of the highest disease burdens in form of water borne diseases (NITI 2018). Unsafe drinking water and poor sanitation results in approximately 10 million cases of diarrhea, more than 7.2 lakh typhoid cases, and 1.5 lakh viral hepatitis cases every year in India (BIS 2012). Accordingly, to eliminate the impact of deteriorated water quality on human health, Bureau of Indian Standards (BIS) has prescribed the limits for water quality parameters in light of World Health Organization (WHO) guidelines. The parameters, which are missing from WHO guidelines and crucial for Indian conditions, have been incorporated from the standards of other countries. This chapter describes the health effect of the water quality parameters, their prescribed limits, and options for their treatment.

27.2 Organoleptic/Physical Parameters

Organoleptic or physical parameters are those parameters which can be detected by the senses. BIS (2012) has prescribed limits for six physical parameters, namely color, odor, pH, taste, turbidity, and total dissolved solids, which can be judged by the sensory organs (Table 27.1). The drinking water supplied to the citizens should not exceed the values mentioned under 'Acceptable limit', however, in absence of alternate source, this may be relaxed to the values mentioned under 'Permissible limit'. However, water sources with values exceeding the 'Permissible limit', will have to be rejected.

Table 27.1 BIS limits for organoleptic parameters (BIS 2012)

Sr. No.	Characteristic	Acceptable limit	Permissible limit
1	Color, Hazen units, Max	5	15
2	Odor	Agreeable	Agreeable
3	pH	6.5–8.5	No relaxation
4	Taste	Agreeable	Agreeable
5	Turbidity, NTU, Max	1	5
6	Total dissolved solids, mg/L, Max	500	2000

27.2.1 Color

Drinking water should ideally be colorless. Color in water may be due dissolved or suspended colloidal particles and indicates the presence of iron, manganese, or organic matter. Iron salts impart brown/red; manganese salts impart black, chromium salts impart yellow, and decaying leaves and macroscopic plants impart brownish-yellow hue to the water (Nathanson 1986). Color is also introduced in the water through mixing of the industrial effluents.

Color is expressed in Pt–Co or Hazen units, and the units are interchangeable. Color is measured by visual comparison of the sample with known concentrations of colored solutions or by spectrophotometer, after adjusting the pH to around 7.6 followed by centrifugation for 1 h to remove turbidity (APHA 2017).

Presence of color in drinking water is undesirable from the aesthetic point of view and is generally not harmful to human health, but indicator of a hazardous situation. A substantial change in color of water should be investigated for the harmful content.

27.2.2 Odor and Taste

These are indicators of the presence of undesirable impurities such as hydrogen sulfide, sewage contamination, excess dissolved solids, corrosion byproducts, disinfection byproducts, petroleum products, etc. A change in the normal appearance, taste, or odor of a drinking-water supply indicates change in the quality of the raw water source or deficiency in the treatment process, demanding immediate curative measures. The limits prescribed for odor and taste are related to consumer acceptability, are of great significance for consumer confidence, and are not directly related to human health (WHO 2017).

Odor and taste are analyzed by the threshold odor test and the flavor profile analysis. Both the tests are subjective tests since both depends on the human perception of the odor and taste in water (APHA 2017).

The tastes and odors can be removed by coagulation-flocculation, aeration, ozonation, or activated carbon. The selection of the method will depend on the taste and odor causing components in the water.

27.2.3 *pH*

pH is defined as negative logarithm of hydrogen ion concentration. The pH of pure water is 7 at 25 °C because hydrogen and hydroxyl ion concentrations are equal, however, natural waters usually have pH values in the range of 4–9, and most are slightly alkaline. Waters in humid areas with highly leached soils in general have lower pH than waters in areas with limestone formations or those in semi-arid or arid regions. Waters originating from forested areas also tend to have a low pH due to presence of high concentrations of humic substances. Although, pH has no direct impact on human health, the prescribed standard is operational parameter.

The pH of water is generally measured electrometrically using pH meter consisting of potentiometer, a glass electrode, a reference electrode, and a temperature compensating device.

If pH deviates from the operational range, acid or alkali dosing is used. The pH is generally reduced by aerating with CO_2.

27.2.4 *Turbidity*

Turbidity in water is due to the presence of suspended and colloidal matter, such as silt, clay, tiny fragments of organic/inorganic matter, and microscopic plants and organisms. These suspended matter scatter and adsorb light rays, reducing the clarity of water and giving it murky or turbid appearance. Turbidity in water is due to the poor source, poor treatment, and disturbance of sediments / biofilms in the distribution system, and ingress of dirty water through the supply line breaks and other faults. High turbidity, generally more than 5 NTU, leads to staining of materials, fittings, and clothes, and also reduces the acceptability of drinking water. Although, most particles that contribute to turbidity have no health significance, their presence may provide a shield to pathogenic micro-organisms during the disinfection process and can lead to waterborne diseases.

Turbidity is measured by nephelometer. A ray of light is passed through the water sample and the scattered light is measured by detectors at 90° to the incident light in comparison to standard suspension in nephelometric turbidity units (NTU).

Turbidity in water is generally removed by coagulation, flocculation and clarification followed by filtration. The water with low turbidity can be treated through slow sand filter, microfiltration, or ultrafiltration membrane for removal of suspended solids.

27.2.5 *Total Dissolved Solids*

Total dissolved solids (TDS) in water includes all dissolved material in solution, whether ionized or not. TDS is numerical sum of all mineral constituents dissolved in water and is expressed in mg/L. TDS in drinking water is contributed from natural sources, sewage, urban runoff and industrial wastewater. Due to the difference in the solubility of minerals in different geological regions, the dissolved solids in water vary from one place to another. Based on TDS contents, water can be classified in to four categories as fresh, brackish, saline and brine water (Table 27.2).

No health-based guideline value for TDS has been proposed due to unavailability of reliable data on possible health effects due to consumption of TDS in drinking water (WHO 2011a). However, the water suppliers should keep in mind that drinking water is palatable up to 600 mg/L, and significantly unpalatable at TDS levels greater than 1000 mg/L. TDS associated with high concentrations of carbonates of Ca and Mg results in scale formation and bitter test, however, TDS associated with high concentration of chlorides and sulfates of cations, particularly Na & K, leads to salty/brackish taste and increased corrosivity.

It is estimated by filtering a known volume of water sample through 2.0 μm (or smaller) nominal pore size filter, evaporating the filtrate to dryness at 180 ± 2 °C, and reporting the weight of solids remaining after evaporation in mg/L.

Dissolved solids in water can be removed by precipitation, ion exchange processes, nanofiltration membranes, reverse osmosis membranes, and thermal distillation processes.

27.3 General Inorganic Parameters

BIS (2012) has prescribed limits for 13 general inorganic parameters, namely ammonia, barium, calcium, magnesium, total hardness, chloride, fluoride, nitrate, sulfate, sulfide, total alkalinity, chloramines, and free residual chlorine which are undesirable in excessive amounts, and 1 inorganic parameter, cyanide, which is toxic (Table 27.3).

Table 27.2 Classification of water based on total dissolved solids

Sr. No.	TDS (mg/L)	Water quality
1	0–1,000	Fresh water
2	1000–10,000	Brackish water
3	10,000–100,000	Saline water
4	>100,000	Brine

Table 27.3 BIS limits for general inorganic parameters (BIS 2012)

Sr. No.	Characteristic	Acceptable limit	Permissible limit
1	Ammonia (as NH_4-N), mg/L, *Max*	0.5	No relaxation
2	Barium (as Ba), mg/L, *Max*	0.7	No relaxation
3	Calcium (as Ca), mg/L, *Max*	75	200
4	Magnesium (as Mg), mg/L, *Max*	30	100
5	Total hardness (as CaCO3), mg/L, *Max*	200	600
6	Total alkalinity (as CaCO3), mg/L, *Max*	200	600
7	Chloride (as Cl), mg/L, *Max*	250	1000
8	Fluoride (as F), mg/L, *Max*	1.0	1.5
9	Nitrate (as NO_3), mg/L, *Max*	45	No relaxation
10	Sulfate (as SO_4), mg/L, *Max*	200	400
11	Sulfide (as H_2S), mg/L, *Max*	0.05	No relaxation
12	Chloramines (as Cl_2), mg/L, *Max*	4.0	No relaxation
13	Free residual chlorine (as Cl_2), mg/L, *Min*	0.2	1.0
14	Cyanide (as CN), mg/L, *Max*	0.05	No relaxation

27.3.1 Ammonia

Ammonia in water comprises of non-ionized (NH_3) and ionized form (NH_4^+), originating from mammalian metabolism, agricultural runoff, and industrial effluents. Less than 0.2 mg/L ammonia is encountered in oxic water, whereas anaerobic water resources may contain upto 3 mg/L NH_3. Presence of ammonia in drinking water indicates possible bacterial, sewage, and animal waste pollution.

The guideline value prescribed by BIS for ammonia is aesthetic based and not health-based, as the toxicological effects are about 200 mg/kg body weight. WHO has not prescribed any guideline value for ammonia. The threshold odor and taste concentration is approximately 1.5 mg/L and 35 mg/L respectively (WHO 2017).

Ammonia is estimated in drinking water by ammonia selective electrode method or phenate method. Ammonia selective electrode uses a hydrophobic gas permeable membrane that separate the sample from the electrode internal solution of ammonium chloride. The ammonium present in the water is converted to ammonia gas by raising the pH above 11, which diffuses through the membrane, resulting in a change in the pH of the internal solution, which is sensed by a pH electrode. The phenate method is a colorimetric method, in which ammonia in water reacts with phenol to form an intense blue compound, indophenol, in the presence of hypochlorite, and sodium nitroprusside as a catalyst.

Ammonia in drinking water can be removed by ion-exchange resin, biological filter (nitrification), air stripping, and reverse osmosis membrane.

27.3.2 *Barium*

Barium (Ba) compounds are present in igneous and sedimentary rocks, and their average abundance in the earth's crust is 390 mg/kg and in soils it is 63 to 810 mg/kg. It is found primarily in barite ($BaSO_4$) or in witherite ($BaCO_3$). They are used in drilling fluid additive, X-ray diagnostic tests, plastics, rubbers, paint, glass, carpets, ceramics, sealants, furniture, fertilizers and pesticides. The average concentration of Ba in streams is 10 mg/L and in groundwaters 0.05–1 mg/L (APHA 2017). Barium in water comes primarily from natural sources as well as industrial emissions and anthropogenic uses, and the concentration is controlled by the solubility of $BaSO_4$, and to some extent by adsorption on hydroxides.

The health effects of the different barium compounds depend on their solubility and the insoluble compounds are not generally harmful. Ingestion of Ba laden water may cause difficulties in breathing, increased blood pressure, changes in heart rhythm, stomach irritation, brain swelling, muscle weakness, and damage to the liver, kidney, heart, and spleen, in people, but are not carcinogen (WQA 2013a; WHO 2017). The U.S. EPA and WHO recommends guideline value of 1 mg/L and 1.3 mg/L, respectively.

Barium concentration in drinking water can be estimated by iodometric titration, atomic absorption spectrometer, and inductively coupled plasma mass spectrometer.

Barium in drinking water can be removed by ion exchange, lime softening, reverse osmosis, and distillation to below 1 mg/L.

27.3.3 *Total Hardness, Calcium, and Magnesium*

Hardness in drinking water is due to dissolved polyvalent metallic ions, predominantly calcium and magnesium cations, and is expressed as milligrams of calcium carbonate per litre. The hardness of drinking-water is important for aesthetic acceptability and it does not pose a health hazard. Hardness in water prevents the lather (foam) formation with the soap, and is important for economic and operational considerations. It is classified as soft water, hard water, moderately hard water and very hard water (Table 27.4).

Drinking-water can be an important contributor of calcium and magnesium to those who are marginal for calcium and magnesium. Typical recommended dietary

Table 27.4 Classification of drinking water based in hardness (WHO 2011a, b)

Sr. No.	Total hardness (mg/L)	Water type
1	<60	Soft water
2	60–120	Moderately hard water
3	120–180	Hard water
4	>180	Very hard water

intake for Ca & Mg is 1000 mg/day and 200–400 mg/day respectively (WHO b), and a glass of milk (200 mL) can met ≈30% Ca and ≈15% Mg requirement (Brink et al. 1992; Gaucheron 2005). The recommended upper intake level for Ca is 2500 mg/day, and the individuals exposed to high concentration are protected by a tightly regulated intestinal absorption and elimination mechanism through the action of 1,25-dihydroxyvitamin D. The excess absorbed calcium is excreted by the kidney in healthy people who do not have renal impairment. Drinking water with both magnesium and sulfate, above 250 mg/L each, can have a laxative effect, although the effect recedes as exposure continues.

The taste threshold for the calcium ion is in the range of 100–300 mg/L, however, some consumers can tolerate water hardness in excess of 500 mg/L. WHO has not established any guideline value considering the fact that the levels found in drinking water does not pose a health hazard to humans (WHO 2011a, b).

Total hardness, calcium and magnesium in drinking water is analyzed by EDTA titrimetric method. These parameters can also be analyzed by Ion Chromatograph.

The removal of hardness from drinking water depends on type of hardness and extent of hardness removal. Lime soda softening, ion exchange resins, nanofiltration membranes, and reverse osmosis membranes are capable of reducing hardness of water.

27.3.4 *Total Alkalinity*

Alkalinity is the buffering capacity of the water to neutralize acids and bases, and is dependent on the presence of certain chemicals in the water like bicarbonate, carbonate, and hydroxides. The alkalinity in the water comes mostly from the rocks and land of the catchment area of the water body. Water with alkalinity levels less than 150 mg/L is more likely to be corrosive, and alkalinity levels greater than 150 may contribute to scaling. The alkalinity of water affects the amount of chemicals required to accomplish effective coagulation and softening during treatment. Alkalinity in drinking water, due to naturally occurring materials such as carbonate and bicarbonate up to approximately 400 mg/L as calcium carbonate, is not a health hazard (USEPA 1976), therefore, WHO has not prescribed any guideline value for alkalinity.

Alkalinity in water is determined by titration with standard sulfuric acid using pH sensitive indicators or pH meter.

Alkalinity in water can be removed by acidification of water and can be added by passing the low alkaline water through calcite filter.

27.3.5 *Chloride*

Chloride ions come into water either from geogenic or anthropogenic sources. The geogenic sources includes chloride bearing minerals like halite, potassium chloride, calcium chloride, limestone, sandstone, shale, etc. Anthropogenic sources of chloride are human sewage, livestock waste, synthetic fertilizer, etc. Chloride is a conservative anion in most aqueous environments and its movement is not retarded by the interaction of water with soils, sediments, and rocks. Further, it is not affected by the microbial action or redox chemistry in environment. Hence, it can be used as an indicator of other type of contamination. High concentration of chloride in water gives a salty taste. Concentration of chloride associated with calcium, sodium and potassium in excess of 250 mg/L is detected by taste, however, no side effects on human health has been observed at the levels found in drinking water, due to which WHO has not proposed any guideline value. Excessive chloride concentration in water makes it corrosive and is important from operational point of view.

Chloride concentration in drinking water can be determined with argentometric titration using potassium chromate indicator or potentiometer. It can also be analyzed by ion chromatograph.

Chloride in drinking water can be removed by reverse osmosis or distillation.

27.3.6 Fluoride

The average abundance of fluoride in earth's crust is 300 mg/kg, and is found at significant levels in a wide variety of minerals, including fluorspar, rock phosphate, cryolite, apatite, mica, hornblende and others. It is found in all-natural waters at some concentration; in seawater, it is about 1 mg/L; in rivers and lakes, it is less than 0.5 mg/L; and in groundwater, its concentration varies depending on the nature of the rocks and the occurrence of fluoride-bearing minerals. Fluoride concentration in water is limited by fluorite solubility, means, the presence of 40 mg/L calcium will limit fluoride concentration to 3.1 mg/L (Hem 1989). Therefore, high fluoride concentrations are expected in groundwaters from calcium poor aquifers and in areas where fluoride-bearing minerals are common (Nanyaro et al. 1984; Gacırı and Davis 1993; Kundu et al. 2001). The highest natural level reported is 2800 mg/L (WHO 2004).

Fluoride in drinking water has a narrow range between intakes that cause beneficial (0.5–1.0 mg/L) and detrimental health effects (>1.0 mg/L), primarily dental and skeletal fluorosis. WHO (2011a) and BIS (2012) have prescribed the maximum allowable limit for fluoride uptake to human's in drinking water as 1.5 mg/L. However, the national standard should be based on the average water intake and intake from other sources.

Fluoride in drinking water can be estimated by fluoride selective electrode, spectrophotometer (SPADNS method), or ion chromatograph.

Fluoride in drinking water can be removed by precipitation, adsorption with aluminum based media, evaporation, or reverse osmosis membrane. Use of reverse osmosis membrane should be avoided for water having Ca and F ions as CaF_2 will lead to precipitation fouling of the membrane.

27.3.7 Nitrate

Nitrate (NO_3^-) concentration in both surface water and groundwater is due to agricultural activities, natural vegetation, wastewater, and human and animal excreta waste. Nitrate concentration in drinking water is a potential health hazard, when present in large quantities. Nitrites are formed by reduction of nitrate in the human body, which combines with hemoglobin in the blood to form methemoglobin that leads to methaemoglobinaemia (blue baby syndrome) in infants. The combination of nitrates with amines, amides, or other nitrogenous compounds through the action of bacteria in the digestive tract results in the formation of nitrosamines, which are potentially carcinogenic. The guideline values prescribed by WHO (2017) for nitrate in drinking water is 50 mg/L as NO_3 considering the fact that no adverse health effects has been observed below this concentration in epidemiological studies.

Nitrate in the drinking water can be estimated by electrometric (nitrate selective electrode), colorimetric (Cd reduction method), or ion chromatographic methods.

Nitrate in drinking water can be removed by biological denitrification, reverse osmosis, or distillation.

27.3.8 Sulfate

Sulfates (SO_4^{-2}) occur naturally in numerous minerals, including barite ($BaSO_4$), epsomite ($MgSO_4 \cdot 7H_2O$) and gypsum ($CaSO_4 \cdot 2H_2O$). Other sources of sulfur in water includes decomposition of organic, plant and animal, matter. Sulfate in drinking-water can cause taste impairment depending on associated cation, ranging from 250 mg/L for sodium sulfate to 1000 mg/L for calcium sulfate. Very high levels of sulfate may cause a laxative effect in unaccustomed consumers, and therefore, the health authorities should be notified of sources of drinking water that contain sulfate concentrations in excess of 500 mg/L (WHO 2011a). Adults generally adapt to high sulfate concentrations within 1 or 2 weeks, however, infants may be more sensitive (USEPA 2003).

The concentration of sulfate found in natural waters is generally not at levels to cause adverse health effects and the existing data do not identify a level of sulfate that is likely to impact the human health, due to which WHO has not proposed any health based guideline (WHO 2017).

Sulfate in the drinking water can be estimated by gravimetric, turbidimetric, or ion chromatographic methods.

Sulfate concentration in drinking water can be reduced by reverse osmosis, ion exchange, or distillation.

27.3.9 *Sulfide*

Hydrogen sulfide (H_2S) gas produces an offensive "rotten egg" odor and unpleasant taste in the water. Its concentration in groundwater and surface water depends on the rock mineralogy, microorganisms present, and influx of effluents. The sulfur/sulfate reducing bacteria uses sulfate and sulfur compounds in decaying plant material, rocks or soil, as an electron acceptor for the degradation of organic compounds, resulting in the production of S^{-2}, CO_2, and CH_4 in anaerobic environment (Muyzer and Stams 2008). In water, H_2S dissociates into monohydrogensulfide (HS^-) and sulfide (S^{2-}) ions, and the concentration of these species depends on pH of water, with hydrogen sulfide concentrations increasing with decreasing pH. H_2S gets rapidly oxidized to SO_4^{-2} in well aerated or chlorinated waters. Further, the taste and odor threshold of H_2S in water is between 0.05 and 0.1 mg/L respectively, due to which the chances of a person consuming a harmful dose of H_2S from drinking water becomes very thin and is the basis for WHO not proposing a health based guideline value. The oral dose of sodium sulfide fatal to humans is around 10–15 g, leading to irritation of mucous membranes, nausea, vomiting, and epigastric pain (WHO 2003a).

Sulfide in drinking water can be estimated by titrimetric iodometric method or spectrophotometric methylene blue method (BIS 1987).

Trace amount of sulfide (0.05–0.3 mg/L) in water can be removed by adsorption on activated carbon and more than 0.3 mg/L by oxidation to sulfate by aeration (<2 mg/L), manganese greensand filter (<10 mg/L), and chlorination (<75 mg/L).

27.3.10 *Chloramines*

Chloramines are a group of compounds (monochloramine, dichloramine and trichloramine) formed from the reaction of chlorine with ammonia. Among the group, monochloramine is the only useful disinfectant and other chloramines are formed occasionally, therefore, WHO has proposed the health based guideline value only for monochloramine. Chloramine is preferred by water treatment operators in place of chlorine, as it is more stable and reduces the formation of Trihalomethanes (THMs). Monochloramine is weaker oxidant than free chlorine and are not very effective in getting rid of taste and odor already present. Higher chloramines, particularly trichloramine, give rise to taste and odor complaints, however, they are little unstable and escapes from the water before reaching the taps. WHO (2017) has prescribed 3 mg/L as the guideline value for monochloramine. IARC has classified chloramine in Group 3 (not classifiable as to its carcinogenicity to humans).

Chloramines in drinking water can be estimated by amperometric titration, *N,N*-diethyl-*p*-phenylenediamine (DPD) ferrous titration, or DPD colorimetric method (APHA 2017).

Chloramines in water can be removed by activated carbon filter with 10 min empty bed contact time. Since chloramines are small, stable molecules with no net charge; distillation, reverse osmosis, and ion exchange resins are not efficient technologies for their removal (WQA 2013b).

27.3.11 Free Residual Chlorine

Chlorine (Cl_2) is the most widely used disinfectant and oxidant in drinking water treatment for killing pathogens and reducing disagreeable tastes and odors in water. Chlorine dissolves rapidly in water to establish an equilibrium with hypochlorous acid (HOCl) and hydrochloric acid (HC1). In dilute solution with pH levels above 4.0, molecular chlorine cease to exist and HOCl becomes the dominant species. Between pH 6.0 and 8.5, hypochlorous acid dissociates to form hypochlorite ion (OCl^-), which becomes dominant above 9.0 pH. The total concentration of Cl_2, HOCl, and OCl^- in water is defined as 'free available chlorine', and the concentration of these species left after reaction with microorganisms, organics, and metals is termed as 'free residual chlorine' (IARC 1991).

Chlorine in drinking water is sensed by most of the consumers at concentrations well below 5 mg/L, and some may even sense as low as 0.3 mg/L. WHO (2017) has prescribed the health-based guideline value of 5 mg/L for chlorine. Ingestion of a small amount of bleach results in the irritation of the esophagus, a burning sensation in the mouth and throat, and spontaneous vomiting, however, specific adverse health related effects have not been observed in humans and animals exposed to chlorine in drinking-water. IARC has concluded that hypochlorites are not classifiable as to their carcinogenicity to humans (Group 3).

Free residual chlorine in water can be estimated by iodometric titration, amperometric titration, DPD ferrous titrimetric method, DPD colorimetric method, or syringaldazine (FACTS) method.

Chlorine in drinking water can be removed by activated carbon filter or by reducing with sodium metabisulfite.

27.3.12 Cyanide

Cyanides (CN^-) compounds are powerful and rapid-acting poisons, can both occur naturally or anthropogenically in drinking waters. Natural sources of CN are certain species of bacteria, fungi, and algae, and a number of food and plant products like almonds, millet sprouts, lima beans, soy, spinach, bamboo shoots, and cassava roots. The anthropogenic sources are discharges from some metal mining and processing industries, some organic chemical industries, vehicle exhaust, burning of municipal waste, landfill leachates, and cyanide-containing pesticides. Cyanides are generally absent in drinking-water and can occur occasionally at very low concentrations. Although, at few occasions spills of cyanide from industries may give rise to concentrations of concern in drinking waters, particularly surface waters.

Cyanide is a well-established toxin, however, the actual dose resulting in acute fatal intoxication to humans has not been established due to the lack of well-conducted studies on sub-lethal toxicity. The lowest recorded lethal dose for humans is 0.56 mg/kg bodyweight. Low exposures to cyanides are not fatal to humans with efficient detoxification system. The major pathway for conversion of cyanide to much less toxic thiocyanate ion is via the intramitochnodrial enzyme rhodanease, a liver enzyme that catalyzes the transfer of sulfur from a donor to cyanide. Urine is the major route of excretion of thiocyanate, normally ranging between 0.85 and 14 mg over a 24 h period (Health Canada 1991). USEPA has allowed 0.2 mg CN per L in the drinking water as the maximum limit. WHO has not established a guideline value for cyanide in drinking water as cyanide concentration in water resources has been observed to be well below the health concern.

Cyanide in drinking water can be estimated by titration with $AgNO_3$, colorimetric method, or CN selective electrode method.

Cyanide in drinking water is reduced adsorption on activated carbon, oxidation by 2-step chlorination, hydrogen peroxide, or persulfate, or reverse osmosis.

27.4 Organic Parameters

BIS (2012) has prescribed limits for 3 organic parameters, namely anionic detergents, mineral oil, and phenolic compounds, which are undesirable in excessive amounts, and 24 parameters, namely polychlorinated biphenyls, polynuclear aromatic hydrocarbons, trihalomethanes (bromoform, bromomonochloromethane, and bromodichloromethane), and pesticides (alachlor, atrazine, Aldrin/dieldrin, alpha HCH, beta HCH, Butachlor, chlorpyriphos, delta HCH, 2,4-dichlorophenoxyacetic acid, DDT, endosulfan, ethion, gamma HCH, isoproturon, malathion, methyl parathion, monocrotophos, and phorate), which are toxic substances (Table 27.5).

27.4.1 Anionic Detergents

Anionic detergents are surface active compounds (fatty acid soaps, linear alkyl benzene sulfonates, alkyl ether sulfates, and alkyl sulfates) consisting of a hydrophobic alkyl chain and a hydrophilic negatively charged head group such as carboxylate, sulfate, phosphate or sulfonate. These are used in consumer products like laundries, car washing, dry cleaning, soaps, dishwashing, and personal care products, and are released to the environment after use, causing pollution of water resources (Brook et al. 2013). High concentration of detergents in drinking-water results in foaming, taste problems, and indicator of wastewater ingress, requiring attention of water treatment operators.

Anionic detergents are not able to penetration through intact skin, however, orally ingested detergents are readily absorbed from the gastrointestinal tract. The adsorbed detergents are metabolized in the liver, and the surfactants along with their metabolites are mostly eliminated through urine and only minor amounts are eliminated with feces. The detergents on contact with the human body may cause swelling and

Table 27.5 BIS limits for organic parameters (BIS 2012)

Sr. No.	Characteristic	Acceptable limit	Permissible limit
1	Anionic detergents (as MBAS), mg/L, *Max*	0.2	1.0
2	Mineral oil, mg/L, *Max*	0.5	No relaxation
3	Phenolic compounds (as C_6H_5OH), mg/L, Max	0.001	0.002
4	Polychlorinated biphenyls, mg/L, *Max*	0.0005	No relaxation
5	Polynuclear aromatic hydrocarbons (as PAH), mg/L, *Max*	0.0001	No relaxation
6	Trihalomethanes, mg/L, *Max*		
6a	Bromoform	0.1	No relaxation
6b	Dibromochloromethane	0.1	No relaxation
6c	Bromodichloromethane	0.06	No relaxation
6d	Chloroform	0.2	No relaxation
7	Pesticides, μg/L, Max		
7a	Alachlor	20.0	No relaxation
7b	Atrazine	2.0	No relaxation
7c	Aldrin/ Dieldrin	0.03	No relaxation
7d	Alpha HCH	0.01	No relaxation
7e	Beta HCH	0.04	No relaxation
7f	Delta HCH	0.04	No relaxation
7 g	Gamma HCH (Lindane)	2.0	No relaxation
7 h	Butachlor	125.0	No relaxation
7i	Chlorpyriphos	30.0	No relaxation
7j	2,4-Dichlorophenoxyacetic acid	30.0	No relaxation
7 k	DDT (o,p & p,p – DDT, DDE, DDD)	1.0	No relaxation
7 l	Endosulfan (alpha, beta, and sulfate)	0.4	No relaxation
7 m	Ethion	3.0	No relaxation
7n	Isoproturon	9.0	No relaxation
7o	Malathion	190.0	No relaxation
7p	Methyl parathion	0.3	No relaxation
7q	Monocrotophos	1.0	No relaxation
7r	Phorate	2.0	No relaxation

irritation of the eyes, skin, and mucous membranes (Madsen et al. 2001). WHO has not prescribed any guideline value for anionic detergents.

Anionic detergents in water samples can be determined spectrophotometrically by reacting the anionic detergents with methylene blue in basic medium, extracting the excess methylene blue into chloroform, and measurement of blue color in $CHCl_3$ at 652 nm.

Anionic detergents in drinking water can be removed by $FeCl_3$ and cationic polyelectrolyte assisted precipitation (Aboulhassan et al. 2006; Kaleta and Elektorowicz 2013), adsorption (Tripathi et al. 2013), electrochemical oxidation (Rabbani et al. 2017), and constructed wetlands (Kruszelnicka et al. 2019). Narkis and Weinber (1989) observed that coagulation-flocculation-sedimentation, followed by sand filtration and adsorption through granular activated carbon is most effective means for removal of detergents.

27.4.2 *Mineral Oil*

Mineral oil (also known as base oil, mineral base oil or lubricant base oil) are highly purified chemical substances prepared from crude petroleum oil. It is used in baby lotions, cold creams, ointments and many other cosmetic and personal care products, due to its ability to reduce water loss from skin. The extent of refining/purification is determined by the Modified Ames test (ASTM 1996) and the IP346 assay (Institute for Petroleum 1985; CONCAWE 1994). Untreated and mildly treated mineral oils are carcinogenic to humans (Group 1) (IARC 2012; WHO 2005a).

Mineral oil in drinking water is analyzed by partition gravimetric method, partition infra-red method, or soxhlet extraction method. BIS prefer partition infra-red method involving acidification and extraction, followed by measuring the absorbance at 2980 cm^{-1} against reference standards.

Mineral oil in drinking water can be removed be adsorption on aluminum hydroxide precipitates, granular activated carbon filters, and ceramic membranes. PVDF UF membranes can also be employed for mineral oil removal from water.

27.4.3 *Phenolic Compounds*

Phenolic compounds are defined as hydroxyl derivatives of benzene and its condensed nuclei. These compounds may enter into the water resources from natural, industrial, domestic and agricultural activities. Alkylphenols may be derived from lignin transformation and tannins. Phenol derivatives are present in coal tar, synthetic resins, plastics, rubber proofing, and dye manufacturing. In agricultural activities, phenolic compounds are employed as herbicides and insecticides. These chemicals, after entering into the water resources, may undergo transformation into other moieties which may be more harmful than the original compound (Davi and Gnudi 1999; Anku et al. 2017). The carcinogenic and mutagenic properties of phenolic compounds makes them a threat to human beings. WHO has prescribed health based guideline value of 0.2 and 0.009 mg/L for 2,4,6-trichlorophenol and pentachlorophenol respectively. IARC has classified both these compounds in Group 2B (possibly carcinogenic to humans).

Phenolic compounds in water samples can be analyzed by distilling the sample, followed by treating the distillate with 4-aminoantipyrine at pH 8.0 in the presence of potassium ferricyanide to form a colored antipyrene dye, extracting the dye with chloroform, and measuring the absorbance at 460 nm (BIS 2003).

Phenolic compounds present in the drinking water can be removed by activated carbon adsorption, electro-fenton oxidation, photocatalysis, and reverse osmosis membrane.

27.4.4 Polychlorinated Biphenyls

Polychlorinated biphenyls (PCBs) are chlorinated hydrocarbon compounds comprising of two benzene rings linked by a single carbon–carbon bond. As the amount of chlorine in these compound increases, the product becomes progressively thicker, up to a solid wax. PCBs are a concern due to their persistence, long range transport ability, bioaccumulation, and negative impact on humans and environment (Huang et al. 2014). They have been used in plasticizers, surface coatings, inks, adhesives, flame retardants, pesticide extenders, paints, and dielectric fluids in transformers and capacitors. Most of the countries have banned production and use of PCBs. PCBs were never manufactured in India and their import was banned in 1998. Indian government has also decided to completely prohibit the use of PCBs in any form by December 2025 (UNIDO 2015).

The orally ingested PCBs are rapidly absorbed from the gastrointestinal tract and accumulate in the liver and adipose tissue. They also cross the placenta, are excreted in milk, and accumulate in the fetus/infant. PCBs in human tissue are believed to be associated with an elevated risk of cancer of the digestive system, particularly the liver, and of malignant melanoma. PCBs exposure is also associated with reduced growth rates, retarded development, neurological effects, and immunological changes (WHO 2003b).

Analysis of PCBs in drinking water involves extraction at neutral pH with methylene chloride using separatory funnel, continuous liquid–liquid extractor, or solid phase extraction, followed by sequential sulfuric acid/potassium permanganate cleanup and extract analysis using gas chromatograph equipped with electron capture detector (ECD), electrolytic conductivity detector (ELCD), or mass detector (MS).

PCBs in drinking water can be removed by phytoremediation, microbial degradation, activated carbon, and biofilm covered activated carbon (Jing et al. 2018).

27.4.5 Polynuclear Aromatic Hydrocarbons

Polynuclear aromatic hydrocarbons (PAHs) contain two or more fused aromatic rings of carbon and hydrogen atoms. Most of the PAHs in the environment originates from the incomplete combustion of fossil fuels, road runoff, industrial wastewater, leaching from creosote-impregnated wood, and petroleum spills, and only a small fraction of PAHs are produced commercially. PAHs have very low solubility in water and high affinity for particulate matter, due to which they are not found in notable concentrations (WHO 2003c; Karyab et al. 2013). Concentrations of individual PAHs in surface and coastal waters are generally around 50 μg/L, in uncontaminated groundwater usually in the range of 0–5 μg/L, and at heavily contaminated sites upto 10 μg/L (Environment Canada 1994).

The major concern regarding PAHs is the potential carcinogenicity related to lung, bladder, as well as skin cancer (IPCS 1998). Since the concentration of all the PAHs except benzo(a)pyrene are at level well below those of health concern, WHO has proposed a drinking-water guideline value of 0.7 μg/L for benzo(a)pyrene.

PAHs in drinking water can be analyzed by extracting with methylene chloride, followed by analysis with HPLC equipped with UV and fluorescence detector or GC equipped with FID/MS.

PAHs in drinking water can be removed by adsorption on activated carbon, biodegradation under aerobic conditions, phytoremediation, advanced oxidation processes, and membrane processes (Mojiri et al. 2019).

27.4.6 Trihalomethanes

Trihalomethanes (THMs) are generated in drinking water as a byproduct of disinfection. Chloroform ($CHCl_3$), bromodichloromethane ($CHBrCl_2$), dibromochloromethane ($CHBr_2Cl$), and bromoform ($CHBr_3$) are the most commonly occurring and regulated THMs. The concentration of THMs in drinking water is directly proportional to chlorine dose, organic matter/humic acid concentration, bromide ion concentration, temperature, and pH (Bond et al. 2012). Chloroform is the principal disinfection by-product in chlorinated drinking water, however, the presence of bromide results in preferential formation of brominated THMs and proportional decrease in chloroform concentrations. The exposure to THMs are primarily from inhalation of indoor air and ingestion of tap water, and may lead to cancer of bladder and colon, and reproductive and developmental toxicity (Krasner et al. 1989; Golfinopoulos et al. 2003; WHO b). IARC has classified chloroform and bromodichloromethane as possible carcinogen to humans (Group 2B), and bromoform and dibromochloromethane not classifiable as to its carcinogenicity to humans (Group 3). The guideline value prescribed by WHO for chloroform, Bromoform, dibromochloromethane, and bromodichloromethane are 0.3 mg/L, 0.1 mg/L, 0.1 mg/L and 0.06 mg/L respectively.

THMs in drinking water can be analyzed by employing liquid/liquid extraction using isooctane/pentane for at least five fold concentration enhancement, followed by analyzing the extract in a gas chromatograph equipped with electron capture detector or mass spectrometer.

THMs in the drinking water can be reduced by removing the precursors through enhanced coagulation and dosing chlorine after filtration. In case some oxidation is required before filtration, an alternative to chlorine like potassium permanganate, peroxide, or ozone may be dosed. THMs can be removed by degassing, adsorption on activated carbon or by reverse osmosis membranes.

27.4.7 Pesticides

Pesticide is a composite term that encompasses all chemicals that are used to kill or control insects, weeds, fungi and other pests, to protect the crops. They are divided into herbicides (protection against weeds), insecticides (against insects), and fungicides (against fungi) based on their usage. Although the use of pesticides has improved the agricultural yield and ensured food security, the quality of yield has significantly affected. Most of the pesticides are inherently toxic, not only to the pests, against which they are used, but also to other organisms. Damage to non-target organisms, perturbation of structure and function of environment, and toxic contamination of environment are few consequences of pesticide use. Long term and rampant use of pesticides resulted in persistence, bio-accumulation and long range transport of these hazardous chemicals. The toxicants affect entire ecological balance and result in severe health hazards to human beings. These changes occur at a very slow pace and the adverse impacts become visible at a stage where it is almost impossible to reverse the trend.

The health effects of pesticides depend on the type of pesticide, concentration in water, duration of exposure, and individual health status. Humans and animals can be exposed to pesticides through contact with the skin, ingestion, or inhalation. Within a human or animal body, pesticides may be metabolized, excreted, stored, or bio-accumulated in body fat. The negative health effects associated with chemical pesticides are dermatological, gastrointestinal, neurological, carcinogenic, respiratory, reproductive, and endocrine effects (Nicolopoulou-Stamati et al. 2016; Sabarwal et al. 2018).

The pesticides in the water samples can be analyzed by extraction of pesticides from water samples with organic solvents, followed by analysis with GC equipped suitable detectors (Electron capture detector for chlorinated pesticides, Nitrogen phosphorus detector for nitrogen or phosphorus containing pesticides, Flame ionization detectors, or Mass spectrometer). Identification of compounds is done by comparing the retention times with those of standard pesticides and the qualitative determination is carried out by comparing the retention time and peak area of the pesticides (EPA 2007; APHA 2017).

Pesticides in drinking water can be reduced by adsorption, oxidation, microbial degradation, membrane processes, or combination of these (Baldauf 1993; Ormad et al. 2008; Ahmad et al. 2010; Marican and Duran-Lara 2017). However, the most efficient process seems to be oxidation before coagulation-flocculation-sedimentation followed by adsorption on activated carbon. Nanofiltration has also emerged as a good alternative for removal of pesticides without removing all the dissolved solids.

27.4.7.1 Alachlor

Alachlor is a chloroacetanilide herbicide. It is used to control the annual grasses and broad leaved weeds in corn, peanuts, soybeans, and other crops. The research data indicates that the groundwater contamination due to alachlor usage takes place after a gap of considerable time. Alachlor and its metabolite 2,6-diethylanilchlor has shown to be mutagenic. IARC has not evaluated alachlor and has recommended for identification of carcinogenic hazards in medium priority (IARC 2019a). The guideline value prescribed by WHO for Alachlor in drinking water is 0.02 mg/L.

27.4.7.2 Atrazine

Atrazine is a chlorotriazine herbicide. It is used to selectively control the annual broadleaf and grassy weeds in fruit orchards, coffee plantation, grasslands, cereal crops, sugar cane, roses and vines. Along with atrazine, the metabolite hydroxy-atrazine is found in water contaminated with atrazine. IARC has concluded that atrazine is not classifiable as to its carcinogenicity in humans (Group 3) and the 2019 Priorities Advisory Group assigned atrazine a medium priority (IARC 2019a). WHO (2017) has prescribed the guideline value of 0.1 and 0.2 mg/L for atrazine and hydroxyatrazine respectively.

27.4.7.3 Aldrin/Dieldrin

Aldrin and Dieldrin are synthetic organochlorine insecticides. They were used to control broad spectrum of soil-dwelling pests, for seed dressings, and for wood protection. Aldrin is rapidly converted to dieldrin under most environmental conditions. The use of both compounds for use in agricultural practices were banned by most of the countries since the early 1970s. These are highly persistent with high affinity for soil and sediments, due to which most dieldrin in the environment is found in the bottom sediments of the water bodies. Further, the solubility of dieldrin in water is very low ($\approx$27 μg/L), and therefore, the concentrations are not high in natural waters. Plants grown on dieldrin contaminated soil, store it in their leaves and roots. Similarly, fish, animals, or humans feeding dieldrin-contaminated materials store it in their fat (WHO 2017; IARC 2019b), and the bio-accumulated dieldrin is typically measured in blood or body tissues. The target organs by dieldrin are the central nervous system and liver. It is excreted in the bile, feces, and breast milk, and can cross the placenta (ASTDR 2002; IARC 2019b). IARC has classified dieldrin in Group 3 (not classifiable as to its carcinogenicity to humans). WHO (2017) has prescribed guideline value of 0.03 μg/L for drinking and other domestic usage for combined concentration of aldrin and dieldrin.

27.4.7.4 Hexachlorocyclohexane (HCH)

HCH consists of eight isomers, but only α-HCH, β-HCH, γ-HCH, and δ-HCH are of commercial significance. All of the isomers are toxic to animals to varying degrees and are persistent in the environment, with only γ-HCH (Lindane) has an appreciable insecticidal activity. Lindane is used as an insecticide on fruit and vegetable crops, for seed treatment, and in forestry. Technical BHC is a mixture of a number of stereoisomers, principally alpha-BHC, beta-BHC, gamma-BHC and delta-BHC, and the proportion of these isomers can vary in different technical mixtures. The Stockholm Convention held in 2009 accepted inclusion of γ-, α-, and β-HCH in Persistent Organic Pollutants (POPs) list requiring the address to these chemicals at global level (Vijgen et al. 2010). However, the convention exempted the use of lindane as human health pharmaceutical. Several countries have already restricted the use of lindane. Lindane is not likely to pose a carcinogenic risk to humans, which was further confirmed by the epidemiological study designed to assess the potential association between breast cancer and exposure to chlorinated pesticides (WHO 2011a). WHO (2017) has prescribed the guideline value 0.002 mg/L for γ-BHC.

27.4.7.5 Butachlor

Butachlor is a chloroacetanilide herbicide for the pre-emergent control of grass and broadleaf weeds in rice and barley. Butachlor is non-irritating to the skin and moderately irritating to the eyes, and almost non-toxic. Further, it has been shown that butachlor is not genotoxic and is not oncogenic (Heydens et al. 2010; Furukawa et al. 2014). WHO has not prescribed any guideline value for butachlor.

27.4.7.6 Chlorpyriphos

Chlorpyriphos is a broad-spectrum organophosphorus insecticide, used for soil treatment (pre-plat and at planting), seed treatment, and foliar spray. They are used to control mosquitoes, flies, various crop pests in soil and on foliage. It is used as an insecticide on grain, cotton, corn, almonds, vegetable crops, ornamental plants and fruit trees (Femia et al. 2013). WHO Pesticide Evaluation Scheme (WHOPES) does not recommend the addition of chlorpyriphos to water for public health purposes, however, its use as an aquatic larvicide for the control of mosquito larvae in some countries can't be ruled out. Chlorpyriphos has low solubility in water. Further, the chlorpyriphos attached to the soil is not leached out because of the strong ionic interaction between the two, however, it gets degraded by microbial action at a slow rate. Chlorpyriphos exposure can result in neurological disorder, persistent developmental disorders and autoimmune disorders. Exposure during pregnancy retards the mental development of children. Chlorpyriphos is not genotoxic and do not pose carcinogenic risk to humans. WHO (2017) has prescribed the guideline value of 0.03 mg/L for drinking and other domestic usage, considering the fact that it is used as a mosquito larvicide in water bodies.

27.4.8 *2,4-Dichlorophenoxyacetic Acid (2,4-D)*

2,4-D is a systemic herbicide for control of broad-leaved weeds, including aquatic weeds. 2,4-D is easily biodegradable in the environment and its residues rarely exceed a few tens of micrograms per kilogram in food (Trivedi and Mandavgane 2018). Few epidemiological studies indicated 2,4-D promoting two forms of cancer in humans: soft tissue sarcomas and non-Hodgkin lymphoma, however, the results were inconsistent and therefore, Joint FAO/WHO Meeting on Pesticide Residues (JMPR) concluded that 2,4-D is neither genotoxic nor carcinogenic. A guideline value of 0.03 mg/L has been proposed for 2,4-D by WHO (2017).

27.4.8.1 DDT and Metabolites

DDT (1,1,1-trichloro-II,2-bis(p-chlorophenyl)ethane) is an organochlorine insecticide. It was widely used across the globe to control insects on agricultural crops and vectors. Nowadays, it is used only in a few countries for controlling vectors transmitting diseases like malaria, typhus, yellow fever, and sleeping sickness. Technical-grade DDT is a mixture of three forms, p,p′-DDT (85%), o,p′-DDT (15%), and o,o′-DDT (trace amounts), and may also contain DDE (1,1-dichloro-II,2-bis(p-chlorophenyl)ethylene) and DDD (1,1-dichloro-II,2-bis(p-chlorophenyl)ethane) as contaminants. DDD was also used to kill pests, and one form, o, p′-DDD, has been used to treat cancer of the adrenal gland. DDT and its metabolites, DDE and DDD, are persistent in the environment for a very long time, potentially for hundreds of years. Most of the DDT in the soil breaks down to DDD and DDE through microbial degradation and therefore, the metabolites should be analyzed instead of DDT. Further, these metabolites may evaporate into the air and deposited in other places. DDT and its metabolites get strongly adsorbed to the soil, and therefore generally remain in the surface layers of soil. Part of DDT and its metabolites may reach the surface water resources along with the soil particles, and only a very small fraction may seep into the groundwater.

IARC (1991) has concluded that there is insufficient evidence in humans and sufficient evidence in experimental animals for the carcinogenicity of DDT (Group 2B). WHO (2017) prescribes a guideline value of 0.001 mg/L for DDT and its metabolites in drinking water to protect the human health, however, it further states that the benefits of DDT use in malaria and other vector control programmes outweigh any health risk from the presence of DDT in drinking-water.

27.4.8.2 Endosulfan

Endosulfan, an organochlorine insecticide, is a mixture of two isomers, α-endosulfan (70%) and β-endosulfan (30%). It is used to control pests on fruit, vegetables, tea, and non-food crops. It is also used to control the tsetse fly and home garden pest, and as a wood preservative. WHO classifies endosulfan in Category 2 (moderately hazardous) and United States Environmental Protection Agency (USEPA) classifies endosulfan as Category 1b (highly hazardous). WHO (2017) has not established any guideline value for endosulfan citing the reason that the concentration of endosulfan encountered in drinking water is well below those of health concern.

27.4.8.3 Ethion

Ethion is an organophosphate pesticide. It used to kill aphids, mites, scales, thrips, leafhoppers, maggots and foliar feeding larvae, on a wide variety of food, fiber and ornamental crops, orchids, lawns and turf. It is also used as a cattle dip for ticks and for buffalo flies. Ethion is absorbed by the skin, as well as by the respiratory and gastrointestinal tracts. Short term exposure symptoms include nausea, vomiting, abdominal cramps, diarrhea, excessive salivation, headache, giddiness, weakness, muscle twitching, difficult breathing, blurring or dimness of vision, and loss of muscle coordination. Ethion is very toxic for humans at a probable oral dose of 50–500 mg/kg, and may lead to death from failure of the respiratory center, paralysis of the respiratory muscles, intense bronchoconstriction, or all three. Long term exposure may damage the nervous system (Pohanish 2015). WHO (2017) has not recommend any guideline value for ethion.

27.4.8.4 Isoproturon

Isoproturon (3-(4-isopropylphenyl)-1,1-dimethylurea) is a selective, systemic phenylurease class herbicide used to control annual grasses and broad-leaved weeds in cereals. It is mobile in soil, however, it gets photodegraded, hydrolyzed and biodegraded. It does not possess significant genotoxic activity, but it causes marked enzyme induction and liver enlargement (EC 2002; WHO 2003). Its use has been banned in EU after 2007 considering potential groundwater contamination and risk to aquatic life. WHO (2017) recommends guideline value 0.009 mg/L for Isoproturon.

27.4.8.5 Malathion

Malathion is an organophosphorus insecticide to control mosquitoes and a variety of insects that attack fruits, vegetables, landscaping plants and shrubs, by acetylcholinesterase inhibition. It is also used to control ticks and insects on pets and lice on human body. Under least favorable conditions (i.e. low pH and little organic content), Malathion in low pH water, with little organics, may persist for years, however, in normal conditions, the half-life is roughly 7–14 days. Malathion travels to the liver and kidneys and affects the nervous system. Generally, the body can break down Malathion and removes it quickly. The health based value for malathion is 0.9 mg/L, and the values encountered in drinking water resources are much lower than this values and hence WHO (2017) considered unnecessary to derive a formal guideline value for malathion in drinking water.

27.4.8.6 Methyl parathion

Methyl parathion is an organophosphorus insecticide and acaricide for use on many crops, in particular cotton. It derives its insecticidal properties from acetylcholinesterase inhibition, and this inhibition is responsible for the toxicity in humans. After application, part of methyl parathion get adsorbed on soil particles and part of it volatilizes to air. Methyl parathion and its breakdown products, generally does not move through soil, and hence their chances of reaching groundwater is minimal. Methyl parathion in environment is degraded by microbes. Half-life of methyl parathion in water ranges from weeks to months. Poisoning with methyl parathion leads sweating, dizziness, vomiting, diarrhea, convulsions, cardiac arrest, respiratory arrest, and, in extreme cases, death, due to cholinergic crisis. The health based value for methyl parathion is 9 μg/L, and the values encountered in drinking water resources are much lower than this values and hence, WHO (2017) considered unnecessary to derive a formal guideline value for malathion in drinking water.

27.4.8.7 Monocrotophos

Monocrotophos is an organophosphate insecticide and acaricide used to protect crops from mites, ticks, leaf hoppers, aphids and other insects (Singh and Walker 2006). Monocrotophos is highly soluble in water because of its hydrophilic nature, and is weakly sorbed by soil particles. Monocrotophos poisoning in humans is characterized by blurred vision, muscular weakness, profuse perspiration, confusion, vomiting, small pupils and even death due to respiratory failure. It also leads to genotoxic, cardiotoxicity, hyperglycaemic and stressogenic effects to non-target living organisms (Kaur and Goyal 2019). Use of monocrotophos has been banned in most of the countries and is unlikely to occur in drinking-water (WHO 2017).

27.4.8.8 Phorate

Phorate is an organophosphate insecticide/nematocide used for the control of sucking, biting, and chewing insects, mites and certain nematodes. Phorate is a restricted use pesticide with no residential application and is among the most poisonous chemicals commonly used for pest control. Phorate can cause cholinesterase inhibition in humans causing nausea, dizziness, confusion, and at very high exposures, respiratory paralysis and death.

Phorate applied to soil is rapidly degraded by microorganisms and interaction with water. It's half-life in field study was observed to be 7.5 days and in aqueous solution, it is two hours at pH 8 and 70 °C (Health Canada 1986), due to which phorate is unlikely to occur in drinking water (WHO 2017).

27.5 Trace Toxic Metal and Metalloid

BIS (2012) has prescribed limits for 8 metal/metalloid, namely aluminum, boron, copper, iron, manganese, selenium, silver, and zinc, which are undesirable in excessive amounts, and 7 metal/metalloid, namely cadmium, lead, mercury, molybdenum, nickel, arsenic and chromium, which are toxic substances (Table 27.6).

Trace metals in the water samples can be analyzed by atomic absorption spectrometer (AAS), inductively coupled plasma optical emission spectrometry (ICP-OES), or inductively coupled plasma mass spectrometer (ICP-MS). For analysis of trace metals by these techniques, the samples are digested in nitric acid and hydrogen peroxide for oxidation/removal of organics in microwave reaction system and centrifuged before injecting in AAS, ICP-OES, or ICP-MS.

Table 27.6 BIS limits for trace toxic metal/metalloid (BIS 2012)

Sr. No.	Characteristic	Acceptable limit	Permissible limit
1	Aluminum (as Al), mg/L, *Max*	0.03	0.2
2	Boron (as B), mg/L, *Max*	0.5	1.0
3	Copper (as Cu), mg/L, Max	0.05	1.5
4	Iron (as Fe), mg/L, *Max*	0.3	No relaxation
5	Manganese (as Mn), mg/L, *Max*	0.1	0.3
6	Selenium (as Se), mg/L, *Max*	0.01	No relaxation
7	Silver (as Ag), mg/L, *Max*	0.1	No relaxation
8	Zinc (as Zn), mg/L, *Max*	5.0	15.0
9	Cadmium (as Cd), mg/L, *Max*	0.003	No relaxation
10	Lead (as Pb), mg/L, *Max*	0.01	No relaxation
11	Mercury (as Hg), mg/L, *Max*	0.001	
12	Molybdenum (as Mo), mg/L, *Max*	0.07	No relaxation
13	Nickel (as Ni), mg/L, *Max*	0.02	No relaxation
14	Total arsenic (as As), mg/L, *Max*	0.01	0.05
15	Total chromium (as Cr), mg/L, *Max*	0.05	No relaxation

27.5.1 Aluminum

Aluminum (Al) is the most abundant metallic element and constitutes about 8% of Earth's crust. High Al concentration in drinking water results in undesirable color and turbidity. The orally ingested aluminum is not acutely toxic to humans, however, it has been hypothesized that aluminum exposure is a risk factor for the development or acceleration of onset of Alzheimer disease in humans. WHO (2017) recommends the guideline value of 0.9 mg/L for aluminum in drinking water.

Aluminum in drinking water can be reduced to around 0.1 mg/L by optimizing pH in the coagulation process, use of polyelectrolyte, and media filtration.

27.5.2 Boron

Boron (B) is a naturally occurring element and is used in many consumer products like fiberglass, borosilicate glass, soaps and detergents, flame retardants, neutron absorbers for nuclear installations, mild antiseptics, cosmetics, pharmaceuticals (as pH buffers), boron neutron capture therapy (for cancer treatment), pesticides, and fertilizers. The concentration of boron in the earth's crust has been estimated to be <10 mg/kg, but in boron-rich areas, boron concentrations may be as high as 100 mg/kg. Boron enters in the environment mainly through the weathering of rocks, boric acid volatilization from seawater, and volcanic activity. Boron concentration in groundwater and surface water is generally small, however, the concentration can be significantly increased depending on the anthropogenic activities and surrounding geology. Boron toxicity results in gastrointestinal tract distress, vomiting, abdominal pain, diarrhea, and nausea (Simonnot et al. 2000; Yazbeck et al. 2005; Health Canada 2020). The guideline value of 2.4 mg/L has been prescribed by WHO (2017) for boron in drinking water.

Boron in drinking water can be removed by reverse osmosis membranes up to some extent but effective removal is achieved by boron specific ion exchange resins like Amberlite PWA10, Purolite S108, Diaion CRB 02 (Simonnot et al. 2000; Kabay et al. 2007).

27.5.3 Copper

Copper (Cu) is an essential nutrient as well as a drinking water contaminant. It is used to make pipes, valves and fittings, and is present in alloys and coatings. Copper concentration above 1 mg/L results in staining of laundry and sanitary ware, and beyond 5 mg/L, copper imparts astringent taste and cause discoloration. The epidemiological studies have not been able to establish the relationship between ingestion of copper in drinking water on the gastrointestinal tract, carriers of the gene for Wilson disease and other metabolic disorders of copper homeostasis (NRC 2000). US EPA (1991) classifies copper as Group D, not classifiable as to human carcinogenicity. The guideline value of 2 mg/L prescribed by WHO (2017) is to protect against gastric irritation.

Copper from drinking water can be removed by lime soda softening process, adsorption on activated carbon, reverse osmosis, electrodialysis, and distillation.

27.5.4 Iron

Iron (Fe) is one of the most abundant metals in Earth's crust, and in fresh waters, it ranges from <0.001 to 90 mg/L (Health Canada 1978). Fe in trace amount is essential for nutrition, however, higher concentrations impart inky flavor, bitter, and astringent taste to water. Taste and odor problems may be caused by filamentous organisms that prey on iron compounds (frenothrix, gallionella and leptothrix). In addition, these bacterial may clog the well screens or may develop in the distribution system. Fe in water also stains the laundry and plumbing fixtures. So, the presence of Fe is unacceptable in drinking water, not due to health reasons but due to aesthetic/operational reasons. No guideline value for iron in drinking-water is proposed by WHO (2017) citing the reason that concentrations found in drinking water do not pose any health effect.

Iron in drinking water can be removed by aeration-flocculation-sedimentation followed by media filtration, manganese greensand filter, iron specific ion exchange resins, calcium carbonate based minerals, constructed soil filter, electrocoagulation, UF membrane, and nanomaterials (Khadse et al. 2015; Khatri et al. 2017).

27.5.5 Manganese

Manganese is one of the most abundant metals in Earth's crust and generally co-exist with iron. It is used in the manufacture of iron and steel alloys, as an oxidant for cleaning, bleaching and disinfection (as potassium permanganate), and as an octane enhancer (as methyl-cyclo-pentadienyl manganese tri-carbonyl) in petrol. Manganese occurs in surface water sources which are anaerobic or anoxic. At levels exceeding 0.1 mg/L, manganese in water supplies causes an undesirable taste in beverages and stains sanitary ware and laundry, and at levels exceeding 0.2 mg/L, manganese will often form a coating on pipes which may slough off as a black precipitate.

Manganese is an essential element for humans, however, consuming drinking water with concentration more than 0.4 mg/L is expected to be associated with adverse effect on learning in children. WHO (2017) has not established a guideline value for manganese, as the health based value (0.4 mg/L) is well above concentrations of manganese, 0.1 mg/L, normally causing acceptability problems in drinking water.

Manganese in drinking water can be removed by chemical precipitation, coagulation, flotation, ion-exchange, oxidation/filtration, electrochemical treatment, adsorption, and membrane filtration (Health Canada 2016; Alvarez-Bastida et al. 2018). Oxidation by $KMnO_4$, followed by precipitation and filtration can achieve a manganese concentration of 0.05 mg/L in drinking water.

27.5.6 Selenium

Selenium (Se), a metalloid, is a member of group VIA and generally present in elemental form or ionic [selenide (Se^{-2}), selenate ($SeO_4{}^{-2}$), or selenite ($SeO_3{}^{-2}$)]. It is widely distributed in the Earth's crust, in association with sulfur containing minerals, at a concentration of 50–90 μg/kg. Its concentration in groundwater and surface water ranges from 0.06 to 400 μg/L. It is an essential element and FAO/WHO recommends daily intake of 6–21, 26, and 30 μg/L Se for infants/children, females, and males respectively. Most water soluble inorganic and organic selenium are efficiently absorbed across the gastrointestinal tract, which is cleared by the liver, transported to peripheral tissues, and distributed to all organs, with highest concentration in kidney, liver, spleen, testes, and skeletal muscle. High intake of Se results in gastrointestinal disturbances, discoloration of the skin, decayed teeth, hair or nail loss, nail abnormalities, changes in peripheral nerves, and several type of cancer (WHO 2011c; Health Canada 2014). A provisional guideline value of 0.04 mg/L has been recommended by WHO (2017).

Se in drinking water can be removed by adsorption on iron/aluminum coated adsorbents, zerovalent iron/iron oxides, ion-exchange resins, and reverse osmosis (Kapoor et al. 1995; Sharma et al. 2019).

27.5.7 Silver

Silver (Ag) is a member of group XI, occurring primarily in the form of the sulfide (argentite Ag_2S) or intimately associated with other metal sulfides, and gold. Its abundance in earth's crust is 0.08 mg/kg. Ag is used in electroplating, as a conductor, in dental alloys, paints, jewelry, silverware, coinage, mirror production, and disinfectant. The average Ag concentration in natural water are 0.2–0.3 μg/L and occasionally above 5 μg/L. Levels in drinking-water treated with silver for disinfection may be above 50 μg/L.

The excess Ag intake results in heavy discoloration of skin and hair and the condition is known as argyria. The estimated acute lethal dose of $AgNO_3$ for humans is minimum 10 g. There are no adequate data to derive a health-based guideline value for silver in drinking-water. WHO (2017) does not specify any guideline value for Ag in drinking water.

Silver present in drinking water can be reduced by reverse osmosis membranes, distillation, and ion exchange resins.

27.5.8 Zinc

Zinc (Zn), a member of group XII, is an essential trace element found in virtually all food and potable water. The concentration of Zn in surface water and groundwater normally does not exceed 0.01 and 0.05 mg/L respectively. Zn concentrations in excess of 3–5 mg/L impart an undesirable astringent taste to water and develop a greasy film on boiling, although, drinking water rarely contains Zn concentration above 0.1 mg/L. The daily requirement of Zn for adult humans is 15–22 mg. Drinking water usually makes a negligible contribution to zinc intake and is not of health concern at levels found in drinking water, therefore, WHO (2017) has not established any guideline value.

Zn present in drinking water can be removed by coagulation, ion exchange, adsorption, and evaporation. Reverse osmosis membranes could also be effective in zinc removal although Zn precipitates may foul the membranes (Alswata et al. 2017; Khulbe and Matsuura 2018).

27.5.9 Cadmium

Cadmium (Cd), a group XII element, and its compounds are widely used in batteries, pigments, electronic components, and nuclear reactors. It is released to the environment through wastewater, leaching of solid waste, and diffuse pollution from fertilizers and local air pollution. Cadmium concentration in natural waters is usually below 1 μg/L and can go up to 100 μg/L. Cadmium accumulates primarily in the kidneys, has a long biological half-life in humans of 10–35 years, and adversely affects the kidney. The absorption of cadmium compounds is dependent on the solubility of the compounds. The epidemiological studies had established the Cd induced carcinogenicity by the inhalation route, and IARC has classified cadmium and cadmium compounds in Group 2A (probably carcinogenic to humans), however, evidence of carcinogenicity by the oral route is lacking (WHO 2011d; Health Canada 2019; Idrees et al. 2018). WHO (2017) has prescribed a guideline value of 0.003 mg/L for Cd in drinking water.

Cadmium in drinking water can be removed by coagulation/filtration, electrocoagulation, ion exchange, adsorption, distillation, or reverse osmosis (Zhao et al. 2002; Yaacoubi et al. 2014; Heffron et al. 2016, 2019).

27.5.10 Lead

Lead (Pb), a group X element, accounts for 13 mg/kg of Earth's crust, and is used principally in the production of lead-acid batteries, solder, alloys, cable sheathing, pigments, rust inhibitors, ammunition, glazes, and plastic stabilizers. The organic lead compounds tetraethyl and tetra methyl lead has also been used extensively as antiknock and lubricating agents in petrol, although their use in many countries including India has largely been phased out. Exposure to lead is associated with a wide range of effects, including various neurodevelopmental effects, mortality (mainly due to cardiovascular diseases), impaired renal function, hypertension, impaired fertility and adverse pregnancy outcomes. The evidence for the carcinogenicity of lead in humans is inconclusive because of the limited number of studies, and therefore, Pb has been placed in Group 2B (possible human carcinogen) of the IARC classification, however, inorganic lead compounds have been placed in Group 2A (namely probable human carcinogen). The provisional guideline value for lead in drinking water prescribed by WHO (2017) is 0.01 mg/L.

Pb in drinking water can be removed by coagulation and filtration, lime softening, electrocoagulation, ion-exchange resins, or reverse osmosis (Heffron et al. 2016; Health Canada 2017).

27.5.11 Mercury

Mercury (Hg), a group XII element, is a toxic element. It is present in Earth's crust at an average concentration of 0.08 mg/kg and cinnabar (HgS) is the most common ore. Mercury and its compounds are used in dental preparations, thermometers, fluorescent and ultraviolet lamps, and pharmaceuticals, fungicides, and chlor-alkali industry. In oxic waters, Hg exists as $Hg(OH)_2$ and $HgCl_2$, and in anoxic sediments, it is immobilized as the sulfide. Mercury content in the surface and groundwater is generally less than 0.5 μg/L, however, concentration up to 5.5 μg/L has been reported in the wells in Izu Oshima Island (Japan). Mercury poisoning results in neurological and renal disturbances. Neurological disturbances is due to organic mercury ingestion and renal disturbance is due inorganic mercury. The guideline value for mercury prescribed by WHO (2017) is 0.006 mg/L.

The treatment technologies that are effective in removing Hg from drinking water are $FeCl_3$ assisted co-precipitation at pH 8, lime softening, adsorption on granular activated carbon, ion-exchange resins, or reverse osmosis.

27.5.12 Molybdenum

Molybdenum (Mo), a group VI element, is an essential trace element for human, animal, and plant health, and exposure to high doses can be detrimental. Its abundance in Earth crust is around 1.5 mg/kg. Mo is present in special steels, electric contacts, filaments, X-ray tubes, spark plugs, induction heating elements, lubricants, fertilizer, and treatment of seeds. The estimated daily requirement of Mo for adults is 0.1–0.3 mg, however, Mo content in the drinking water usually do not exceed 10 μg/L (WHO 2011a, e; Smedly and Kinniburgh 2017). Therefore, WHO (2017) did not recommend any guideline value.

Molybdenum in drinking water can be removed by co-precipitation with ferric iron, adsorption on modified biochar and activated carbon, or ion-exchange resins (LeGendre and Runnells 1975; Verbinnen et al. 2012; Zhang et al. 2015; Lian et al. 2018).

27.5.13 Nickel

Nickel (Ni), a group X element, is an essential metal for several animal species, micro-organisms and plants, and toxicity symptoms can occur when too little or too much nickel is taken up. The average abundance of nickel in the earth's crust is 1.2 mg/L, in soils it is 2.5 mg/L, in streams it is 1 μg/L and in groundwater it is <0.1 mg/L. Nickel is obtained chiefly from pyrrhotite and garnierite. Nickel is released to the environment from the burning of fossil fuels and waste discharge from electroplating industries. In general concentration of nickel in water resources is generally below 0.02 mg/L. IARC has included inhaled nickel compounds in Group 1 (carcinogenic to humans) and metallic nickel in Group B (possibly carcinogenic), however, there is a lack of evidence of a carcinogenic risk from oral exposure to nickel. The guideline value for nickel prescribed by WHO (2017) is 0.07 mg/L.

Nickel in drinking water can be removed by chemical coagulation-sedimentation-filtration, adsorption, or electrocoagulation (WHO 2005c; Kalkan et al. 2012; Heffron et al. 2016; Thanh et al. 2018).

27.5.14 Arsenic

Arsenic (As), a group XV metalloid, is usually present in natural waters at concentrations of less than 1–2 μg/L. However, in groundwaters, the concentrations can be significantly elevated depending on the aquifer geology. Arsenic can exist in four valences –3, 0, +3, and +5, out of which arsenite (As^{+3}) is dominant in reducing conditions and arsenate (As^{+5}) is dominant in oxic environment. The acute toxicity of arsenic compounds in humans is predominantly a function of their rate of removal from the body. Arsine is considered to be the most toxic form, followed by the arsenites, the arsenates and organic arsenic compounds. Signs of chronic arsenicism, including dermal lesions such as hyperpigmentation and hypopigmentation, peripheral neuropathy, skin cancer, bladder and lung cancers and peripheral vascular disease, have been observed in populations ingesting arsenic-contaminated drinking-water. WHO (2017) has recommended 0.010 mg/L as the provisional guideline value for drinking water.

As in drinking water can be removed by oxidation followed by adsorption, coagulation-flocculation, electrocoagulation, nanomaterials, or reverse osmosis (Ng et al. 2004; Thirunavukkarasu et al. 2005; Nicomel et al. 2015; Heffron et al. 2016; Hao et al. 2018).

27.5.15 Chromium

Chromium (Cr), a group VI element, is widely distributed in Earth crust with an average concentration of 100 mg/kg. In water, chromium occurs in two oxidation states, Cr (III) and Cr (VI). Chromium (III) is an essential human dietary element and is found in many vegetables, fruits, meats, grains, and yeast. Cr enters the environment by industrial processes discharge. Further, chromium enters the environment by leakage, poor storage, or inadequate industrial waste disposal practices. IARC has classified chromium (VI) in Group 1 (human carcinogen) and chromium (III) in Group 3 (not classifiable as to its carcinogenicity to humans). WHO (2017) prescribes provisional guideline value of 0.05 mg/L for Cr.

Cr in drinking water can be removed by adsorption, ion exchange, electrocoagulation, biosorption, nanomaterial, or reverse osmosis membranes (Rengraj et al. 2001; Shah et al. 2011; Heffron et al. 2016; Mitra et al. 2017; Chen et al. 2018; Liu et al. 2018).

27.6 Bacteriological Quality

The greatest risk to public health from microbes in water is associated with consumption of drinking-water that is contaminated with human and animal excreta, although other sources and routes of exposure may also be significant. Infectious diseases caused by pathogenic bacteria, viruses and parasites are the most common and widespread health risk associated with drinking-water. Microbiological contamination is most likely to arise from the entry of fecal matter to waters. Specific waterborne disease outbreaks attributed to contaminated groundwater include cholera, salmonellosis, shigellosis, infectious hepatitis, gastroenteritis, and amoebic dysentery.

Microbiological examination of water samples is conducted to determine the sanitary quality and degree of contamination with wastes. The examination involves

Table 27.7 BIS limits for bacteriological parameters (BIS 2012)

Sr. No.	Characteristic	Requirement
1	Total coliform bacteria	Shall not be detectable in any 100 mL sample
2	*E. coli* or thermotolerant coliform bacteria	Shall not be detectable in any 100 mL sample

detection and enumeration of indicator organisms, rather than pathogens. The coliform group of bacteria is the principal indicator of suitability of a water for domestic, industrial, or other uses. Escherichia coli (*E. coli*) is the major species in the fecal coliform group. Of the five general groups of bacteria that comprise the total coliforms, only *E. coli* is generally not found growing and reproducing in the environment. Consequently, *E. coli* is considered to be the species of coliform bacteria that is the best indicator of fecal pollution and the possible presence of pathogens. Fecal coliforms may enter surface water by a number of ways, from contaminated soil runoff from storm water, from vegetation and insects, wash from cities, or from direct sewage pollution by man or animals. BIS (2012) has prescribed absence of total coliform and *E. coli* bacteria in the drinking water (Table 27.7). WHO (2017) also recommends that total coliforms and *E. coli* (or, alternatively, thermotolerant coliforms) should be absent in the drinking water and their presence indicate inadequate treatment or breaches in distribution system integrity.

Coliforms in the drinking water can be eliminated by disinfection with chlorine, chloramine, chlorine dioxide, hydrogen peroxide, ozone, silver nanoparticles, and ultraviolet radiation (EPA 2011; Percival et al. 2014; Nguyen et al. 2017).

27.7 Radioactive Substances

Water used for drinking may contain radioactive substances and are harmful to human health. Minute traces of radioactivity are normally found in all drinking water, which may vary from place to place depending on the radiochemical composition of the soil and rock strata through which the raw water has passed. Several naturally occurring and human made sources present throughout the environment results in radioactivity of water, and it is quite expensive as well as time consuming to identify individual radionuclides. Therefore, measurement of total radioactivity in the form of alpha and beta emitters is conducted, followed by investigation of the concentration of individual radionuclides, in case the alpha and beta emissions exceed the guideline value. The acceptable limits prescribed by BIS (2012) for radioactive substances in drinking water is provided in Table 27.8.

WHO (2017) recommends guideline levels of 0.5 Bq/L for gross alpha activity and 1.0 Bq/L for gross beta activity.

The alpha radioactivity in water sample is measured by alpha counter having ZnS-Ag scintillation counter after evaporating the sample to near dryness. Similarly,

Table 27.8 BIS limits for parameters concerning radioactive substances (BIS 2012)

Sr. No.	Characteristic	Acceptable limit	Permissible limit
1	Alpha emitters, Bq/L, *Max*	0.1	No relaxation
2	Beta emitters, Bq/L, *Max*	1.0	No relaxation

beta radioactivity in the water sample is measured by Geiger Muller counting system after evaporating the sample to near dryness.

The radioactive material in the drinking water can be removed by aeration, adsorption, and membrane processes (RO/NF) (Huikuri et al. 1998; Montana et al. 2013; Baeza et al. 2017; Khulbe and Matsuura 2018).

27.8 Summary and Conclusion

The quality of water is very important and a determining factor of its usefulness. Significance of water quality parameters can be judged from the fact that unsafe water is responsible for around 88% of the deaths from diarrheal diseases. Accordingly, BIS has prescribed the limiting values for 66 drinking water quality parameters important from human health, aesthetic reasons, and operational issues. The water supply departments should prioritize microbial safety of drinking water on the top, followed by chemical constituents causing adverse health effects, organoleptic parameters, and operational parameters at the end. The aesthetic or organoleptic parameters, although do not impact human health, and are significant in the sense that they decide the acceptability of water for drinking. The contribution of drinking water to the total exposure of radioactivity is very small, however, in certain areas with prominent radioisotopes in their geological formations, radiological parameters become important.

All the prescribed BIS parameters should be analyzed at least once when the treatment system is initially operationalized. During continuous operation, it is recommended to analyze the microbial parameters every day, the operational and organoleptic parameters weekly, and other parameters annually. However, depending upon the geology and source, the critical parameters should be analyzed more frequently.

References

Aboulhassan MA, Souabi S, Yaacoubi A, Baudu M (2006) Removal of surfactant from industrial wastewaters by coagulation flocculation process. Int J Environ Sci Tech 3(4):327–332

Ahmad T, Rafatullah M, Ghazali A, Sulaiman O, Hashim R, Ahmad A (2010) Removal of pesticides from water and wastewater by different adsorbents: a review. J Environ Sci Health, Part C 28(4):231–271

Alswata AA, Ahmad MB, Al-Hada NM, Kamari HM, Hussein MZB, Ibrahim NA (2017) Preparation of Zeolite/Zinc Oxide Nanocomposites for toxic metals removal from water. Res Phys 7:723–731

Alvarez-Bastida C, Martínez-Miranda V, Solache-Ríos M, Linares-Hernández I, Teutli-Sequeira A, Vázquez-Mejía G (2018) Drinking water characterization and removal of manganese from water. J Environ Chem Eng 6(2):2119–2125

Anku W W, Mamo MA, Govender PP (2017) Phenolic compounds in water: sources, reactivity, toxicity and treatment methods. In: Soto-Hernandez M, Palma-Tenango M, del Rosario Garcia-Mateos M (eds) Phenolic compounds—natural sources, importance and applications. IntechOpen. https://doi.org/10.5772/66927

APHA (2017) Standard methods for the examination of water and wastewater, 23rd edn. American Public Health Association, American Water Works Association, and Water Environment Federation.

ASTM (1996) Determining carcinogenic potential of virgin base oils in metalworking fluids (E 1687-95), West Conshohocken, PA

ATSDR (2002) Toxicological profile for aldrin/dieldrin. Agency for Toxic Substances and Disease Registry, Atlanta (GA), USA

Baeza A, Salas A, Guillén J, Muñoz-Serrano A, Ontalba-Salamanca MÁ, Jiménez-Ramos MC (2017) Removal naturally occurring radionuclides from drinking water using a filter specifically designed for drinking water treatment plants. Chemosphere 167:107–113

Baldauf G (1993) Removal of pesticides in drinking water treatment. Acta Hydrochim Hydrobiol 21(4):203–208

BIS (1987) Methods of sampling and test (physical and chemical) for water and wastewater, Part 29: Sulfide (IS: 3025–1986, Reaffirmed 2003). Bureau of Indian Standards, New Delhi

BIS (2003) IS 3025 (Part 43) (1992, Reaffirmed 2003): methods of sampling and test (physical and chemical) for water and wastewater, Part 43: Phenols (First Revision). Bureau of Indian Standards, New Delhi

BIS (2012) Indian Standard—Drinking Water Specification (2nd Revision), IS 10500: 2012. Bureau of Indian Standards, New Delhi

Bond T, Goslan EH, Parsons SA, Jefferson B (2012) A critical review of trihalomethane and haloacetic acid formation from natural organic matter surrogates. Environ Technol Rev 1(1):93–113

Brink EJ, Dekker PR, Beresteijn ECHV, Beynen AC (1992) Bioavailability of magnesium and calcium from cow's milk and soya-bean beverage in rats. Br J Nutr 68:271–282

Brook I, Malchi T, Nir S (2013) Removal of anionic detergents from water and treatment of gray water by micelle–clay composites. Desalin Water Treat 53(8):2184–2192

Chen Y, An D, Sun S, Gao J, Qian L (2018) Reduction and removal of Chromium VI in water by powdered activated carbon. Materials 11(2):269

CONCAWE (1994) The use of the Dimethyl Sulfoxide (DMSO) extract by the IP346 method as an indicator of the carcinogenicity of lubricant base oils and distillate aromatic extracts. Report Number 94/51.

Davi ML, Gnudi F (1999) Phenolic compounds in surface water. Water Res 33(14):3213–3219

EC (2002). Isoproturon (SANCO/3045/99-final). European Commission, Health and Consumer Protection Directorate General

Environment Canada (1994) Priority substances list assessment report: polycyclic aromatic hydrocarbons. Ministry of Supply and Services, Government of Canada, Ottawa, Ontario

EPA (2007) Method 1699: pesticides in water, soil, sediment, biosolids, and tissue by HRGC/HRMS. US Environmental Protection Agency, Washington DC

EPA (2011) Water treatment manual: disinfection. Environmental Protection Agency, Ireland

Femia J, Mariani M, Zalazar C, Tiscornia I (2013) Photodegradation of chlorpyrifos in water by UV/H_2O_2 treatment: toxicity evaluation. Water Sci Technol 68(10):2279–2286

Furukawa S, Harada T, Thake D, Latropoulos MJ, Sherman JH (2014) Consensus diagnoses and mode of action for the formation of gastric tumors in rats treated with the Chloroacetanilide herbicides alachlor and butachlor. Toxicol Pathol 386–402

Gaciri SJ, Davies TC (1993) The occurrence and geochemistry of fluoride in some natural waters of Kenya. J Hydrol 143:395–412

Gaucheron F (2005) The minerals of milk. Reprod Nutr Dev 45(4):473–483

Gloxhuber C, Künstler K (1992) Anionic surfactants: biochemistry, toxicology, dermatology, 2nd edn, vol 43. Marcel Dekker, Inc., New York, United States

Golfinopoulos SK, Nikolaou AD, Lekkas TD (2003) The occurrence of disinfection by-products in the drinking water of Athens, Greece. Environ Sci Pollut Res 10:368–372

Hao L, Liu M, Wang N, Li G (2018) A critical review on arsenic removal from water using iron-based adsorbents. RSC Adv 8(69):39545–39560

Health Canada (1978) Guidelines for Canadian drinking water quality: guideline technical document—iron (Updated November 1987). Government of Canada, Ontario

Health Canada (1986) Guidelines for Canadian drinking water quality: guideline technical document—Phorate. Government of Canada, Ontario

Health Canada (1991) Guidelines for Canadian drinking water quality: guideline technical document—cyanide, chap 28. Government of Canada, Ontario

Health Canada (2014) Guidelines for Canadian drinking water quality: guideline technical document—Selenium. Water and Air Quality Bureau, Healthy Environments and Consumer Safety Branch, Health Canada, Ottawa, Ontario

Health Canada (2016) Manganese in drinking water: document for public consultation. Government of Canada, Ontario

Health Canada (2017) Lead in drinking water: document for public consultation. Government of Canada, Ontario

Health Canada (2019) Cadmium in drinking water: guideline technical document for public consultation. Government of Canada, Ontario

Health Canada (2020) Boron in drinking water: guideline technical document for public consultation. Government of Canada, Ontario

Heffron J, Marhefke M, Mayer BK (2016) Removal of trace metal contaminants from potable water by electrocoagulation. Sci Rep 6(1):28478

Hem JD (1989) Study and interpretation of the chemical characteristics of natural water. Water supply paper 2254. US Geological Survey, Washington

Heydens WF, Lamb IC, Wilson AGE (2010) Chloracetanilides. In: Krieger R (ed) Hayes' handbook of pesticide toxicology, 3rd edn. Academic Press, USA, pp 1753–1769

Huang Y, Li J, Xu Y, Xu W, Cheng Z, Liu J, Wang Y, Tian C, Luo C, Zhang G (2014) Polychlorinated biphenyls (PCBs) and hexachlorobenzene (HCB) in the equatorial Indian Ocean: temporal trend, continental outflow and air–water exchange. Mar Pollut Bull 80:194–199

Huikuri P, Salonen L, Raff O (1998) Removal of natural radionuclides from drinking water by point of entry reverse osmosis. Desalination 119(1–3):235–239

IARC (1991) Chlorinated drinking-water; chlorination by-products; some other halogenated compounds; cobalt and cobalt compounds (IARC monographs on the evaluation of carcinogenic risks to humans, No. 52). International Agency for Research on Cancer, Lyon, France

IARC (2012) A review of human carcinogens (Volume 100F: Chemical agents and related Occupations). IARC Monographs on the Evaluation of Carcinogenic Risks to Humans, Lyon, France

IARC (2019a) IARC monographs on the identification of carcinogenic hazards to humans: report of the advisory group to recommend priorities for the IARC monographs during 2020–2024. International Agency for Research on Cancer, Lyon, France

IARC (2019b) IARC monographs on the identification of carcinogenic hazards to humans: pentachlorophenol and some related compounds, vol 117. International Agency for Research on Cancer, Lyon, France

Idrees N, Tabassum B, Abd Allah EF, Hashem A, Sarah R, Hashim M (2018) Groundwater contamination with cadmium concentrations in some West U.P. Regions, India. Saudi J Biol Sci 1365–1368

Institute for Petroleum (1985) IP Standards for petroleum and its products. Part I—Methods Anal Test 2:346.1

IPCS (1996) Linear alkylbenzene sulfonates and related compounds. Environmental Health Criteria. World Health Organization, Geneva, Switzerland, p 169

IPCS (1998) Selected non-heterocyclic polycyclic aromatic hydrocarbons. World Health Organization, International Programme on Chemical Safety (Environmental Health Criteria 202), Geneva.

Jing R, Fusi S, Kjellerup BV (2018) Remediation of polychlorinated biphenyls (PCBs) in contaminated soils and sediment: state of knowledge and perspectives. Front Environ Sci 6:79

Kabay N, Sarp S, Yuksel M, Arar Ö, Bryjak M (2007) Removal of boron from seawater by selective ion exchange resins. React Funct Polym 67(12):1643–1650

Kaleta J, Elektorowicz M (2013) The removal of anionic surfactants from water in coagulation process. Environ Technol 34(8):999–1005

Kalkan E, Nadaroglu H, Demir N (2012) Experimental study on the nickel (II) removal from aqueous solutions using silica fume with/without apocarbonic anhydrase. Desalin Water Treat 44(1–3):180–189

Kapoor A, Tanjore S, Viraraghavan T (1995) Removal of selenium from water and wastewater. Int J Environ Stud 49(2):137–147

Karyab H, Yunesian M, Nasseri S, Mahvi A, Ahmadkhaniha R, Rastkari N, Nabizadeh R (2013) Polycyclic aromatic hydrocarbons in drinking water of Tehran, Iran. J Environ Health Sci Eng 11(1):25

Kaur R, Goyal D (2019) Toxicity and degradation of the insecticide monocrotophos. Environ Chem Lett 17:1299–1324

Khadse GK, Patni PM, Labhasetwar PK (2015) Removal of iron and manganese from drinking water supply. Sustain Water Res Manag 1:157–165

Khatri N, Tyagi S, Rawtani D (2017) Recent strategies for the removal of iron from water: a review. J Water Process Eng 19:291–304

Khulbe KC, Matsuura T (2018) Removal of heavy metals and pollutants by membrane adsorption techniques. Appl Water Sci 8:19

Krasner SW, McGuire MJ, Jacangelo JG, Patania NL, Reagan KM, Aieta EM (1989) The occurrence of disinfection by-products in us drinking water. J Am Water Works Assoc 81(8):41–53

Kruszelnicka I, Ginter-Kramarczyk D, Wyrwas B, Idkowiak J (2019) Evaluation of surfactant removal efficiency in selected domestic wastewater treatment plants in Poland. J Environ Health Sci Eng 17:1257–1264

Kundu N, Panigrahi MK, Tripathy S, Munshi S, Powell MA, Hart BR (2001) Geochemical appraisal of fluoride contamination of groundwater in the Nayagarh district of Orissa, India. Environ Geol 41:451–460

LeGendre GR, Runnells DD (1975) Removal of dissolved molybdenum from waste waters by precipitates of ferric iron. Environ Sci Technol 9(8):744–749

Lian JJ, Huang YG, Chen B, Wang SS, Wang P, Niu SP, Liu ZL (2018) Removal of molybdenum(VI) from aqueous solutions using nano zero-valent iron supported on biochar enhanced by cetyltrimethyl ammonium bromide: adsorption kinetic, isotherm and mechanism studies. Water Sci Technol 2017(3):859–868

Liu W, Yang L, Xu S, Chen Y, Liu B, Li Z, Jiang C (2018) Efficient removal of hexavalent chromium from water by an adsorption–reduction mechanism with sandwiched nanocomposites. RSC Adv 8(27):15087–15093

Madsen T, Boyd HB, Nylen D, Pedersen AR, Petersen GI, Simonse F (2001) Environmental and health assessment of substances in household detergents and cosmetic detergent products (Environment project No. 615). Centre for Integrated Environment and Toxicology (CETOX), Horsholm, Denmark

Marican A, Durán-Lara EF (2017) A review on pesticide removal through different processes. Environ Sci Pollut Res 25(3):2051–2064
Mitra S, Sarkar A, Sen S (2017) Removal of chromium from industrial effluents using nanotechnology: a review. Nanotech Environ Eng 2:11
Mojiri A, Zhou JL, Ohashi A, Ozaki N, Kindaichi T (2019) Comprehensive review of polycyclic aromatic hydrocarbons in water sources, their effects and treatments. Sci Total Environ 133971
Montaña M, Camacho A, Serrano I, Devesa R, Matia L, Vallés I (2013) Removal of radionuclides in drinking water by membrane treatment using ultrafiltration, reverse osmosis and electrodialysis reversal. J Environ Radioact 125:86–92
Muyzer G, Stams AJM (2008) The ecology and biotechnology of sulpate-reducing bacteria. Nat Rev Microbiol 6(6):441–454
Nanyaro JT, Aswathanarayana U, Mungure JS (1984) A geochemical model for the abnormal fluoride concentrations in waters in parts of northern Tanzania. J Afr Earth Sci 2:129–140
Narkis N, Weinber H (1989) Removal of surfactants from effleunts. Tenside, Surfactants, Deterg 26(6):400–405
Nathanson (1986) Basic environmental technology: Water supply, waste disposal, and pollution control. Wiley
Ng K-S, Ujang Z, Le-Clech P (2004) Arsenic removal technologies for drinking water treatment. Rev Environ Sci Bio/Technol 3(1):43–53
Nguyen MT, Allemann L, Ziemba C, Larivé O, Morgenroth E, Julian TR (2017) Controlling bacterial pathogens in water for reuse: treatment technologies for water recirculation in the blue diversion autarky toilet. Front Environ Sci 5
Nicolopoulou-Stamati P, Maipas S, Kotampasi C, Stamatis P, Hens L (2016) Chemical pesticides and human health: the urgent need for a new concept in agriculture. Front Public Health 4:148
Nicomel N, Leus K, Folens K, Van Der Voort P, Du Laing G (2015) Technologies for arsenic removal from water: current status and future perspectives. Int J Environ Res Public Health 13(1):62
NITI (2018) Composite water management index: A tool for water management. National Institute for Transforming India (NITI) Aayog, New Delhi
NITI (2019) Composite water management index. National Institute for Transforming India (NITI) Aayog, New Delhi
NRC (2000) Copper in drinking water. National Research Council (US) Committee on Copper in Drinking Water, National Academy Press, Washington, D.C
Ormad MP, Miguel N, Claver A, Matesanz JM, Ovelleiro JL (2008) Pesticides removal in the process of drinking water production. Chemosphere 71(1):97–106
Percival SL, Yates MV, Williams DW, Chalmers RM, Gray NF (2014) Microbiology of waterborne diseases: microbiological aspects and risks, 2nd edn. Academic Press, London, UK
Pohanish RP (2015) Sittig's handbook of pesticides and agricultural chemicals, 2nd edn. William Andrew Publishing, Norwich, NY, USA
Rabbani D, Mostafaii GR, Eskandari V, Dehghani R, Atoof F (2017) Study of electrochemical process effect on detergent removal from polluted water and fish bioassay test of the effluent. Int Arch Health Sci 4:53–57
Rengaraj S, Yeon K-H, Moon S-H (2001) Removal of chromium from water and wastewater by ion exchange resins. J Hazard Mater 87(1–3):273–287
Sabarwal A, Kumar K, Singh RP (2018) Hazardous effects of chemical pesticides on human health-cancer and other associated disorders. Environ Toxicol Pharmacol 62:103–114
Shah BA, Shah AV, Singh RR, Patel NB (2011) Reduction of Cr (VI) in electroplating wastewater and investigation on the sorptive removal by WBAP. Environ Prog Sustain Energy 30(1):59–69
Sharma VK, Sohn M, McDonald TJ (2019) Remediation of selenium in water: a review. In: Ahuja S (ed) Advances in water purification techniques: meeting the needs of developed and developing countries. Elsevier, The Netherlands
Simonnot MO, Castel C, NicolaÏ M, Rosin C, Sardin M, Jauffret H (2000) Boron removal from drinking water with a boron selective resin: is the treatment really selective? Water Res 34(1):109–116

Singh BK, Walker A (2006) Microbial degradation of organophosphorus compounds. FEMS Microbiol Rev 30(3):428–471

Smedley PL, Kinniburgh DG (2017) Molybdenum in natural waters: a review of occurrence, distributions and controls. Appl Geochem 84:387–432

Thanh DN, Novák P, Vejpravova J, Vu HN, Lederer J, Munshi T (2018) Removal of copper and nickel from water using nanocomposite of magnetic hydroxyapatite nanorods. J Magn Magn Mater 456:451–460

Thirunavukkarasu OS, Viraraghavan T, Subramanian KS, Chaalal O, Islam MR (2005) Arsenic removal in drinking water—impacts and novel removal technologies. Energy Sources 27(1–2):209–219

Tripathi SK, Tyagi R, Nandi BK (2013) Removal of residual surfactants from laundry wastewater: a review. J Dispersion Sci Technol 34(11):1526–1534

Trivedi NS, Mandavgane SA (2018) Fundamentals of 2, 4 dichlorophenoxyacetic acid removal from aqueous solutions. Sep Purif Rev 47(4):337–354

UNIDO (2015) Independent mid-term evaluation report: environmentally sound management and final disposal of PCB'S in India. United Nations Industrial Development Organization

US EPA (1991) Maximum contaminant level goals and national primary drinking water regulations for lead and copper; final rule. US Environmental Protection Agency. Fed Reg 56(110):26460–26564

USEPA (2003) Drinking water advisory: consumer acceptability advice and health effects analysis on sulfate. U.S. Environmental Protection Agency, Health and Ecological Criteria Division, Washington, DC

USEPA (1976) Quality criteria for water EPA 440-9-76-023). U.S. Environmental Protection Agency, Washington, D.C.

Verbinnen B, Block C, Hannes D, Lievens P, Vaclavikova M, Stefusova K, Gallios G, Vandecasteele C (2012) Removal of molybdate anions from water by adsorption on zeolite-supported magnetite. Water Environ Res 84(9):753–760

Vijgen J, Abhilash PC, Li YF, Lal R, Forter M, Torres J, Singh N, Yunus M, Tian C, Schaffer A, Weber R (2010) Hexachlorocyclohexane (HCH) as new stockholm convention POPs—a global perspective on the management of Lindane and its waste isomers. Environ Sci Pollut Res 18(2):152–162

WHO (2003a) Hydrogen sulfide in drinking water: Background document for development of WHO guidelines for drinking water quality (WHO/SDE/WSH/03.04/07). World Health Organization, Geneva, Switzerland

WHO (2003b) Polychlorinated biphenyls: human health aspects (Concise international chemical assessment document 55). World Health Organization, Geneva, Switzerland

WHO (2003c) Polynuclear aromatic hydrocarbons in drinking water: Background document for development of WHO guidelines for drinking water quality (WHO/SDE/WSH/03.04/59). World Health Organization, Geneva, Switzerland

WHO (2003d) Isoproturon in drinking-water: Background document for development of WHO guidelines for drinking water quality (WHO/SDE/WSH/03.04/37). World Health Organization, Geneva, Switzerland

WHO (2004) Fluoride in drinking water: Background document for development of WHO guidelines for drinking water quality (WHO/SDE/WSH/03.04/96). World Health Organization, Geneva, Switzerland

WHO (2005a) Petroleum products in drinking water: background document for development of WHO guidelines for drinking water quality (WHO/SDE/WSH/05.08/123). World Health Organization, Geneva, Switzerland

WHO (2005b) Trihalomethanes in drinking-water: Background document for development of WHO guidelines for drinking water quality (WHO/SDE/WSH/05.08/64). World Health Organization, Geneva, Switzerland

WHO (2005c) Nickel in drinking-water: Background document for development of WHO guidelines for drinking water quality (WHO/SDE/WSH/05.08/55). World Health Organization, Geneva, Switzerland
WHO (2011a) Guidelines for drinking-water Quality, 4th edn. Geneva, Switzerland
WHO (2011b) Hardness in drinking-water: Background document for development of WHO Guidelines for Drinking-water Quality (WHO/HSE/WSH/10.01/10/Rev/1). World Health Organization, Geneva, Switzerland
WHO (2011c) Selenium in drinking-water: Background document for development of WHO Guidelines for Drinking-water Quality (WHO/HSE/WSH/10.01/14). World Health Organization, Geneva, Switzerland
WHO (2011d) Cadmium in drinking-water: Background document for development of WHO guidelines for drinking-water quality (WHO/SDE/WSH/03.04/80/Rev/1). World Health Organization, Geneva, Switzerland
WHO (2011e) Molybdenum in drinking-water: Background document for development of WHO Guidelines for drinking-water quality (WHO/SDE/WSH/03.04/11/Rev/1). World Health Organization, Geneva, Switzerland
WHO (2016) Quantitative microbial risk assessment: application for water safety management (ISBN 978-92-4-156537-0). World Health Organization, Geneva, Switzerland
WHO (2017) Guidelines for drinking-water quality: fourth edition incorporating the first addendum. World Health Organization, Geneva, Switzerland
WQA (2013a) Barium fact sheet. Water Quality Association, Illinois, USA
WQA (2013b) Chloramine fact sheet. Water Quality Association, Illinois, USA
WWAP (2019) The United Nations World water development report 2019: leaving no one behind. UNESCO, Paris
Yaacoubi H, Zidani O, Mouflih M, Gourai M, Sebti S (2014) Removal of cadmium from water using natural phosphate as adsorbent. Procedia Eng 83:386–393
Yazbeck C, Kloppmann W, Cottier R, Sahuquillo J, Debotte G, Huel G (2005) Health impact evaluation of boron in drinking water: a geographical risk assessment in Northern France. Environ Geochem Health 27(5–6):419–427
Zhang X, Ma J, Lu X, Huangfu X, Zou J (2015) High efficient removal of molybdenum from water by Fe2(SO4)3: effects of pH and affecting factors in the presence of co-existing background constituents. J Hazard Mater 300:823–829
Zhao X, Höll WH, Yun G (2002) Elimination of cadmium trace contaminations from drinking water. Water Res 36(4):851–858

Chapter 28
Water Quality Assessment of Upper Ganga Canal for Human Drinking

Tesfamariam Abreha Bahita, Sabyasachi Swain, Deen Dayal, Pradeep K. Jha, and Ashish Pandey

Abstract Water is fundamental need for existence of life. The deterioration in quality of drinking water may lead to severe impacts on human health. Thus, the quality assessment drinking water sources is of paramount concern. This study presents a comprehensive evaluation of water quality of Upper Ganga Canal (UGC) in Roorkee, Haridwar, Uttarakhand, India. The water samples were collected every month from 18 sites of the UGC from November 2014 to October 2015 and assessed at seasonal level (winter, summer and monsoon) for 15 important physicochemical parameters and 10 toxic trace metals. The results were compared with guidelines prescribed by four authentic standards viz., BIS (Bureau of Indian Standards: Drinking water specifications, 2012), EPA (Parameters of water quality: interpretation and standards. Environmental Protection Agency, Ireland, 2001), ICMR (Manual of Standards of Quality for Drinking Water Supplies, 2012) and WHO (Guidelines for drinking-water quality: recommendations, 2004). The arithmetic weightage-based Water Quality Index (WQI) was also computed to evaluate the pollution status of the canal. The results reveal that the UGC water is not suitable for human drinking as the physicochemical parameters and toxic trace metals exceeded the permissible limits of the standards considered, at numerous sites. The parameters also possessed strong seasonal variation in concentration. The overall water quality index was beyond the

T. A. Bahita · S. Swain · D. Dayal · A. Pandey (✉)
Department of Water Resources Development and Management, Indian Institute of Technology Roorkee, Roorkee, Uttarakhand 247667, India
e-mail: ashish.pandey@wr.iitr.ac.in

T. A. Bahita
e-mail: tesfamariam.abreha@yahoo.com

S. Swain
e-mail: sabyasachiswain16@gmail.com

D. Dayal
e-mail: deemishra26@gmail.com

P. K. Jha
Department of Mechanical and Industrial Engineering, Indian Institute of Technology Roorkee, Roorkee, Uttarakhand 247667, India
e-mail: pkjhafme@iitr.ac.in

A. Pandey et al. (eds.), *Climate Impacts on Water Resources in India*, Water Science and Technology Library 95, https://doi.org/10.1007/978-3-030-51427-3_28

permissible limits of human drinking at all the sites throughout all the seasons. Therefore, the water samples of the UGC are polluted and unfit for human drinking purpose. The exorbitant concentration of these parameters may be attributed to disposal of industrial effluents, domestic sewages and other human activities.

28.1 Introduction

Water is indispensable for existence of all living organisms, including human beings. Utilization of this water for various activities depends on its physical, chemical and biological characteristics (Budzinski et al. 1997; Tomar 1999; Lamotte et al. 2002; Lau et al. 2004; Kundu 2012). Thus, the availability of adequate water supply in terms of its quality and quantity is essential for the existence of life (Oki and Kanae 2006; Swain et al. 2018a). Safety of drinking water is a preliminary requirement for good health (Aarab et al. 2004; Soclo et al. 2008). The availability of freshwater is already a thought-provoking concern in various parts of globe (Bhuiyan and Ray 2017; Swain et al. 2017; Kumari 2018; Mehdi et al. 2018; Pavlidis and Tsihrintzis 2018). The situation is predicted to be further detrimental in the future, driven by urbanization, population explosion and climate change (Mohsin et al. 2013; Verma et al. 2016; Swain et al. 2018b).

The major resources of freshwater include rivers, lakes, reservoirs, streams etc., which are also the sources of canal waters (Michael 2008; Raju and Singh 2017; Tiwari et al. 2017; Wuijts et al. 2018). At present days, water pollution is largely a problem due to rapid urbanization and industrialization (Srivastava et. al. 2011; Li et al. 2017; Sharma et al. 2017; Swain et al. 2019). The pollution of water is on the increase and has become a global challenge in recent times (Goel 2006; Yazdi and Moridi 2017). The industry, agriculture and man activities have proven to be the main cause as well as victim of this issue. Therefore, the quality of water should be properly studied and monitored.

Canal system is one of the important requirements of modern civilization for the sustainable utilization of water in river basins (Matta 2014) and they may be polluted by various means. The UGC system around Roorkee is originated from the River Ganga and plays a significant role in the Northern India as a source of water for agriculture/ irrigation, navigation, drinking etc. The UGC system consists of main canal of 291 and 428 km long distribution channels. The water from the canal is utilized by people from Delhi, Ghaziabad, Noida, etc. for drinking purposes (NRCD 2009; Sharma and Singh 2011; Mishra et al. 2013). The water demand for drinking purposes is huge in its associated districts, which is growing day-by-day. Looking to the aforementioned issues, the present investigation has been carried out to assess the physicochemical parameters and trace metals relevant to water quality of the Upper Ganga Canal, Roorkee, for human drinking purpose.

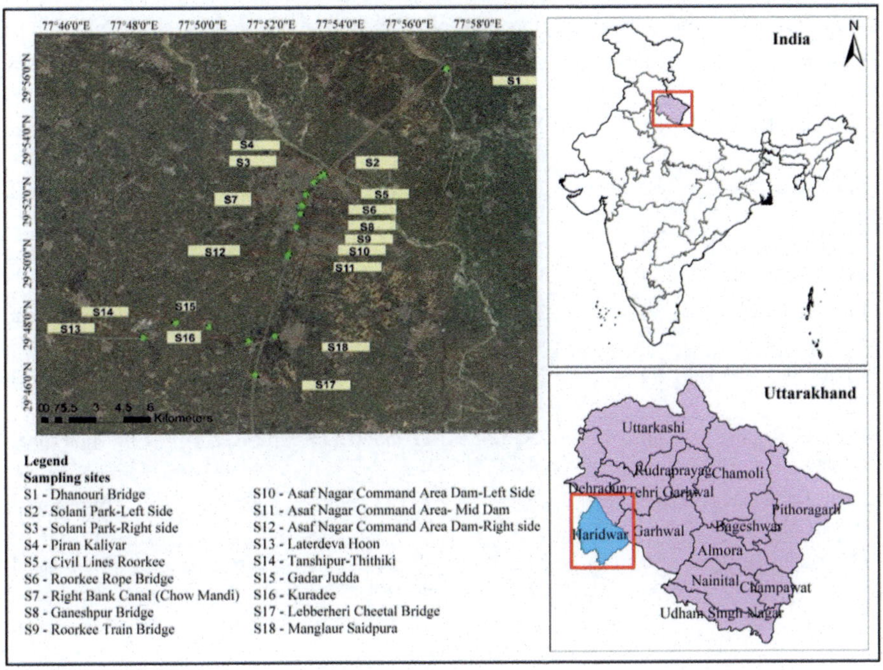

Fig. 28.1 Locations of sampling sites in the study area. *Source* Google Earth

28.2 Study Area

The study was carried out at the Upper Ganga Canal (UGC) around Roorkee City. Roorkee is located geographically at latitude 29° 87° N′ and longitude 77° 88′ E in the District of Haridwar, (Uttarakhand), India. It has an average elevation of 268 m. In the present study, to evaluate the pollution status of UGC and assess its suitability for drinking water, eighteen stations viz. S1, S2, …, S17 and S18, were selected as sampling stations. All the 18 sites are within the district of Haridwar in Uttarakhand, as shown in Fig. 28.1. Out of these, 12 sites are in main canal and 6 sites are in the branch canals of the UGC. The details of location (latitude, longitude, elevation) of the sites are presented in Table 28.1.

28.3 Methodology

In this study, water samples were collected from 18 sites i.e., S1, S2, S3,…, S17 and S18, from a depth of 20–30 cm. The time of collection of the samples were between 6 A.M. and 12 A.M. in the morning and 1 P.M. to 5 P.M. in the afternoon hours. The one-liter polyethylene plastic bottles were thoroughly cleaned by 8M

Table 28.1 Description of the sampling Sites of UGC

Description	Site code	Latitude	Longitude	Elevation (m)	Type of site location
Dhanouri Bridge	S1	29° 56′ 0.70″ N	77° 57′ 12.25″ E	261	Main canal
Solani Park-Left Side	S2	29° 52′ 42.55″ N	77° 53′ 36.74″ E	261	Main canal
Solani Park-Right side	S3	29° 52′ 41.44″ N	77° 53′ 30.75″ E	263	Main canal
Piran Kaliyar	S4	29° 55′ 15.72″ N	77° 56′ 01.20″ E	261	Main canal
Civil Lines Roorkee	S5	29° 52′ 26.39″ N	77° 53′ 20.20″ E	267	Main canal
Roorkee Rope Bridge	S6	29° 52′ 04.34″ N	77° 53′ 05.33″ E	271	Main canal
Right Bank Canal (Chow Mandi)	S7	29° 51′ 43.68″ N	77° 52′ 57.95″ E	269	Main canal
Ganeshpur Bridge	S8	29° 51′ 27.4″ N	77° 52′ 56.18″ E	266	Main canal
Roorkee Train Bridge	S9	29° 51′ 01.15″ N	77° 52′ 48.34″ E	267	Main canal
Asaf Nagar Command Area Dam-Left Side	S10	29° 50′ 06.66″ N	77° 52′ 35.50″ E	265	Main canal
Asaf Nagar Command Area-Mid Dam	S11	29° 50′ 01.10″ N	77° 52′ 32.63″ E	267	Main canal
Asaf Nagar Command Area Dam-Right side	S12	29° 50′ 02.74″ N	77° 52′ 35.19″ E	263	Main canal
Laterdeva Hoon	S13	29° 47′ 24.55″ N	77° 48′ 17.36″ E	259	Branch canal
Tanshipur-Thithiki	S14	29° 47′ 53.24″ N	77° 49′ 15.60″ E	262	Branch canal
Gadar Judda	S15	29° 47′ 32.36″ N	77° 51′ 05.58″ E	263	Branch canal
Kuradee	S16	29° 47′ 16.7″ N	77° 51′ 22.53″ E	262	Branch canal
Lebberheri Cheetal Bridge	S17	29° 46′ 10.04″ N	77° 51′ 34.64″ E	262	Branch canal
Manglaur Saidpura	S18	29° 47′ 26.35″ N	77° 52′ 09.41″ E	265	Branch canal

HNO_3, followed by repeated washing with de-ionized water, before collecting the samples. The samples were collected at monthly interval for a period of one-year (from November, 2014 to October, 2015) with the aim of understanding the seasonal (winter, summer, and monsoon) variation. The water samples were preserved in deep refrigerator (ice box) at 4° C to prevent the possible alteration of parameters. The water samples preserved in ice box were then analyzed in the laboratory. The physicochemical parameters namely pH, water temperature, electrical conductivity, Dissolved Oxygen and Total Dissolved Solids of the water were measured In-Situ at every monitoring station whereas other physicochemical parameters and toxic trace metals were analyzed in the laboratory. The analytical methods and instruments employed during the determination of the physicochemical parameters pertinent to quality of human drinking water, are presented in Table 28.2. The procedures and methods of analysis were as per guidelines of American Public Health Association (APHA 2005) and Guide Manual: Water and Wastewater Analysis (Gautam 2008).

The metal constituents in water, which are important parameters for water quality, were measured after digestion with a concentrated HCO_3 using the GBC-AVANTA flame atomic adsorption spectrophotometer (AAS) model. The details of the analytical methods and instruments used to determine the toxic trace metals are also presented in Table 28.2, as per guidelines of American Public Health Association (APHA 2005). Before measuring the parameters, the instrument GBC AVANTA A.A.S needs to be calibrated at the specified wavelengths for each of the parameters. It can be observed that, there are a total of 10 parameters, which are regarded as the toxic trace metals.

In order to assess the suitability for human drinking, these quality parameters should be compared with the reliable standards/guidelines of water quality for drinking purposes. The results were compared to the guidelines by Bureau Indian Standards (IS 10500: 2012), Indian Council Medical Research ICMR (2012), World Health Organization (WHO 2004) and Environmental Protection Agency (EPA 2001), to assess suitability for human drinking in all sampling sites of the UGC, Roorkee. The allowable limits of various parameters were obtained from the aforementioned standards.

28.3.1 Water Quality Index (WQI)

WQI method has been widely used by various scientists (Akkoyunlu and Akiner 2012) for decision making on quality of water. The concept of Water Quality Index (WQI) was developed by Horton (1965) and modified by Brown et al. (1970), who assigned weightage to different parameters. The weighted arithmetic WQI classify the water quality based on the degree of purity by using the standard variables. (Mishra et al. 2013). The weights for different quality parameters are supposed to vary inversely with respect to their corresponding recommended standards. In the present study, for computing WQI, twenty-seven important physicochemical and toxic trace metals parameters were chosen, collected and analyzed monthly to

Table 28.2 Analytical methods and instruments employed for determination of the physicochemical parameters and toxic trace metals for human drinking water

Parameter	Units	Methods employed for Estimation	Instrument/equipment/apparatus
Water temperature (T_w)	°C	Electronic	Portable Single-multi-parameters water quality meters (Model OR900P)
pH	–		
Electrical conductivity (EC)	μS/cm		
Total dissolved solids (TDS), dissolved oxygen (DO)	mg/L		
Total alkalinity (TA) as mg of $CaCO_3$/L	mg/L	Titration by standard H_2SO_4, 0.02N	Digital Burette
Total hardness (TH), Calcium (Ca^{2+}), Magnesium (Mg^{2+}) (expressed as mg of $CaCO_3$/L)	mg/L	Titration by EDTA	
Sodium (Na^+), Potassium (K^+)	mg/L	Flame photometry	Flame PHOTOMETER (Mode eL-378)
Chloride (Cl^-)	mg/L	Argentometric Titration using Standard $AgNO_3$, 0.0141N	Burette titration
Nitrate (NO_3)	mg/L	Ultraviolet screening	Ultraviolet visible (UV–VIS) spectrophotometer (DN 5000 HACH)
Chemical oxygen demand (COD)	mg/L	UV-spectrophotometry using COD digester	
Boron (B)	mg/L	Carmine method	
Sulfates (SO_4^{2-})	mg/L	Turbidimetry	Turbidimeter (Model HACH 2100AN)
Aluminium (Al), Arsenic (As), Cadmium (Cd), Chromium (Cr), Copper (Cu), Iron (Fe), Mercury (Hg), Manganese (Mn), Lead (Pb), Zinc (Zn)	mg/L	Flame Atomic Absorption Spectrophotometry (AAS)	A.A. Spectrophotometer (GBC-AVANTA)

evaluate the usability of water from UGC, Roorkee, for human drinking. All the eighteen samples were evaluated individually for water quality analysis. The WQI was calculated using the drinking water standards as approved by the Bureau of Indian Standards (IS 10500: 2012). The calculated WQI values were compared to the water quality rating for canal as per Table 28.3 which has already been used in several prior studies (Amadi et al. 2010; Goher et al. 2014; Sun et al. 2016; Zahedi 2017).

The weighted arithmetic index method was used for the calculation of WQI of the UGC as provided in the following steps.

The quality rating (q_n) was obtained using the expression given below (Eq. 28.1):

$$q_n = 100\left[\frac{(V_n - V_{id})}{(S_n - V_{id})}\right] \tag{28.1}$$

where, q_i is quality rating of ith parameter for a total of n water quality parameter, V_n is actually estimated valued of the nth parameter at a given sampling site, S_n is standard permissible limit value of the nth parameter and, V_{id} is the ideal value of nth parameter in pure water. These can be obtained from the authentic Standards or Guidelines.

All V_{id} values are equal to zero for drinking water parameters barring pH and DO, whose ideal values are 7 and 14.6 mg/L respectively (Al-Mashagbah 2015; Amadi et al. 2010; Goher et al. 2014).

The relative unit weight was computed by the following expression (Eq. 28.2):

$$W_n = \frac{k}{S_n} \tag{28.2}$$

where, W_n is Relative unit weight for the nth parameters, S_n represents standard value for nth parameter and, $K = 1$ for Constant proportionality.

Finally, the overall WQI was computed by the weighted Arithmetic Index as given (Eq. 28.3).

$$WQI = \frac{\sum q_n W_n}{\sum W_n} \tag{28.3}$$

where, q_n = quality rating, W_n = relative unit weight.

Table 28.3 Water quality rating as per weighted arithmetic WQI of canal (Goher et al. 2014)

WQI value	Rating of water quality	Grading
0–25	Excellent	A
26–50	Good	B
51–75	Poor	C
76–100	Very poor	D
Above 100	Unsuitable for drinking	E

Generally, WQI is defined for a specified and purposive water use. In the present study, WQI is computed to assess the quality for human consumption (drinking) and the allowable limit of WQI for drinking water is considered as 100 (Chauhan and Singh 2010; Goher et al. 2014; Akter et al. 2016). The calculated WQI values of the UGC, Roorkee water were established from the various twenty-five parameters. The calculated WQI values of the UGC, Roorkee water were established from the various twenty-five parameters.

28.4 Results and Discussion

In this study, 15 physicochemical parameters were measured and compared with multiple reliable standards viz., BIS (2012), EPA (2001), ICMR (2012) and WHO (2004), for permissible limits of human drinking. The results obtained for all the 18 sites during winter, summer and monsoon seasons for UGC water and comparisons with the permissible limits of aforementioned standards are presented in Tables 28.4, 28.5 and 28.6, respectively.

According to WHO (2004), a direct relationship between human health and the pH is so closely associated with other aspects of water quality. Deviation from ideal value of pH will show alkalinity or acidity with organoleptic consequences (EPA 2001) and adversely affects the mucous membrane (IS 10500: 2012). There is no evidence of severe physiological reactions occurring in persons consuming drinking-water supplies that have Electrical conductivity (EC) and Total Dissolved Solids (TDS) in excess, but their higher values cause insipid taste of water. Moderate concentration of alkalinity and hardness is desirable in most water supplies to stabilize the corrosive effects of acidity and enhance palatability. The alkalinity and hardness are mainly imparted by the concentration of Calcium and Magnesium ions, which in excess, don't have severe repercussions on human health (Mohsin et al. 2013). However, inadequacy of these may cause poor blood clotting, rickets, bone fracture and cardiovascular diseases. The cations like Sodium and Potassium are present in small to appreciable amounts in all natural water and are necessary for functioning of living organisms. However, excess consumption of these may cause nausea, hypertension and weakness in human body (EPA 2001). Similarly, the anions like Chlorine, Sulphate and Nitrate, when in excess, may cause metabolic disorders in human body (Ellina et al. 2017; Adimalla et al. 2018). DO and COD, although are not primary causes of health hazards, are important parameters for overall water quality (Parween et al. 2017).

Out of 15 physicochemical parameters considered, pH, TDS, Sodium (Na^+) and Nitrate (NO_3^-) concentrations were within the permissible limits of all the standards at all sampling sites during winter, summer and monsoon seasons. Among the rest, although there was not a single parameter which exceeded the permissible limits of all standards at all the sites throughout the three seasons, but at some sites, the concentrations were beyond limit. Moreover, 4 authentic standards were considered and for some parameters, their permissible limits are different. Hence, at some cases,

Table 28.4 Observed values of the physicochemical parameters for winter season

Sites	pH	EC	TDS	TA	TH	Ca	Mg	Na	K	Cl	SO_4	NO_3	DO	COD	B
S1	8.3	200	128	76.25	57.8	42	15.75	2.83	1.4	16	94.5	4.3	9.72	25.17	0.453
S2	8.4	187	120	76.25	64	46	21.5	3.2	1.55	10.75	96.75	3.88	9.69	39.01	0.568
S3	8.4	181	121	83.5	67.5	41.5	22.5	2.85	1.5	14.25	97.25	4.65	9.67	32.98	0.665
S4	7.1	1316	826	257.5	219	121	98	42.3	14.65	117.5	89	1.69	2.16	202.06	0.936
S5	6.8	984	637	260.5	364.3	335.25	29	54.8	31.18	345	93.5	2.36	1.54	491.09	0.988
S6	8.3	193	124	79.75	64.3	43.25	21	4.45	2.43	14.25	100.75	5.83	9.75	260.88	0.688
S7	6.9	985	636	215.25	220.8	137.5	83.25	43	22.25	148	176.25	2.63	2.11	28.89	0.947
S8	8.3	199	127	75.25	60.3	37.75	22.5	3.95	1.45	11.5	98	5.98	9.87	22.39	0.828
S9	8.3	185	118	73	61.8	39	22.75	2.78	1.55	16.25	101.25	3.9	9.77	31.59	0.831
S10	8.3	182	116	78.5	63.3	41	22.25	3.18	2.55	15	102	3.78	9.69	16.74	0.831
S11	8.2	183	117	89.75	60.8	44.25	22.25	3.35	1.88	15.25	100	4.15	9.64	26.3	0.832
S12	8.3	195	125	83.75	64	36.25	24.5	3.18	1.8	14	104.5	4.03	9.7	18.48	0.831
S13	8.3	186	119	84.75	58.8	36	25.25	3.18	1.98	18	100	2.98	9.51	22.42	0.836
S14	8.3	238	152	82.75	56.5	35.5	21	3.25	1.9	15.75	103.25	3.35	9.74	16.1	0.946
S15	8.4	243	156	81.5	66.8	44	22.75	3.5	2.68	24	115	4.9	9.64	48.53	1.531

(continued)

Table 28.4 (continued)

Sites	pH	EC	TDS	TA	TH	Ca	Mg	Na	K	Cl	SO_4	NO_3	DO	COD	B
S16	8.3	242	155	86.5	55.5	33	28.25	3.53	1.95	11.75	103.75	3.08	9.85	27.49	0.836
S17	8.3	241	154	86.25	58.5	33	27.5	3.28	1.95	14.75	103	3	9.68	22.14	0.833
S18	8.3	195	125	86.75	59.8	34.25	28	3.55	2.88	14.75	116	3	9.5	30.49	0.884
BIS	6.5–8.5	2000	2000	600	600	200	100	200	10	1000	400	100	6	10	1
ICMR	5.5–8.5	1000	1500	400	500	75	30	200	12	250	200	50	5	40	2
WHO	6.5–8.5	1400	1500	200	500	75	30	200	12	250	250	50	5	10	0.5
EPA	5.5–8.5	1000	–	400	200	–	–	200	–	250	200	50	–	40	2

Sources Bureau of Indian Standards (IS 10500: 2012), Indian Council of Medical Research ICMR (2012), World Health Organization (WHO 2004) and Environmental Protection Agency (EPA 2001)

Table 28.5 Observed values of the physicochemical parameters for summer season

Sites	pH	EC	TDS	TA	TH	Ca	Mg	Na	K	Cl	SO_4	NO_3	DO	COD	B
S1	7.8	198	127	52.75	64.5	43	21.5	1.18	1.65	15.75	219.75	4.5	8.77	2.98	0.525
S2	7.7	201	129	62.53	67.8	43.5	28	1.9	2.23	16.75	143.75	6.25	9.05	3.09	0.612
S3	7.8	202	129	72.5	71.5	44.2	23.5	2.8	1.68	17	126	4.15	9.02	3.55	0.652
S4	7.4	1226	653	100.85	202.3	132.5	69.75	12.4	20.25	119	209.5	4.69	2.51	218.73	1.252
S5	7.1	1237	713	177.25	247.8	173.75	74	16.8	13	169	205.25	5.06	1.52	295.98	1.352
S6	7.6	209	134	66.5	67.8	42.5	25.25	2.45	1.48	16.75	86.5	3.63	8.95	170.75	0.663
S7	7.3	1029	679	110.5	224.5	136.25	88.25	13.32	28.08	141.25	193.75	6.54	2.24	18.86	1.253
S8	7.8	208	133	65.25	65.8	42	23.75	1.05	1.23	16.5	105.25	3.73	9.02	3	0.664
S9	7.8	180	115	67.75	65	45.5	19.5	1.48	1.4	16.25	117	3.88	9.12	8.73	0.725
S10	7.8	195	124	69.75	66.3	42.25	24	2.9	1.85	17	92	3.9	9.26	20.88	0.828
S11	7.7	164	105	71.25	65.3	42	23.25	3.25	1.23	17.5	85.5	4.08	9.22	21.21	0.827
S12	7.7	180	115	70.25	65.3	42.25	23	3.33	1.5	17.5	82.25	3.5	9.28	16.61	0.828
S13	7.6	243	155	74.25	66.3	43.75	22.5	1.98	1.65	17.5	85.5	3.85	9.28	29.26	0.872
S14	7.7	161	103	70.75	65.5	44.25	21.25	2.45	1.93	17.75	79.75	3.6	9.12	6.98	0.923
S15	7.7	249	160	90.25	76	53.75	22.25	1.9	4.75	22.75	94.25	9.88	9.38	15.85	0.846
S16	8	168	108	88.55	66.8	41.75	25	2.08	1.45	19.75	84	3.8	8.76	4.55	0.854
S17	7.8	196	125	72.5	64.8	43.25	21.5	2.08	1.45	18	76.25	4.05	8.76	6.15	0.835
S18	7.7	170	108	70.25	65.5	42	22.5	2.08	1.65	16.2	87	4.15	8.9	5.24	0.845
BIS	6.5–8.5	2000	2000	600	600	200	100	200	10	1000	400	100	6	10	1
ICMR	5.5–8.5	1000	1500	400	500	75	30	200	12	250	200	50	5	40	2
WHO	6.5–8.5	1400	1500	200	500	75	30	200	12	250	250	50	5	10	0.5
EPA	5.5–8.5	1000	–	400	200	–	–	200	–	250	200	50	–	40	2

Table 28.6 Observed values of the physicochemical parameters for monsoon season

Sites	pH	EC	TDS	TA	TH	Ca	Mg	Na	K	Cl	SO_4	NO_3	DO	COD	B
S1	7.3	166	106	58.75	78	49.75	28.25	2.2	2.85	22.25	110.5	1.08	8.56	12.97	0.671
S2	7.2	198	127	63	78.3	50.75	27.55	2.63	2.75	22.55	114.75	1.5	8.56	12.81	0.727
S3	7	153	98	63	77	51.75	25.25	2.35	2.55	23.5	118.25	1.63	8.52	20.88	0.707
S4	7.1	1020	734	335	184	124.45	59.55	49.13	16.35	103.5	139.75	3.2	2.51	96	0.911
S5	7.3	667	427	179.25	178.7	113.75	64.9	57.68	10.48	95.75	136.75	3.83	1.88	272.54	0.912
S6	7.3	205	131	62.75	70.3	52	18.3	3.18	2.95	23.65	119.25	1.1	8.4	205.19	0.708
S7	7.5	1064	681	208.75	157.8	111.75	46.05	48.2	18.15	120.75	144.55	5.33	2.19	19.23	0.952
S8	7.2	224	143	63.25	73	53.25	19.75	3.03	2.93	24.75	120.75	1.33	8.56	24.5	0.735
S9	7.1	209	133	61.75	77.3	54	23.3	2.85	3.03	24.8	122.25	1.33	8.52	11.54	0.747
S10	7.4	213	136	62.25	86.8	54.25	32.55	3.28	3.3	24.85	127.25	1.18	8.44	12.8	0.818
S11	7.3	256	168	61.75	82.5	55.5	27	3.18	3.6	24.95	130.25	1.18	8.61	30.94	0.816
S12	7.4	215	142	62.25	86	54.75	31.25	3.3	3.28	23.75	128.25	1.15	8.52	11.22	0.816
S13	7.3	246	158	61.75	78.8	56	22.8	3.4	3.3	25.25	132.75	1.03	8.28	41.16	0.815
S14	7.2	250	160	61.75	72.5	55	17.5	3.45	3.38	24.95	131.75	1.2	8.44	48.26	0.837
S15	7.1	240	153	138.25	118.8	68	50.8	10.95	11.05	39.5	136.2	1.43	8.18	58.66	0.867
S16	7.4	200	127	78.75	92.3	57	35.3	3.05	3.4	27.25	136	1	8.52	13.02	0.925
S17	7.5	213	136	63.35	87.5	57.21	30.29	3.6	3.4	26.25	138.75	1.08	8.36	7.8	0.814
S18	7.4	219	140	64.25	83.8	55.25	28.55	3.58	3.53	24.95	129.25	1.08	8.48	38.47	0.827
BIS	6.5–8.5	2000	2000	600	600	200	100	200	10	1000	400	100	6	10	1
ICMR	5.5–8.5	1000	1500	400	500	75	30	200	12	250	200	50	5	40	2
WHO	6.5–8.5	1400	1500	200	500	75	30	200	12	250	250	50	5	10	0.5
EPA	5.5–8.5	1000	–	400	200	–	–	200	–	250	200	50	–	40	2

concentration of a parameter may be acceptable as per limits of one standard, but not for the others. The measured EC values were within the limits of BIS and WHO for all the seasons, however they exceeded the limits of ICMR and EPA at S4 during winter season, at S4, S5 and S7 during summer season and, at S4 and S7 during monsoon season. Similarly, Total Alkalinity (TA) was within limits as per all the standards considered in summer season, but was beyond the limits of only WHO at S4, S5 and S7 in winter season and, S4 as well as S7 in monsoon season. The parameter of Total Hardness (TH) was exceeding the allowable limits of only EPA at S4, S5 and S7 during winter and summer seasons. The concentration of divalent cations i.e. Ca^{2+} and Mg^{2+} were above the allowable limits of only ICMR and WHO for S4, S5 and S7 throughout all the seasons with an exception of Mg^{2+} at S5 during winter season. The concentration of Potassium ion (K^+) at S5 and S15 during monsoon season was above the permissible limits of only BIS. However, for S4 and S7 during monsoon season; and for S4, S5 and S7 during winter as well as summer season, the K^+ ion was above permissible limits of all the standards. The anionic concentration of Chlorine (Cl^-) was within limits per all sites except S5 during winter season. Similarly, Sulphate ion ($SO_4{}^{2-}$) exceeded the permissible limits of ICMR and EPA at two sites (S4 and S5) in summer season. The concentration of Boron was within the limits of ICMR and EPA, but exceeded the WHO limits at all sites except S1. Based on the BIS guidelines, the Boron was above allowable limits only for S15 in winter, S4, S5, S7 in summer and at no sites in monsoon season. The concentration of Chemical Oxygen Demand (COD) was found to be undesirable for many sites throughout all the seasons. Unlike all the parameters, higher concentration is desirable for dissolved oxygen (DO). However, DO level was much lower than the permissible limits of all the standards at S4, S5 and S7 in all the three seasons, although it was at desirable state for all other sites.

The purpose of considering multiple standards is to evaluate the quality of water for human drinking at utmost accuracy. Since the quality of drinking water directly influences human health, a proper quality assessment was of paramount concern. From the comparison of measured values of the physicochemical parameters with respect to the authentic standards, it can be inferred that the quality of UGC water is not suitable for human drinking at multiple sites. The concentration of physicochemical parameters also possessed a strong seasonal variation. Moreover, it can also be observed that, most of the parameters possessed exorbitant concentration at sites S4, S5 and S7 compared to other sites.

Apart from the physicochemical parameters, there were 10 toxic trace metals viz., Aluminum (Al), Arsenic (As), Cadmium (Cd), Chromium (Cr), Copper (Cu), Iron (Fe), Mercury (Hg), Manganese (Mn), Lead (Pb) and Zinc (Zn), which were measured and compared with permissible limits of the reliable standards i.e. BIS (2012), EPA (2001), ICMR (2012) and WHO (2004), for human drinking. A comparison of the results obtained for all the 18 sites during winter, summer and monsoon seasons for UGC water with the permissible limits of standards are presented in Tables 28.7, 28.8 and 28.9 respectively.

Most of the trace metals under investigation for water quality assessment of UGC are essential for human body, but in little amounts. The consumption of water with

Table 28.7 Observed values of the toxic trace metals for winter season

Sites	Al	As	Cd	Cr	Cu	Fe	Hg	Mn	Pb	Zn
S1	0.532	0.002	0.001	0.021	0.047	1.151	0.0021	0.035	0.012	0.567
S2	0.134	0.003	0.001	0.027	0.037	1.973	0.0029	0.043	0.020	0.621
S3	0.131	0.003	0.002	0.028	0.031	2.962	0.003	0.056	0.021	0.072
S4	0.152	0.014	0.005	0.042	0.037	3.922	0.0192	0.213	0.057	2.054
S5	0.165	0.013	0.009	0.030	0.020	3.911	0.0181	0.214	0.050	2.064
S6	0.236	0.004	0.001	0.028	0.067	2.972	0.0035	0.045	0.021	0.801
S7	0.153	0.011	0.003	0.037	0.063	3.918	0.0211	0.213	0.055	2.074
S8	0.136	0.003	0.002	0.029	0.051	2.978	0.0022	0.07	0.024	0.822
S9	0.136	0.004	0.002	0.029	0.036	2.971	0.0011	0.055	0.024	0.861
S10	0.166	0.002	0.002	0.029	0.059	2.973	0.0028	0.097	0.024	0.986
S11	2.327	0.002	0.006	0.041	0.051	2.973	0.0034	0.098	0.024	1.013
S12	0.166	0.002	0.001	0.025	0.059	2.974	0.0039	0.097	0.024	1.013
S13	0.136	0.005	0.019	0.032	0.057	2.983	0.0021	0.102	0.029	1.023
S14	0.186	0.004	0.003	0.087	0.061	2.982	0.0022	0.102	0.029	1.023
S15	0.186	0.002	0.005	0.081	0.071	2.983	0.0011	0.105	0.03	1.034
S16	1.184	0.002	0.002	0.080	0.064	2.983	0.0012	0.106	0.029	1.034
S17	1.168	0.003	0.001	0.030	0.077	2.976	0.0023	0.102	0.024	1.012
S18	2.157	0.002	0.006	0.039	0.092	2.986	0.0021	0.106	0.029	1.021
BIS	0.2	0.01	0.003	0.05	1.5	0.3	0.001	0.3	0.05	15
ICMR	–	0.05	0.01	–	1.5	1	0.001	0.5	0.5	5
WHO	0.2	0.05	0.003	0.05	1	0.3	0.001	0.5	0.01	5
EPA	0.2	0.05	0.05	0.05	0.05	0.2	0.001	0.05	0.05	3

high concentration of trace metals may have dire consequences for human health. The excessive concentration of trace metals imparts toxicity in water, which is harmful for human body. The health issues that may arise due to excessive consumption of toxic trace metals are carcinogenic problems, skin irritation, digestive disturbances, neurological damage, mutagenic effects, dehydration, abdominal pain, nausea and dizziness etc. (Štambuk-Giljanović 1999; Sharma et al. 2014; Izah et al. 2016; Mahato et al. 2016; Borah et al. 2017; Su et al. 2017; García-Ávila et al. 2018; Wang et al. 2018).

From the Tables (28.7, 28.8 and 28.9), it is evident that all the trace metals are not in desirable condition. The concentration of Aluminum exceeded limits of all the standards for the sites S1, S6, S11, S16, S17 and S18 during winter season; all the sites except S1 in summer season and, all the sites except S1, S2, S3 and S6 in monsoon season. The concentration of Arsenic was within the limits of ICMR, BIS and WHO, however it was above BIS limits at S4, S5 and S7 during all seasons. The trace metals viz., Mercury (Hg) and Iron (Fe) exceeded the permissible limits of all the standards throughout the year at all the sites. Zinc concentration was within

Table 28.8 Observed values of the toxic trace metals for summer season

Sites	Al	As	Cd	Cr	Cu	Fe	Hg	Mn	Pb	Zn
S1	0.134	0.002	0.002	0.021	0.023	1.151	0.0021	0.103	0.022	0.57
S2	0.234	0.003	0.002	0.024	0.035	2.329	0.0029	0.083	0.024	0.762
S3	0.335	0.003	0.003	0.024	0.065	2.337	0.0030	0.116	0.024	0.788
S4	1.152	0.013	0.008	0.092	0.073	2.429	0.0192	0.179	0.042	2.986
S5	1.165	0.013	0.008	0.086	0.036	2.336	0.0181	0.172	0.099	3.141
S6	0.336	0.004	0.003	0.026	0.176	2.556	0.0035	0.171	0.026	0.798
S7	1.153	0.011	0.008	0.098	0.04	2.922	0.0199	0.113	0.048	2.606
S8	0.446	0.003	0.003	0.028	0.051	2.322	0.0022	0.106	0.029	0.888
S9	0.456	0.004	0.004	0.038	0.049	2.416	0.0012	0.104	0.030	0.888
S10	0.548	0.002	0.004	0.038	0.049	2.416	0.0028	0.127	0.030	0.988
S11	0.547	0.002	0.004	0.038	0.092	2.413	0.0034	0.157	0.030	0.988
S12	0.548	0.002	0.004	0.046	0.053	2.413	0.0039	0.138	0.030	0.988
S13	0.766	0.005	0.007	0.048	0.042	2.414	0.0021	0.108	0.033	0.889
S14	0.757	0.004	0.007	0.049	0.045	3.007	0.0022	0.132	0.034	1.122
S15	0.859	0.002	0.007	0.047	0.044	3.016	0.0012	0.133	0.035	1.122
S16	0.735	0.002	0.007	0.046	0.052	2.413	0.0012	0.131	0.034	1.223
S17	0.548	0.003	0.004	0.039	0.052	2.412	0.0023	0.137	0.03	1.111
S18	0.843	0.002	0.006	0.048	0.053	2.414	0.0021	0.154	0.034	1.121
BIS	0.2	0.01	0.003	0.05	1.5	0.3	0.001	0.3	0.05	15
ICMR	–	0.05	0.01	–	1.5	1	0.001	0.5	0.5	5
WHO	0.2	0.05	0.003	0.05	1	0.3	0.001	0.5	0.01	5
EPA	0.2	0.05	0.05	0.05	0.05	0.2	0.001	0.05	0.05	3

limits for all the standards at all sites during winter, summer and monsoon season, barring S5 during summer season, where it exceeded the permissible limits of EPA only. The concentration of Chromium was beyond the permissible limits of all the standards at S14, S15 and S16 in winter season, whereas for S4, S5 and S7 during summer as well as monsoon seasons. Based on the permissible limits of BIS, ICMR and WHO, Manganese was within the permissible limits at all sites. However, except S1, S2 and S6 in winter season and except S13, S14, S15 and S16 in monsoon season, Manganese exceeded the EPA limits in all other sites. Similarly, Lead concentration was within the allowable limits of ICMR, but beyond that of WHO at all the sites during all three seasons. However, based on BIS and EPA guidelines, S4, S5 and S7 in winter; only S5 in summer and no station in monsoon season were above the permissible limits. The concentration of Copper was well within the limits by the guidelines of BIS, ICMR and WHO, but it exceeded the limits of EPA at many sites in all the seasons. Similarly, Cadmium concentration was also beyond the allowable limits at different sites as per the standards considered.

Table 28.9 Observed values of the toxic trace metals for monsoon season

Sites	Al	As	Cd	Cr	Cu	Fe	Hg	Mn	Pb	Zn
S1	0.131	0.002	0.001	0.02	0.020	1.299	0.0021	0.062	0.011	0.37
S2	0.142	0.003	0.001	0.024	0.022	1.363	0.0029	0.086	0.015	0.551
S3	0.142	0.003	0.001	0.025	0.023	1.377	0.0030	0.115	0.016	0.588
S4	1.106	0.017	0.004	0.055	0.056	2.424	0.0192	0.149	0.042	1.873
S5	1.142	0.013	0.005	0.054	0.069	2.244	0.0181	0.135	0.041	1.834
S6	0.152	0.004	0.002	0.026	0.027	1.387	0.0035	0.088	0.015	0.592
S7	1.175	0.011	0.004	0.053	0.063	2.438	0.0222	0.145	0.045	1.954
S8	0.204	0.003	0.002	0.026	0.028	1.767	0.0022	0.112	0.017	0.677
S9	0.205	0.004	0.002	0.026	0.030	1.777	0.0011	0.114	0.019	0.678
S10	0.324	0.002	0.003	0.029	0.038	1.823	0.0028	0.115	0.021	0.688
S11	0.324	0.002	0.003	0.029	0.041	1.815	0.0034	0.116	0.021	0.688
S12	0.324	0.002	0.003	0.029	0.04	1.828	0.0039	0.115	0.021	0.688
S13	0.326	0.005	0.004	0.035	0.035	2.019	0.0021	0.020	0.023	0.776
S14	0.329	0.004	0.004	0.037	0.035	2.088	0.0022	0.021	0.024	0.756
S15	0.329	0.002	0.005	0.037	0.038	2.142	0.0011	0.022	0.024	0.777
S16	0.326	0.002	0.005	0.036	0.039	2.089	0.0012	0.021	0.022	0.738
S17	0.326	0.003	0.004	0.030	0.043	1.839	0.0023	0.119	0.016	0.688
S18	0.326	0.002	0.004	0.034	0.039	2.158	0.0021	0.115	0.024	0.758
BIS	0.2	0.01	0.003	0.05	1.5	0.3	0.001	0.3	0.05	15
ICMR	–	0.05	0.01	–	1.5	1	0.001	0.5	0.5	5
WHO	0.2	0.05	0.003	0.05	1	0.3	0.001	0.5	0.01	5
EPA	0.2	0.05	0.05	0.05	0.05	0.2	0.001	0.05	0.05	3

Tables 28.7, 28.8 and 28.9, it is clear that UGC water contains high undesirable concentration of trace metals at numerous sites. It was also observed that, like the physicochemical parameters, the toxic trace metals possess significant seasonal variation. Further, the concentrations of all the trace metals, in general, were at much higher relative to other sites. Hence, it can be inferred that the UGC water is not suitable for human drinking, at current pollutant level.

Table 28.10, it can be inferred that, the site S7 possessed highest value of WQI throughout all the seasons, which is indicative of highest pollution amongst all the sites. On the other hand, S9 was the least polluted site in all the seasons, based on WQI values. However, the limiting value for the water to be suitable for human drinking is 100. The minimum WQI value among all the sites is beyond this threshold for all the seasons. This reflects that the surface canal water at all the sampling sites is unsuitable for human drinking purpose. Moreover, it can be observed that, except S4, S5 and S7, the WQI values are less than the average of 18 samples. For these three sites, the WQI values are in the range of 1287.59–1518.1, showing the extent

Table 28.10 WQI results for 18 samples of UGC, Roorkee, during winter, summer and monsoon seasons

Site code	WQI from only physicochemical parameters			WQI from only Toxic trace metals			Overall WQI considering both physicochemical parameters and trace metals		
	Winter	Summer	Monsoon	Winter	Summer	Monsoon	Winter	Summer	Monsoon
S1	48.02	52.36	60.49	159.0	173.0	158.1	158.93	172.85	158.03
S2	57.04	57.93	63.79	220.7	228.9	214.5	220.57	228.78	214.39
S3	63.9	60.87	61.21	230.2	243.4	222.1	230.08	243.25	221.91
S4	93.94	114.85	88.76	1369.8	1387.1	1356.1	1368.6	1385.83	1354.89
S5	96.43	122.22	91.35	1325.1	1351.2	1288.7	1323.9	1349.96	1287.59
S6	64.76	60.76	62.56	263.9	278.8	262.7	263.69	278.57	262.46
S7	94.83	116.31	93.29	1486.99	1519.5	1486.01	1485.64	1518.1	1484.66
S8	74.39	61.69	63.75	178.7	193.2	176.9	178.59	193.09	176.82
S9	74.71	65.45	64.38	104.7	127.7	104.3	104.64	127.65	104.24
S10	74.80	72.71	71.79	219.9	242.3	228	219.79	242.16	227.84
S11	74.43	72.11	70.40	295.6	282.7	268.4	295.39	282.46	268.18
S12	74.96	72.04	71.42	291.5	316.6	302.1	291.32	316.35	301.88
S13	75.64	74.52	70.54	302.4	220.2	190.2	302.18	220.02	190.04
S14	82.58	78.76	70.86	191.2	228.2	197.2	191.11	228.07	197.11

(continued)

Table 28.10 (continued)

Site code	WQI from only physicochemical parameters			WQI from only Toxic trace metals			Overall WQI considering both physicochemical parameters and trace metals		
	Winter	Summer	Monsoon	Winter	Summer	Monsoon	Winter	Summer	Monsoon
S15	124.11	73.31	75.46	133.4	155.0	130.7	133.35	154.91	130.61
S16	75.31	76.49	79.3	118.7	160.3	136.1	118.64	160.21	136.06
S17	75.32	73.82	72.02	184.3	208.7	198.3	184.17	208.6	198.20
S18	79.20	73.82	72.04	208.8	213.6	190.6	208.71	213.51	190.51
Min.	**48.02**	**52.36**	**60.49**	**104.67**	**127.71**	**104.28**	**104.64**	**127.65**	**104.24**
Max.	**124.11**	**122.22**	**93.29**	**1486.99**	**1519.46**	**1486.01**	**1485.64**	**1518.1**	**1484.66**
Avg.	**78.83**	**77.73**	**72.86**	**443.83**	**458.87**	**435.07**	**443.48**	**458.51**	**434.71**
SD	**16.83**	**20.28**	**10.03**	**459.72**	**463.95**	**455.81**	**459.28**	**463.52**	**455.38**

of pollution. It will cause severe health hazard, even leading to death, if used for drinking purpose.

28.5 Conclusions

The Upper Ganga Canal water was evaluated for human drinking purpose. A total of 18 sampling sites were located in Upper Ganga Canal (main and branch canal), from which surface water samples were collected for each month from Nov 2014 to Oct 2015. Based on the analysis, the following conclusions are drawn from the present study:

1. Based on guidelines and Standards by BIS (IS 10500: 2012), ICMR (2012), WHO (2004) and EPA (2001), it was found that the physicochemical parameters viz., Ca, K, B, DO and COD were above the permissible limits for human drinking purpose.
2. Most of the trace metals in UGC water possess higher concentration compared to the permissible limits of the standards/guidelines considered for human drinking purposes. The higher concentration of these metals imparts toxicity to the water and hence, is not utilizable for drinking purposes.
3. During all the seasons and at all the sites, the results of the overall WQI values were above the threshold of usability for human drinking purposes, which was mainly due to the excessive concentrations of trace metals.

Acknowledgement First author is thankful for the financial support received by the Government of Ethiopia through the Embassy of the Federal Democratic Republic of Ethiopia, New Delhi, India during the period of study. The scholarship provided by Debre Berhan University, Ministry of Education of the FDRE (Ethiopia) is greatly acknowledged. The authors are thankful to the Department of Water Resources Development and Management, Department of Chemistry, Institute Instrumentation Centre, and Environmental Engineering Laboratory in the Department of Civil Engineering, IIT Roorkee for providing required facilities.

References

Aarab N, Champeau O, Mora P, Daubeze M, Garrigues P, Narbonne JF (2004) Scoring approach based on fish biomarkers applied to French river monitoring. Biomarkers 9(3):258–270

Adimalla N, Li P, Venkatayogi S (2018) Hydrogeochemical evaluation of groundwater quality for drinking and irrigation purposes and integrated interpretation with water quality index studies. Environ Process 5(2):363–383

Al-Mashagbah AF (2015) Assessment of surface water quality of King Abdullah canal, using physico-chemical characteristics and water quality index, Jordan. J Water Resour Protect 7(04):339–352

Akkoyunlu A, Akiner ME (2012) Pollution evaluation in streams using water quality indices: a case study from Turkey's Sapanca Lake Basin. Ecol Ind 18:501–511

Akter T, Jhohura FT, Akter F, Chowdhury TR, Mistry SK, Dey D, Barua MK, Islam MK, Rahman M (2016) Water quality index for measuring drinking water quality in rural Bangladesh: a cross-sectional study. J Health, Popul Nutr 35(1):4

Amadi AN, Olasehinde PI, Okosun EA, Yisa J (2010) Assessment of the water quality index of Otamiri and Oramiriukwa Rivers. Phys Int 1(2):116–123

American Public Health Association (APHA) (2005) Standard Methods for the examination of water and wastewater, 21st edn. American Public Health Association, Washington DC, p 1220p

Bhuiyan C, Ray PKC (2017) Groundwater quality zoning in the perspective of health hazards. Water Resour Manage 31(1):251–267

BIS (2012) Bureau of Indian Standards: drinking water specifications. IS 10500:2012, New Delhi, India

Borah P, Paul A, Bora P, Bhattacharyya P, Karak T, Mitra S (2017) Assessment of heavy metal pollution in soils around a paper mill using metal fractionation and multivariate analysis. Int J Environ Sci Technol 14(12):2695–2708

Brown RM, McClelland NI, Deininger RA, Tozer RG (1970) A water quality index: do we dare? Water Sewage Works 117:339–343

Budzinski H, Jones I, Bellocq J, Pierard C, Garrigues PH (1997) Evaluation of sediment contamination by polycyclic aromatic hydrocarbons in the Gironde estuary. Mar Chem 58(1–2):85–97

Chauhan A, Singh S (2010) Evaluation of Ganga water for drinking purpose by water quality index at Rishikesh, Uttarakhand, India. Rep Opin 2(9):53–61

Ellina G, Papaschinopoulos G, Papadopoulos BK (2017) Fuzzy inference systems: selection of the most appropriate fuzzy implication from available lake water quality statistical data. Environ Process 4(4):923–935

EPA (2001) Parameters of water quality: interpretation and standards. Environmental Protection Agency, Ireland

García-Ávila F, Ramos-Fernández L, Pauta D, Quezada D (2018) Evaluation of water quality and stability in the drinking water distribution network in the Azogues city. Ecuador Data Brief 18:111–123

Gautam SP (2008) Guide manual: water and wastewater analysis. Hydrology Project-II, Ministry of Water Resources (MoWR)

Goel PK (2006) Water pollution: cause, effects and control, revised 2nd edn. New Age International (P) Ltd

Goher ME, Hassan AM, Abdel-Moniem IA, Fahmy AH, El-sayed SM (2014) Evaluation of surface water quality and heavy metal indices of Ismailia Canal, Nile River, Egypt. Egypt J Aquat Res 40(3):225–233

Horton RK (1965) An index number system for rating water quality. J Water Pollut Control Fed 37(3):300–306

ICMR (2012) Manual of standards of quality for drinking water supplies. Indian Council of Medical Research, Spl. Rep. No. 44: 27

Izah SC, Chakrabarty N, Srivastav AL (2016) A review on heavy metal concentration in potable water sources in Nigeria: human health effects and mitigating measures. Exposure Health 8(2):285–304

Kumari P (2018) Distribution of metal elements in capillary water, overlying water, sediment, and aquatic biota of three interconnected ecosystems. Environ Process 5(2):385–411

Kundu S (2012) Assessment of surface water quality for drinking and irrigation purposes: a case study of Ghaggar River system surface waters. Bull Environ Pharmacol Life Sci 1(2):1–6

Lamotte M, de Violet P, Garrigues P, Hardy M (2002) Evaluation of the possibility of detecting benzenic pollutants by direct spectrophotometry on PDMS solid sorbent. Anal Bioanal Chem 372(1):169–173

Lau PS, Wong HL, Garrigues P (2004) Seasonal variation in antioxidative responses and acetylcholinesterase activity in *Perna viridis* in eastern oceanic and western estuarine waters of Hong Kong. Cont Shelf Res 24(16):1969–1987

Li P, Feng W, Xue C, Tian R, Wang S (2017) Spatiotemporal variability of contaminants in lake water and their risks to human health: a case study of the Shahu Lake tourist area, northwest China. Exposure Health 9(3):213–225

Mahato MK, Singh PK, Tiwari AK, Singh AK (2016) Risk assessment due to intake of metals in groundwater of East Bokaro Coalfield, Jharkhand, India. Exposure Health 8(2):265–275

Matta G (2014) Water quality assessment of Ganga Canal system. J Adv Sci Res 5(4):19–24

Mehdi B, Schulz K, Ludwig R, Ferber F, Lehner B (2018) Evaluating the importance of non-unique behavioural parameter sets on surface water quality variables under climate change conditions in a mesoscale agricultural watershed. Water Resour Manage 32(2):619–639

Michael AM (2008) Irrigation theory and practice, 2nd edn. Vikas Publishing House. Pvt. Ltd., New Delhi, India

Mishra N, Khare D, Shukla R, Singh L (2013) A study of temperature Variation in Upper Ganga Canal Command India. Adv Water Resour Protect (AWRP) 1(3):1–7

Mohsin M, Safdar S, Asghar F, Jamal F (2013) Assessment of drinking water quality and its impact on residents health in Bahawalpur city. Int J Humanit Soc Sci 3(15):114–128

NRCD (2009) Status paper on River Ganga: state of environment and water quality. National River Conservation Directorate, Ministry of Environment and Forests, Government of India

Oki T, Kanae S (2006) Global hydrological cycles and world water resources. Science 313(5790):1068–1072

Parween M, Ramanathan AL, Raju NJ (2017) Waste water management and water quality of river Yamuna in the megacity of Delhi. J Int Environ Sci Technol 14(10):2109–2124

Pavlidis G, Tsihrintzis VA (2018) Environmental benefits and control of pollution to surface water and groundwater by agroforestry systems: a review. Water Resour Manage 32(1):1–29

Raju A, Singh A (2017) Assessment of groundwater quality and mapping human health risk in central Ganga Alluvial Plain, Northern India. Environ Process 4(2):375–397

Sharma N, Singh R (2011) History of irrigation in Uttar Pradesh. Indian National Committee on Irrigation and Drainage (INCID), New Delhi

Sharma P, Meher PK, Kumar A, Gautam YP, Mishra KP (2014) Changes in water quality index of Ganges river at different locations in Allahabad. Sustain Water Qual Ecol 3:67–76

Sharma RK, Yadav M, Gupta R (2017) Water quality and sustainability in India: challenges and opportunities. In: Chemistry and water, pp 183–205

Soclo HH, Budzinski H, Garrigues P, Matsuzawa S (2008) Biota accumulation of polycyclic aromatic hydrocarbons in Benin coastal waters. Polycyclic Aromat Compd 28(2):112–127

Srivastava A, Kumar R, Gupta V, Agarwal G, Srivastava S, Singh I (2011) Water quality assessment of Ramganga river at Moradabad by physico-chemical parameters analysis. VSRD-TNTJ 2(3):119–127

Štambuk-Giljanović N (1999) Water quality evaluation by index in Dalmatia. Water Res 33(16):3423–3440

Su H, Kang W, Xu Y, Wang J (2017) Assessment of groundwater quality and health risk in the oil and gas field of Dingbian County, Northwest China. Exposure Health 9(4):227–242

Sun W, Xia C, Xu M, Guo J, Sun G (2016) Application of modified water quality indices as indicators to assess the spatial and temporal trends of water quality in the Dongjiang River. Ecol Ind 66:306–312

Swain S, Patel P, Nandi S (2017) Application of SPI, EDI and PNPI using MSWEP precipitation data over Marathwada, India. In: 2017 IEEE international geoscience and remote sensing symposium (IGARSS). IEEE, pp. 5505–5507

Swain S, Verma MK, Verma MK (2018a) Streamflow estimation using SWAT model over Seonath river basin, Chhattisgarh, India. In: Hydrologic modeling. Springer, Singapore, pp 659–665

Swain S, Nandi S, Patel P (2018b) Development of an ARIMA Model for monthly rainfall forecasting over Khordha District, Odisha, India. In: Recent findings in intelligent computing techniques. Springer, Singapore, pp 325–331

Swain S, Mishra SK, Pandey A (2019) Spatiotemporal characterization of meteorological droughts and its linkage with environmental flow conditions. AGUFM H13O-1959

Tiwari K, Goyal R, Sarkar A (2017) GIS-based spatial distribution of groundwater quality and regional suitability evaluation for drinking water. Environ Process 4(3):645–662

Tomar M (1999) Quality assessment of water and wastewater. CRC press

Verma MK, Verma MK, Swain S (2016) Statistical analysis of precipitation over Seonath river basin, Chhattisgarh, India. Int J Appl Eng Res 11(4):2417–2423

Wang L, Ma L, Yang Z (2018) Spatial variation and risk assessment of heavy metals in paddy rice from Hunan Province, Southern China. Int J Environ Sci Technol 15(7):1561–1572

WHO (2004) Guidelines for drinking-water quality: recommendations, vol 1. World Health Organization

Wuijts S, Driessen PP, Van Rijswick HF (2018) Governance conditions for improving quality drinking water resources: the need for enhancing connectivity. Water Resour Manage 32(4):1245–1260

Yazdi J, Moridi A (2017) Interactive reservoir-watershed modeling framework for integrated water quality management. Water Resour Manage 31(7):2105–2125

Zahedi S (2017) Modification of expected conflicts between drinking water quality index and irrigation water quality index in water quality ranking of shared extraction wells using multi criteria decision making techniques. Ecol Ind 83:368–379

Printed by Printforce, the Netherlands